Engineering as a Global Profession

Engineering as a Global Profession

Technical and Ethical Standards

Michael Davis

ROWMAN & LITTLEFIELD
Lanham • Boulder • New York • London

Published by Rowman & Littlefield
An imprint of The Rowman & Littlefield Publishing Group, Inc.
4501 Forbes Boulevard, Suite 200, Lanham, Maryland 20706
www.rowman.com

Selection, Editorial Matter, Preface, and Chapter 20 Copyright © 2021 by The Rowman & Littlefield Publishing Group, Inc.

All rights reserved. No part of this book may be reproduced in any form or by any electronic or mechanical means, including information storage and retrieval systems, without written permission from the publisher, except by a reviewer who may quote passages in a review.

British Library Cataloguing in Publication Information Available

Library of Congress Cataloging-in-Publication Data

Names: Davis, Michael, 1943– author.
Title: Engineering as a global profession : technical and ethical standards / Michael Davis.
Description: Lanham : Rowman & Littlefield, [2021] | Includes bibliographical references and index.
Identifiers: LCCN 2021030026 (print) | LCCN 2021030027 (ebook) |
  ISBN 9781538155042 (cloth) | ISBN 9781538155059 (epub)
  ISBN 9781538155066 (pbk)
Subjects: LCSH: Engineering ethics. | Engineering—Vocational guidance.
Classification: LCC TA157 .D3196 2021 (print) | LCC TA157 (ebook) |
  DDC 620.0023—dc23
LC record available at https://lccn.loc.gov/2021030026
LC ebook record available at https://lccn.loc.gov/2021030027

For Deb: my lawyer wife, an engineer's daughter, and early supporter of this work

Contents

Permissions		ix
Preface		xiii
PART I: DISTINGUISHING ENGINEERING FROM OTHER PROFESSIONS		**1**
1	Profession	3
2	Engineering—From Chicago to Shantou	23
3	Why Architects Are Not Engineers	37
4	Distinguishing Chemists from Engineers	55
5	Will Software Engineering Ever Be Engineering?	67
6	Engineering and Business Management: The Odd Couple	73
PART II: THE STUDY OF ENGINEERING AS A PROFESSION		**89**
7	Methodological Problems in the Study of Engineering	91
8	Profession as a Lens for Studying Technology	107
PART III: PROFESSIONAL RESPONSIBILITY OF ENGINEERS		**121**
9	"Ain't Nobody Here But Us Social Forces": Constructing the Professional Responsibility of Engineers	123
10	Engineering Ethics, Individuals, and Organizations	151

11	"Social Responsibility" and "Social Justice" for Engineers?	161
12	Macro-, Micro-, and Meso-Ethics	167
13	Doing "the Minimum"	185
14	Re-inventing the Wheel: "Global Engineering Ethics"	191
15	In Praise of Emotion in Engineering	201
PART IV: ENGINEERING'S GLOBALISM		**217**
16	The Whistle Not Blown: WV, Diesels, and Engineers	219
17	Three Nuclear Disasters and a Hurricane: Reflections on Engineering Ethics	235
18	Ethical Issues in the Global Arms Industry: A Role for Engineers	255
19	Temporal Limits of Engineers' Planning	269
20	Epilogue: A Research Agenda	287
Index		303

Permissions

This volume contains the following previously published material:

Chapter 1. Reprinted/adapted by permission from Springer Nature Customer Service Centre GmbH: Springer Nature, *Philosophia* 37, 211–225. Davis, Michael, "Is Engineering a Profession Everywhere?" © June 2009.

Chapter 2. Davis, Michael. "Defining Engineering—From Chicago to Shantou," Monist 92, 3 (July 2009): 325–339 by permission of Oxford University Press.

Chapter 3. Reprinted/adapted by permission from Springer Nature Customer Service Centre GmbH: Springer Nature, Davis, Michael, "Distinguishing Architects from Engineers: A Pilot Study in Differences between Engineers and other Technologists," in *Philosophy and Engineering: An Emerging Agenda*, edited by Ibo van de Poel and David Goldberg, 15–30 © 2010.

Chapter 4. Davis, Michael. "Do the Professional Ethics of Chemists and Engineers Differ?" HYLE 8 (Spring 2002): 21–34.

Chapter 5. Davis, Michael. "Will Software Engineering Ever Be Engineering?" *Communications of the ACM* 54 (November 2011): 32–34.

Chapter 6. Reprinted/adapted by permission from Springer Nature Customer Service Centre GmbH: Springer Nature, Davis, Michael, "Engineers and Business Managers: The Odd Couple," in *The Engineering-Business Nexus – Symbiosis, Tension and Co-Evolution*, edited by Steen Hyldgaard Christensen et al. Philosophy of Engineering and Technology, v. 32, 25–38 © 2019.

Chapter 7. Reprinted/adapted by permission from Springer Nature Customer Service Centre GmbH: Springer Science + Business Media B.V, Davis, Michael, "Engineering as Profession: Some Methodological Problems in its Study," pp. 65–79 in *Engineering, Development and Philosophy: American, Chinese, and European Perspectives*, edited by Steen Hyldgaard Christensen, et al. © 2015.

Chapter 8. Davis, Michael. "Profession as a Lens for Studying Technology," in *Methods for the Ethics of Technology*, edited by Sven-Ove Hansson, 83–96. London: Rowman and Littlefield International, 2016.

Chapter 9. Reprinted/adapted by permission from Springer Nature Customer Service Centre GmbH: Springer Nature, *Science and Engineering Ethics* 18, 13–34. Davis, Michael, "'Ain't no one here but us social forces': Constructing the professional responsibility of engineers" © Winter 2012.

Chapter 10. Reprinted/adapted by permission from Springer Nature Customer Service Centre GmbH: Springer Nature, *Science and Engineering Ethics* 12, 223–231. Davis, Michael, "Engineering Ethics, Individuals, and Organizations" © April 2006.

Chapter 11. Davis, Michael. "Social Responsibility and 'Social Justice' for Engineers?" *Engineering Dimensions* (March/April 2013): 39–41.

Chapter 12. Davis, Michael. "Engineers and Sustainability: An Inquiry into the elusive distinction between Macro-, Micro-, and Meso-Ethics," *Journal of Applied Ethics and Philosophy* 2 (2010): 12–20.

Chapter 13. Reprinted/adapted by permission from Springer Nature Customer Service Centre GmbH: Springer Nature, *Science and Engineering Ethics* 7, 283. Davis, Michael, "Case Study in Engineering Ethics—'Doing the Minimum'" © April 2001.

Chapter 14. Reprinted/adapted by permission from Springer Nature Customer Service Centre GmbH: Springer Nature, Davis, Michael, "In Praise of Emotion in Engineering," in *Philosophy and Engineering: Exploring Boundaries, Expanding Connections*, edited by Diane P. Michelfelder, Byron Newberry, Qin Zhu, 181–194 © 2017.

Chapter 15. Reprinted/adapted by permission from Springer Nature Customer Service Centre GmbH: Springer Nature, Davis, Michael, "'Global

Engineering Ethics': Re-inventing the Wheel?" in *Engineering Ethics for a Globalized World*, edited by Colleen Murphy, et al., 69–78 © 2015.

Chapter 16. Davis, Michael. "A Whistle Not Blown: WV, Diesels, and Engineers" in *Next Generation Ethics: Engineering a Better Society*, edited by Ali Abbas, 217–229. Cambridge University Press, 2019.

Chapter 17. Davis, Michael. "Three Nuclear Disasters and a Hurricane: some reflections on engineering ethics," *Journal of Applied Ethics and Philosophy* 4 (September 2012): 1–10.

Chapter 18. Davis, Michael. "Ethical Issues in the Global Arms Industry: A Role for Engineers" in *Ethical Dilemmas in the Global Defense Industry*, edited by Claire Finkelstein and Kevin Govern (forthcoming) by permission of Oxford University Press.

Chapter 19. Reprinted/adapted by permission from Springer Nature Customer Service Centre GmbH: Springer Nature, *Science and Engineering Ethics* 25, 1609–1624. Davis, Michael, "Temporal Limits on What Engineers Can Plan" © October 2019.

Preface

I am, and ever will be, a white-socks, pocket-protector, nerdy engineer, born under the second law of thermodynamics, steeped in steam tables, in love with free-body diagrams, transformed by Laplace and propelled by compressible flow.

—Neil Armstrong, astronaut, National
Press Club, February 22, 2000

This book is primarily a contribution to the philosophy of engineering. Unlike engineering ethics, the philosophy of engineering is an academic field that even few philosophers had heard of until recently. While *a* philosophy of engineering is a controversial view about how engineering should be practiced, *the* philosophy of engineering is the attempt to reach a justified common understanding of what engineers are, what they do, how they do it, why they do it, and why their doing it in that way is reasonable. For some, "reasonable" means required by reason; for others, allowed by reason; for the remainder, recommended by reason, that is, more than allowed but less than required. For the purposes of this book, "reasonable" means at least allowed by reason.

Insofar as successful, the philosophy of engineering should be uncontroversial, simply bringing engineering into focus. Nonetheless, insofar as successful, the philosophy of engineering should help social scientists (including historians) do a better job with the empirical study of engineers, help government make better policy concerning engineering, help universities improve engineering education, and help engineers better understand what they do. The profits of philosophy, even the philosophy of engineering, can be large, various, and valuable.

The philosophy of engineering is (or, at least, should be) the engineering counterpart of the philosophy of law, the philosophy of medicine, or the philosophy of science. After a particular "philosophy-of" succeeds, philosophers tend to lose interest, abandoning the field to the appropriate sciences, legislators, or practitioners—until new difficulties in understanding arise. Since the philosophy of law, the philosophy of medicine, and the philosophy of science are still among philosophy's active fields, "we" (you, me, and the other readers of this book) can be reasonably sure that the philosophers working in the field still do not understand law, medicine, or science well enough to be satisfied. No doubt, I shall not do much better with engineering. But some understanding is better than none, and no one succeeds without trying. This book is (in part at least) an invitation to join the attempt.

1. PHILOSOPHY OF ENGINEERING TODAY

The philosophy of law, medicine, and science have each been significant academic fields since at least early in the twentieth century. That is not true of the philosophy of engineering. Though George Sinclair (1977), Taft Broome (1985), David Bella (1987), Paul Durbin (1987), Carl Mitcham (1994), and a few others, have called for undertaking the philosophy of engineering and even pointed to work, especially in engineering ethics, already done that could count as contributions to the philosophy of engineering, the philosophy of engineering remained a prairie almost empty to the horizon until the twenty-first century.[1]

The academic fields upon which the philosophy of engineering might draw for information, insight, or problems are not much better off. The sociology of engineering tends in practice to be more about technology (what happens, or should happen, to what engineers *and others* produce) rather than about engineering as such (how *engineers* in particular do what they do, with whom, how, when, and why). Works like that of Kidder (1981), Hacker (1990), and Vaughn (1996) remain rare. As Bruce Sinclair said long ago, "we still know very little of the vast majority of American engineers" (Sinclair, 1986). We know even less of engineers elsewhere.

Only the *history* of engineering seems to have found its subject. And, even there, the work of Edward Constant, Edwin Layton, David McCollough, David Noble, Charles Perrow, Bruce Seeley, Eugene Ferguson, Walter Vincenti, and a few others remains exploratory. In general, the manager, the inventor, and the politician, together with what engineers make, tend to cast a shadow over the engineers with whom they work.

Why is that? Why do philosophers, sociologists, and even historians give so little attention to engineers? Some explanations can be rejected out of hand. The explanation cannot be that engineers are too rare to be worth study. Excluding those engineers, computer scientists, mathematicians, medical researchers, psychologists, and the like who are mere technicians, there are in the United States today, or at least seem to be, more engineers than natural scientists. That is true even after controlling for advanced degrees.

Nor can the explanation be that what engineers do is unimportant compared to what scientists do. What engineers do may (almost by definition) be less important for *theoretical* knowledge than what scientists do, but what engineers do for *everyday* life is more important. Like accountants, lawyers, nurses, physicians, and psychiatrists, engineers belong to an occupation that typically seeks to do rather than to know (seeking to know only so that they may do).

Nor does the "mere" *practical* importance of engineering explain why philosophers, historians, and sociologists have so long neglected engineers. They have devoted much more work to other practical occupations, including law, medicine, and even psychiatry. Nor can this neglect of engineering be explained by any simple prejudice of the liberal arts against engineering's (supposed) origin in the manual arts, its continuing involvement in the dirt and sweat of factories, mines, and construction, or even its isolation in a college, school, or institute separate from the liberal arts. On the one hand, until recently, philosophers have given more attention to the *ethics* of engineering than to the ethics of science. Where was the prejudice of the liberal arts then? On the other hand, sociologists of science have recently begun to study the scientific counterpart of engineering's tacit knowledge, use of models, and reliance on the physical world (for example, "the golden hands" of some researchers). Historians of science now ask about the place of drawings, graphs, and physical surroundings in the thinking of "scientists" (some of whom—like Leonardo da Vinci—might better be counted as engineers than scientists if counted as either).

Still, even in technological universities like my own, even in rare courses like History of Engineering, the engineers tend to disappear into their products or to be mixed up with their employers or the scientists, managers, inventors, technicians, and manual laborers with whom they work.

While the present state of the philosophy of engineering may not entirely explain the present state of the history or sociology of engineering, it seems to be an obvious contributing factor. Certainly, the present state of the philosophy of engineering seems to be affecting how scholars, policymakers, and even engineers think about engineering. For example, historians, sociologists, and engineers, as well as philosophers, seem to disagree about when to start the history of engineering—whether with the invention of the first tool (the beginning of technology), the construction of Egypt's pyramids, England's Stonehenge, or another impressive ancient building (the beginning of large technological systems), the founding of the first *corps de génie* in France in 1677, the rise of science-based industry in the 1800s, or another event or period. If we consider the sort of things engineers have "always" created, we tend to date the beginning of engineering quite early, for example, in the first age of tools, even before *homo sapiens* loped across Africa's savannahs. If instead we look for people performing much the same social function as today's engineers, we tend to start much later (1500 BC, perhaps), with the experts who managed sieges and designed, built, and maintained fortifications. If we look for people *called* "engineers," we tend to emphasize "industrial lore," "rules of thumb," and other "hands-on knowledge" passed from one generation of engineers to the next. We then have trouble distinguishing engineers from the millwrights, carpenters, mechanics, electricians, plumbers, draftsmen, foremen, and other "technicians" with whom engineers share much implicit knowledge, history, and outlook. We begin the story of engineering with England's first industrial revolution sometime in the 1700s. However, if we emphasize modern (scientific) training, we must start the story in the late 1600s and then have trouble distinguishing engineers from the chemists, mathematicians, and physicists with whom they mixed. And so on.

What we emphasize when writing the history of engineering may, in turn, affect what we think is necessary for an engineer's education and, indeed, whom we want to count as an engineer. The present undeveloped state of the philosophy of engineering may, for example, explain (at least in part) why different agencies of the U.S. government give substantially different tallies of engineers. For example, according to the U.S. National Science Board (2016), there were about 1,600,000 college graduates employed in engineering occupations in 2013. In contrast, the US Census for 2015 counted 2,129,782 "engineers," almost a third more than the Science Board counted only a year or two later. Were there approximately 2.1 million engineers in the United States in 2015, approximately 1.6 million, or some other number? How are we to tell?

The philosophy of engineering might also help us answer such questions as whether "software engineering," "genetic engineering," "geo-engineering," "climate engineering," "financial engineering," "re-engineering," or "social engineering" is (or is not) engineering (strictly speaking)—and why the answer to that question might matter for social policy and for understanding engineering.

Our ignorance of engineering is not primarily empirical. We already know much about engineering (dates of inventions, educational programs at various times, degrees handed out each year, issues debated within societies consisting of engineers so called, and so on). Our ignorance is, it seems, primarily conceptual. We do not know what to do with many of the facts we already have. For example, we know how many students in the U.S. graduate with a BS in marine engineering each year and how many sailors are licensed as marine engineers. But we do not know how many of these are engineers (strictly so called). We are rather vague even about what facts we would need to collect to answer that question. In short, we lack even the rudiments of what the philosophy of engineering should provide.

2. FOUNDING THE PHILOSOPHY OF ENGINEERING

That, anyway, is how matters seemed to stand when a small group of engineers, philosophers, and scientists met at the Massachusetts Institute of Technology (MIT) on October 19-20, 2006, to establish what soon became the Workshop on Philosophy and Engineering (WPE). Among those present were Stephanie J. Bird (Neuroscience, MIT), Michael Davis (Philosophy, IIT), David Goldberg (Entrepreneurial Engineering, University of Illinois, Urban-Champaign), Billy Koen (Mechanical Engineering, University of Texas-Austin), Carl Mitcham (Philosophy, Colorado School of Mines), Steven Nichols (Mechanical Engineering, University of Texas-Austin), and Ibo van de Poel (Philosophy, Delft University of Technology). Taft Broome (Civil Engineering, Howard University), who was then visiting MIT for the year while on sabbatical leave, led the effort; Joel Moses (Computer Science, MIT), managed arrangements and found the money to pay for them. The rest of us were invitees who had come when called. We arrived with various rough ideas about philosophy of engineering. We left with a plan to start an organization of philosophers, engineers, and social scientists to continue the discussion we had begun.[2]

WPE held one successful conference at the TUDelft the next year and another at the Royal Academy of Engineering, London, 2008.[3] Those successes brought in scholars with a somewhat different understanding of the field, one in which technology had a larger place. In 2009, WPE became the Forum on Philosophy, Engineering, and Technology (fPET) and, in 2010, held its first biennial meeting for "philosophers, engineers, and other researchers and practitioners related to both fields" (that is, philosophy of technology and philosophy of engineering) at the Colorado School of Mines.[4] Later meetings have so far been held at: the Graduate University of the Chinese Academy of Sciences, Beijing, in 2012; Virginia Tech, Blacksburg, in 2014; Friedrich-Alexander University, Erlangen-Nuremberg, in 2016; and the University of Maryland, College Park, in 2018. The Covit19 epidemic postponed the meeting originally planned for the Universidad de Valparaíso, Valparaíso, Chile, in 2020. Almost from its inception, then, the philosophy of engineering was an undertaking at once interdisciplinary and international.

From these meetings flowed several volumes of collected papers, some published in the Springer series "Philosophy of Engineering and Technology" (see, especially, Murphy et al., 2015; Christensen, et al., 2015; and Mitchelfelder et al., 2017). Meanwhile, *Science and Engineering Ethics* has published an increasing number of contributions to the philosophy of engineering. Meanwhile, Gary Downey continued to publish an interdisciplinary journal *Engineering Studies*, the first volume of which appeared in 2000. Its mission is in part to "1. to advance critical analysis in historical, social, cultural, political, philosophical, rhetorical, and organizational studies of engineers and engineering."[5]

There is, then, reason to think that the philosophy of engineering may be stepping from the twin shadows cast by engineering ethics and the philosophy of technology. Yet, it also seems clear from what has so far been published that the philosophy of engineering is still far from reaching the common understanding it pursues. There are, I think, at least two reasons for that:

First, philosophers still do not understand engineering well enough to guide sociology and history to the study of engineering proper. In large part at least that is because "we" (philosophers, sociologists, historians, and even engineers) have not properly organized what we already know about engineering as a social and intellectual enterprise. We need to identify areas about which the philosophy of engineering still lacks understanding and then figure out how much of that lack of understanding is due to conceptual confusion, how much is due to lack of information and, of course, how much is due to both. We need to develop a research agenda for the philosophy of engineering. That research agenda should seek answers to such questions as those this book will focus on, especially: What is a global profession? What counts as evidence that engineering is a global profession? How do we get that evidence? What, if anything, should count as decisive proof?

Second, philosophers lack a fruitful research agenda because they (most at least) have, I think, been drawing on the wrong models for their agenda. The philosophy of engineering has typically been conceived in one of two ways, either (1) as the attempt to understand engineering as an intellectual enterprise on the model of science (whichever model of science is chosen) or (2) as the attempt to understand engineering as a social enterprise on the model of technology (whichever model of technology is chosen). The attraction to those ways of conceiving of philosophy of engineering is understandable for at least two reasons. First, would-be philosophers of engineering have generally come from philosophy of science or philosophy of technology. They naturally brought their favored model of science or technology with them. Second, at least one of those models should have proved fruitful if engineering were (more or less) just a special sort of science (like anthropology or chemistry) or an activity defined by what it produces (that is, if engineering could be defined as *the* producer of technology of a certain sort, for example, the producer of complex systems). But if, as I shall argue here, engineering is best understood as a distinct *profession*, defined—as other professions are— primarily by the special (historically contingent) way it sets tasks and tries to accomplish them, the best place to look for a helpful model of philosophy of engineering would seem to be the philosophy of professions, especially the most developed of those philosophies, the philosophy of law, the field in which I began to work in the 1970s.

The philosophy of law has traditionally attempted to understand law as a reasonable undertaking. The philosophy of law (or, at least, the part relevant here) focuses on what the legal profession (including judges) thinks or does, not on law as abstract knowledge. Until the 1950s, philosophy of law in the English-speaking world was primarily the domain of law professors—and was known as "jurisprudence." Then, in 1961, H. L. A. Hart published *The Concept of Law*, a work at once philosophically sophisticated and very much engaged with legal practice. Since Hart was an "amphibian," a lawyer by training and a philosopher by adoption, his work opened jurisprudence to other philosophers, especially those with legal training. Hart attempted to revive legal positivism (the separation of law and morality) as the best way to understand law. Other philosophers either soon attempted to defend other theories of law against Hart's criticism or attempted to revise their preferred theory of law (especially, natural law, that is, law as including morality in one way or another) as Hart had for legal positivism. By the end of the 1970s, philosophers or philosopher-lawyer amphibians had published many important works in philosophy of law—on causation in the law, professional responsibility, legal responsibility, legal obligation, contract, the relation of law and morality, and so on.

The only other philosophy-of as developed as philosophy of law, the philosophy of medicine, is (strictly speaking) not a philosophy of a profession. It has tended to ignore the individual medical professions in favor of understanding "bioethics" as a social or intellectual enterprise larger than a profession and quite different, one that includes osteopaths, nurses, hospital administrators, medical researchers, biomedical engineers, and so on, without distinction. In general, bioethics ignores differences in education, academic degrees, ethical codes, and licensing when analyzing problems of "medical ethics."

How far can a model drawn from the philosophy of law take the philosophy of engineering? We shall see.

3. CONTENTS OF THIS BOOK

Beginning with a definition of engineering as a profession to be found in *Thinking Like an Engineer* (Oxford, 1998) and *Profession, Code, and Ethics* (Ashgate, 2002), where I amend the definition, this book concentrates on a question that the last two decades seems to have made critical: Is engineering one *global* profession (like medical doctors) or many analogous national or regional professions (like lawyers)? This book brings together, organizes, updates, and edits to reduce redundancy, nineteen articles and chapters published since 2001, including one article hard for English-speaking philosophers to access because it was published in Chinese, and several other publications easy for philosophers to miss because they are unlikely to be included in an index that a philosopher would search. The book also includes this Preface and an Epilogue, both new. The Epilogue sketches a research agenda this book's argument suggests.

Engineering as a Global Profession has four main parts (along with the Preface, Epilogue, and Index). Part I's first chapter defends a still-controversial definition of "profession" after explaining why defining "profession" is a philosophical problem. Chapter 2 offers an unusual definition of "engineering" (engineering as a profession in the still-controversial sense of "profession" defended in chapter 1). The four succeeding chapters then test these two definitions by using them to distinguish engineers from (respectively) architects, chemists, software engineers, and technical managers. Part II explains how to use "profession" (as defined in Part I) to study engineering (strictly so called), disposes of some methodological problems, and explains what would have to be true if engineering is (more or less) the same profession around the globe. It is up to history or sociology (certain scholars following certain procedures) to confirm or disconfirm these "truths." Part III provides an analysis of the professional responsibility of engineers so defined, distinguishes engineering's professional responsibility from "social responsibility," considers the content and justification of engineering's professional responsibilities, and then disposes of some criticisms of how those responsibilities are taught. Part IV applies the method of profession to several important examples of professional responsibility—including nuclear power, the VW diesel scandal, and global arms sales. The examples illustrate the way individual recommendations of engineers are, or at least should be, shaped by engineering's status as a global profession. While science and technology studies (STS) have increasingly taken an "empirical turn," much of STS research is still so unclear about the *professional* responsibility of engineers that it tends to avoid the subject, leaving engineering ethics without the empirical research needed to teach it better. I therefore hope that this book will suggest ways in which philosophers, social scientists, and others studying engineering (as a social institution) might improve their empirical research.

The Epilogue both picks up the discussion of empirical research where Part IV ended and takes it in a different direction. The method of reflection used in Part IV is particularly suited to philosophers. It is the sort of research one can perform in an armchair. The data is already present in newspapers, one's own experience, online, in every-day conversation, and so on. In contrast, the research discussed in the Epilogue cannot be performed in an armchair (even metaphorically). It requires going into the "field" to collect data, whether by surveys, interviews, or observation (methods typical of social science rather than philosophy) or by examining old textbooks or the contents of archives (a method typical of history rather than philosophy), and so on. Of course, once the data has been collected, the researcher must return to an armchair (or desk) to consider what the data means. The Epilogue's examples of empirical research by philosophers have two purposes. The first is to give examples of research social scientists should be doing. The other purpose is to show philosophers of engineering what they can do until the social scientists take over.

4. ACKNOWLEDGMENTS

This book is more than the sum of its parts. The chapters have been arranged to make a single argument for the claim that engineering is a global profession. All the chapters (except the new Epilogue) have been revised at least a little to standardize terminology, avoid unnecessary repetition, update arguments, and otherwise turn a collection of independent papers published over two decades into one coherent, timely argument. I have also corrected mistakes when I have noticed them. Nonetheless, the chapters remain close descendants of the articles that preceded them.

Chicago, Illinois
October 2020

NOTES

1. For an early important example of philosophy of engineering (though the work of a mechanical engineer), see Koen, 1985.
2. My thanks to Carl Mitcham for providing a planning document he had "squirreled away" and to Stephanie Bird for letting me compare my memory of events with hers. Bo Pong Li, though not present at this meeting, was a presence there because of his recently published *Introduction to the Philosophy of Engineering: I Create, Therefore I Am* (2002). Though available only in Chinese at that time, his book seemed (to those who knew of it) to signal the growing importance of philosophy of engineering outside Europe and the Americas.
3. Royal Academy (2010) and Royal Academy (2011) include papers from the WPE meetings and are available for download at http://www.raeng.org.uk/policy/engineering-ethics/philosophy (Accessed July 1, 2018). Oosterlaken (2010) reviewed the volume that came out of the 2007 meeting.
4. The official announcement for that first conference was still posted at http://www.fpet2016.org/home (accessed May 15, 2020).
5. https://www.tandfonline.com/action/journalInformation?show=aimsScope&journalCode=test20 (accessed June 1, 2020).

REFERENCES

Bella, David A. "Organizations and Systematic Distortion of Information," *Journal of Professional Issues in Engineering* 113 (October 1987): 360–370.

Broome, Jr., Taft H. "Engineering the Philosophy of Science," *Metaphilosophy* 16 (January 1985): 47–56.

Christensen, Steen Hyldgaard, et al., editors, *Engineering, Development and Philosophy: American, Chinese, and European Perspectives* (Springer Science + Business Media B.V, 2015).

Durbin, Paul T. "Toward a Philosophy of Engineering and Science in R & D Settings," in Paul Durbin, ed., *Technology and Responsibility* (Reidel: Dortrecht-Holland, 1987), pp. 309–327.

Hacker, Sally, *Doing it the Hard Way* (Unwin: Boston, 1990).

Kidder, Tracy, *The Soul of a Machine* (Little, Brown, and Company: Boston, 1981).

Koen, Billy Vaughn, *Definition of the Engineering Method* (American Society of Engineering Education: Washington, DC, 1985).

McCullough, David, *The Great Bridge: The Epic Story of the Building of the Brooklyn Bridge*. (Simon&Schuster: New York, New York, 1972).

Michelfelder, Diane, Byron Newberry, Qin Zhu, editors, *Philosophy and Engineering: Exploring Boundaries, Expanding Connections*, Springer, 2017.

Mitcham, Carl, *Thinking through Technology: The Path between Engineering and Philosophy* (University of Chicago Press: Chicago, 1994).

Murphy, Colleen, et al., editors, *Engineering Ethics for a Globalized World* (Springer, 2015).

National Science Board, "Chapter 3: Science and Engineering Labor Force." *Science and Engineering Indicators 2016*, 3–5. https://nsf.gov/statistics/2016/nsb20161/#/report/chapter-3/highlights

Oosterlaken, Ilse (2010) "Ibo van de Poel and David E. Goldberg (eds), *Philosophy and Engineering. An Emerging Agenda*, Springer, The Netherlands, 2007," *Science and Engineering Ethics*.

Royal Academy of Engineering, *Philosophy of Engineering: Volume 1 of the proceedings of a series of seminars held at The Royal Academy of Engineering* (London, June 2010).

Royal Academy of Engineering, *Philosophy of Engineering: Volume 2 of the proceedings of a series of seminars held at The Royal Academy of Engineering* (London, October, 2011).

Sinclair, George, "A Call for a Philosophy of Engineering," *Technology and Culture* 18 (October 1977): 685–689.

United States Census. 2015. "Full-time, Year-Round workers and Median Earnings in the Past 12 Months by Sex and Detailed Occupation: 2015." https://www.census.gov/people/io/files/Median_earnings_2015_final.xlsx

Vincenti, Walter G. *What Engineers Know and How They Know It* (Johns Hopkins University Press: Baltimore, 1990).

Part I

DISTINGUISHING ENGINEERING FROM OTHER PROFESSIONS

Chapter 1

Profession

The purpose of this chapter is to present a definition (interpretation, understanding, or conception) of the term "profession" useful for the study of professions. I regard the question "What is a profession?" as concerned with a fact—but, I admit, an odd fact. Unlike "dog" or "barking," "profession" is not a simple descriptive term. Like "law," "democracy," and indeed "engineering," "profession" is a term connected with institutions that interpret it and are, in turn, shaped by those interpretations. Right now, engineering's status as a profession seems to be disputed in several important countries, including France and Japan.[1] What counts as a profession may affect engineering in those countries, for example, by helping to decide whether a course in professional ethics will be required for the first degree or licensing made a condition of practice. Providing the appropriate conception of "profession" belongs not to lexicography or language analysis but to political philosophy (in the broad sense that includes legal and social philosophy). Lexicography and language analysis can only give us the concept, the most general guide to how to use a word properly. For most purposes, we need a more specific guide, one that abandons certain (proper) senses of the word to have a tool more useful for one purpose. We need what is now commonly called "a conception" (rather than the concept).

1. OCCUPATION, DISCIPLINE, AND PROFESSION

"Profession" has several senses in English. It can be a mere synonym for "occupation"—an occupation being any typically full-time activity defined in part by a discipline by which one can (and people typically do) earn a living. (A discipline is an easily recognizable body of knowledge, skill, and judgment useful for a certain activity.) It is in this sense that we may, without irony or metaphor, speak of someone being a "professional athlete," "professional beggar," or "professional thief," provided the person in question has mastered the relevant *discipline* well enough to make a living by it.

While the *discipline* of engineering has not existed throughout history, its roots certainly go back several centuries—disappearing into the older disciplines of stonemason, siege master, chief builder, and so on. The *occupation* of engineering, on the other hand, did not exist before the nineteenth century. Before then, almost everyone called "engineer" in the sense that interests us, that is, almost everyone sharing the discipline of engineers (mathematics, physics, chemistry, mechanical drawing, and so on)—was a military officer. Being "gentlemen," officers did not work for a living (though we moderns might think otherwise when we look at how they spent their time). While they may have received an income as an officer, they fulfilled (or at least were supposed to fulfill) the duties of their office not because they were paid for it but because doing so was proper whatever money they received for it. Those who claim that every civilization has had engineers (in the sense that interests us) must explain why the history of the word "engineer" is the history of the discipline that corresponds to that French word (and French curriculum).[2] The occupation follows the discipline.

I shall not use "profession" in this broad sense (occupation) or even in the somewhat more limited sense (also common in English) of honest occupation: "Plumbing is a profession; prostitution is not." Our concern is "profession" in the sense in which engineers say, for example, "Engineering is a profession; plumbing is not." Our concern is a special kind of honest occupation, one that we can compare with other similar occupations (architecture, law, medicine, and so on).

There are at least three approaches to conceptualizing profession in this special-kind-of-honest-occupation sense. One, what we may call "the sociological," has its origin in the social sciences. Its language tends to be statistical. The statement of the conception, a definition of sorts, does not claim to give necessary or sufficient conditions for some occupation to be a profession but merely to state what is true of "most professions," "the most important professions," "the most developed professions," or the like. Every sociologist concerned with professions seems to have a list of professions that the definition must capture. Law and medicine are always on the list; the clergy, often; and other occupations commonly acknowledged as professions, such as engineering, sometimes. The arguments for inclusion or exclusion tend to be quite weak.[3]

We may distinguish three traditions in the sociology of professions (what we may call): the economic, the political, and the anthropological. Though individual sociologists often mix their elements, distinguishing them as "ideal types" should help us to think about them more clearly, even in their less ideal (i.e., mixed) forms. What I believe to be wrong with all three ideal types, a failure to understand how central ethics is to profession, remains even when the types mix.

The economic tradition interprets professions as primarily a means of controlling market forces for the benefit of the professionals themselves, that is, as a form of monopoly, guild, or labor union. The economic tradition has at least two branches: Marxist and free market. Among recent sociologists in the Marxist tradition, the best is still Magali Sarfatti Larson (*The Rise of Professionalism*, 1977); among sociologists in the free-market tradition, Andrew Abbott (*The System of Professions*, 1988) is a good example. For sociologists in the economic tradition (whether Marxist or free market), it is the would-be members of a profession who, by acting together under favorable conditions, create their monopoly (or, at least, some approximation of one). Successful professions have high income, workplace autonomy, control of who can join, high social status, and so on; less successful professions lack some or most of these powers (more or less). Morality, if relevant at all, is relevant merely as a means to monopoly, a way of making a "trademark" (the profession's name) more attractive to potential employers. The success in question may be independent of what participants in events sought. The economic tradition delights in discovering "the invisible hand" at work, for example, attempts to serve one's own interest that in fact serve the public interest instead. Like the monopoly itself, signs of a profession's success may be embedded in law but need not be. What matters for the economic tradition is market arrangements ("economic realities"), not (mere) law.

For the political tradition, however, the law is crucial. Often associated with Max Weber, the political tradition interprets profession as primarily a legal condition, a matter of (reasonably effective) laws that set standards of (advanced) education, require a license to practice, and impose discipline upon practitioners through formal (governmental) structures. "Professional ethics"—and, indeed, even ordinary moral standards—are, if distinguished at all, treated as just another form of regulation. To be a profession (in this sense) is to be an occupation bureaucratized in a certain way. For the political tradition, it is the society (government) that creates professions out of occupations, and the society (the public) that benefits (whoever else may benefit as well). The political tradition substitutes society's very visible hands for the invisible hand of economics. The members of a profession have little or no part in making their profession. A recent work in this tradition is Robert Zussman's *Mechanics of the Middle Class* (1985).

The third tradition, the anthropological, is often associated with Emile Durkheim or Talcott Parsons. It interprets professions as primarily cultural facts, the natural expression of a certain social function under certain conditions. Neither the professionals nor society can have much to say about whether a certain occupation is a profession. Professions are a function of special knowledge used in a certain way, a community created by a common occupation requiring advanced study. Its ethics are as much a natural product of that community as anything else about it. Among recent sociologists, the best of those working in the anthropological tradition seems to be Eliot Freidson (in, e.g., *Professionalism: The Third Logic*, 2001).[4]

Distinguishing these three traditions helps make the point that the sociological approach has yet to yield a single definition of profession and, more importantly, is unlikely to. Sociology's way of developing definitions, that is, abstracting from a (short) list of undisputed cases something common to most or all, is unlikely to yield a single conception—or, at least, is unlikely to yield one until sociologists agree on a list of cases sufficiently long to exclude most candidate definitions. Today, only two professions appear on all sociological lists (law and medicine). That is much too few to derive a definition both widely accepted and narrow enough to be useful. Whatever the utility of a particular sociological definition for a particular line of social research, no such definition is likely to seem definitive to more than a minority of sociologists. Why sociologists continue to generate definitions as they have is an intriguing question, but one best left to the history (or sociology) of sociology. We may ignore it here.[5]

What we cannot ignore is that few, if any, of these definitions would rule out an immoral profession, for example, a profession of torturers. Let us assume that there is enough employment for torturers to form an occupation. (Whatever our moral objections to torture, we must admit that torturers might be useful enough to make a living—much as prostitutes and dealers in illicit drugs are.) Nothing in the *economic* conception of profession as such rules out certain persons winning a monopoly over torture—with resulting high income, workplace autonomy, control of who can join them, and so on. Similarly, nothing in the *political* conception as such rules out laws requiring torturers to be educated in certain ways, to pass certain tests, to be licensed after meeting certain conditions, and to be subject to having their license revoked should they prove incompetent, careless, or otherwise unsatisfactory. Last, there is nothing in the *anthropological* conception as such to rule out the special knowledge of how to torture that defines an occupational community, a profession of torturers. Because there is nothing in the sociological approach as such to require professions to be moral undertakings, there is nothing in it to rule out a profession of torturers. Individual sociologists are, of course, free to define profession to exclude torturers (since none of the usual lists of undisputed cases includes any profession that routinely torturers). But sociologists are equally free to define professions as requiring a doctorate—because law, medicine, and other professions on a typical list of undisputed cases require a doctorate to practice.

The sociological approach offers a wilderness of possibilities, but little help choosing among them. So, for example, some sociologists have equated professions with consulting occupations (sometimes also called "free professions" or "liberal professions"), excluding from professional status (or at least from "full professional status") most engineers, journalists, nurses, teachers, professors, and others who work as employees in large organizations (groups that have not only long claimed to be professions but have been accepted as such by physicians, lawyers, and others that these sociologist recognized as belonging to "true professions"). When physicians and lawyers themselves recently began to be absorbed into large organizations in the United States, much was written about their "de-professionalizing," though those professions otherwise continued much as before. That, I think, is enough to make clear how unattractive the sociological approach (and the resulting definitions) should be, even though that approach (unfortunately) continues to dominate discussion of what professions are—their *claim* to rest on empirical methods trumping the obvious fact that they do not.

2. PHILOSOPHICAL APPROACHES

Two other approaches that conceptualize profession are philosophical. They offer necessary and sufficient conditions for an occupation to count as a profession. While a philosophical conception (i.e., a definition resulting from applying a philosophical approach) may leave the status of a few would-be professions unsettled, we should at least be able to use it to explain (in a satisfying way) why those few are neither clearly professions nor clearly not professions. Philosophical conceptions are sensitive to counterexample in a way that sociological conceptions are not. Philosophers cannot use the standard defense of sociologists confronted with a counterexample: "I said 'most', not 'all'."

One of the philosophical approaches to conceptualizing profession (what I shall call), the Cartesian, answers the question, "What do *I* think a profession is?" It attempts to piece together in a coherent way the contents of one person's mind. There may be as many Cartesian conceptions of profession as there are people who ask themselves what they mean by "profession." The Cartesian approach has no procedure for decisively mediating between one individual's definition and another's. (That, indeed, is one reason to call this approach Cartesian: its tendency to be solipsistic.) The differences between Cartesian definitions can be startling. For example, some are as indifferent to morality as any sociological definition. My favorite claims that the *mafia* is a profession (Sanders, 1993).[6]

The other philosophical approach to conceptualizing profession is (more or less) Socratic. It answers the question, "What do *we*—philosophers and (self-described) professionals—('really') think a profession is?" Such a conception must be worked out through a conversation, a uniting of Cartesian *I*'s into a public *we* (i.e., through something like a typical Socratic dialogue). A member of a profession (so called) says what she means by "profession." Philosophers, or other members of her profession or another, test the definition with counter-examples, consider the consequences of adopting the definition, and otherwise examine it in the way philosophers typically do. Any problem so discovered should be fixed by revising the definition in a way that seems to fix the problem. The definition is again examined. And so the process continues until everyone participating in the conversation is satisfied that no problem remains. It is this critical conversation that underwrites the claim that the resulting definition is "what we *really* think a profession is" (i.e., what we think it is after enough thinking of the right sort). The definition is "proved" only so long as no one, whether original participant or new arrival, dissents anew.

In principle, either philosophical approach might yield a definition much like one that a sociological approach has. The definition yielded by an approach depends in part on the approach but in part too on the context to which it is applied. If, for example, professional status were a statistical phenomenon, then even a philosophical method could not (without error) generate anything stronger than a statistical definition of "profession." The advantage of both philosophical approaches is that they should find the necessary and sufficient conditions *if they exist*. The philosophical approaches do not necessarily settle for a merely statistical definition. The Socratic approach has an additional advantage. It automatically treats a profession as a self-conscious institution. Since it includes professionals as equal members in the inquiry, it allows "us" (philosophers as well as professionals) to understand professions from the "inside," that is, as the professionals themselves understand them. For sociologists, what members of a profession think about their profession, or professions in general, is at best another datum, rather as the zoologist understands the leopard's roar than like the opinion of a fellow human with direct access to the practice under study. Cartesian philosophers are, in this respect, much like sociologists. There is no necessary connection between the definition a Cartesian philosopher might reach and what those it covers (the professionals) think. For a practitioner of the Socratic approach, on the other hand, the perspective of members of this or that profession is crucial (though no more crucial than the philosophers' perspective). The conversation between philosophers and professionals does not end until the professionals as well as the philosophers are satisfied with the definition.

The conversation need not, however, end with a definition that all the groups at first admitted to be "professions." The conversation may lead some participants to withdraw their claim to belong to a profession. In my experience, MBAs drop their claim for professional status as the definition moves away from the sociological and they begin to appreciate what the physicians, lawyers, engineers, nurses, and so on have in mind. There is nothing canonical about the original list of professions (as there is in the sociological approach). The Socratic approach nonetheless provides a procedure for resolving disputes, something neither the sociological nor the Cartesian does. Individual insights must be incorporated into a single definition on which everyone agrees. The Socratic approach concludes only when there is no live alternative to its preferred definition, a definition that necessarily excludes individual mistakes and even widespread but indefensible prejudices (such as, e.g., that MBAs must be professionals because they hold an advanced degree or that masons cannot be because their education typically stops before college). In this respect, the resulting definition is a product of reason rather than individual or social psychology (what people happen to think).

After many years of applying this method, I have reached the following definition: *A profession is a number of individuals in the same occupation voluntarily organized to earn a living by openly serving a moral ideal in a morally permissible way beyond what law, market, morality, and public opinion would otherwise require.* For convenience, I shall hereafter refer to this definition as "the Socratic conception" (or "the Socratic definition") though it is technically only a conception (or definition) developed using the Socratic approach and the identical conception might have been developed using the sociological or Cartesian approach.

3. THE SOCRATIC CONCEPTION EXPLAINED

My purpose in this section is to explain the Socratic conception enough to forestall otherwise likely objections—to provide what amounts to a partial proof of the definition, one sufficient to satisfy most philosophers and professionals that this book might be on the right track. I will leave it to the reader to appreciate most of the differences between this definition and the sociological ones discussed above, but I shall point out a few particularly striking differences as I explain the definition.

I can offer only a partial proof here because a full proof would, in part, include the conversations that led to the conception (something space forbids me to reproduce here even if I had kept a record) but in part too because a full proof must include not only the reaction of living philosophers and members of the professions to the conception once stated but also the reaction of those who may hereafter encounter the definition when all of us are dead. The proof is always hostage to the future. It stands so long as philosophers and members of the professions, upon fair consideration, say (something like), "Yes, that's it." They cannot do that here. Since philosophy is not a profession (indeed, not even an occupation, most philosophers making their living as professors), philosophers can only object to this conception as university professors (if they consider that occupation a profession) or on the usual technical grounds on which philosophers would object to this or that conception of "law" or "democracy." Their objections would renew the discussion—but they cannot, by themselves, settle the question one way or the other. The Socratic approach does not give philosophers or professionals the last word (though it does give members of the two groups different roles).

According to the Socratic definition, a profession is a group undertaking. A profession is like an army, family, crowd, or other plurality. There can be no profession with just one member (though there can be a last member, as the profession dies out). This is one respect in which members of a profession differ from mere experts, artists, entrepreneurs, or other knowledgeable, skillful, inventive, or judicious people. Such people can be one of a kind.

The group forming a profession must share an occupation (though its members may be a subset of the occupation rather than the whole).[7] Whether the occupants of a certain collection of jobs constitutes one occupation, two, or several is as much a matter of decision as of fact—much as is the amount of hair one must have on one's head to defend against a charge of baldness. To decide, we need to know how similar the skills in question, how much movement between jobs of different descriptions, how similar the work of occupants of different jobs, how different from neighboring occupations is the "occupation" in question, how important the differences are, and so on. There is usually room for argument—and, often, room even for more than one good answer. For example, for the purpose, say, of membership in the Institute for Electrical and Electronic Engineers, computer scientists may count as belonging to the same occupation as electrical engineers. But, for another purpose, say, the study of engineering ethics or curriculum, computer scientists may be too different from electrical engineers (since computer scientists have their own code of ethics and distinct curriculum). Though occupations do have flexible boundaries, their boundaries are not infinitely flexible. Law and medicine will (in all probability) never be one profession; nor will engineering and journalism. The underlying disciplines are just too different.[8]

According to the Socratic definition, the group in question (the would-be profession) must organize to work in a morally permissible way. If there is no morally permissible way to carry on the occupation, there can be no profession. There can, therefore, be no profession of thieves or torturers (since theft and torture are—almost always—morally impermissible). Morality thus limits what can be a profession. Some professions ("professional thief" and "professional torturer") are conceptually impossible. Of course, if morality changes over time, or from country to country, then some professions may be possible in one time or place and not in another.

The moral permissibility of a profession's occupation is one way that, according to the Socratic definition, profession is conceptually connected with morality. There are two others. One concerns "moral ideals." A moral ideal is a state of affairs "everyone" (every reasonable person at her most reasonable) recognizes as a significant good. (That the state of affairs in question is a good is shown by everyone's wanting it to be (at least everyone at her most reasonable). Its status as a *significant* good is shown by everyone's willingness to help realize it in at least minor ways if others do the same. For most professions, stating the distinctive moral ideal (roughly) is easy: physicians have organized to cure the sick, comfort the dying, and protect the healthy from disease; lawyers, to help people obtain justice within the law; accountants, to represent financial information in ways both useful and accurate; architects, to build durable, convenient, and beautiful structures; and so on. Health, a comfortable death, justice within the law, accurate financial information, beauty, and the like are goods we all recognize as significant. The moral ideal engineering serves is also easy to state (roughly): the design, construction, maintenance, improvement, and disposal of safe, efficient, and useful physical systems (or, in brief, the material progress of humanity).

"Moral ideal" is a term of art. Though "ideal" has its usual sense (a good state of affairs hard to achieve fully but worth approaching even a little), "moral" does not. We may, I think, understand morality as consisting of those standards of conduct every reasonable person (at his or her most reasonable) wants everyone else to follow even if that means having to do the same. The "moral" in "moral ideal" resembles morality so defined insofar as it involves something every reasonable person wants enough to do something for it (help, endorse, or at least allow it) if enough others do the same. It differs from morality only in not involving a standard of conduct as such but merely an outcome (important enough to be worth doing something for it). The analogy is, I think, close enough to justify the term "moral," especially since such ideals routinely have an important place in moral discussions (as other ideals, such as the perfect chess opening or perfect orchid, do not). But, for those who think otherwise, I am happy to let them substitute another term. What is important is the conception that "moral ideal" names, not the name itself.

"Moral ideal" is, I should add, not a mere synonym for "public service." The ideals I just listed are all easily understood as forms of public service. But some are not. For example, the natural sciences typically seek to "understand nature" (different sciences focusing on different parts of nature). They seek to understand nature without necessarily claiming to serve anyone but other scientists. An understanding of nature is nonetheless a moral ideal if, but only if, *all* of us (at our most reasonable) are interested in understanding nature, even parts of it, such as distant galaxies, the understanding of which does us no good (or, at least, no good beyond satisfying curiosity). That scientists do not seek to serve us all ("the public") is consistent with their in fact serving us all. Not the intentions of scientists but "human nature" (what interests us at our most reasonable) determines whether the ideal that scientists seek to serve is a moral ideal and therefore whether a certain science can be a profession.

Perhaps I can be a morally decent person without actively serving any moral ideal, but an occupation cannot be a profession unless it serves one. A profession serves its chosen moral ideal by setting (and following) appropriate standards for carrying on its occupation that go beyond what law, market, morality, and public opinion would otherwise require.[9] At least one of those standards must be *special*, that is, something not imposed by law, market (ordinary) morality, or public opinion. Otherwise, the occupation (the candidate profession) would remain nothing more than an honest way to earn a living. So, for example, what distinguish the professional soldier from the mere mercenary (however expert and honest) are the special standards of a professional soldier. To be a (good) mercenary, one need only competently carry out the terms of one's (morally permissible) contract of employment, but to be a (good) professional soldier, one must do more, for example, serve one's country honorably even when the contract of employment, ordinary morality, law, and public opinion do not require it.

Some philosophers might object to building morality into a conception of profession. "We should," they might say, "leave the moral status of profession to be settled by argument. Settling a moral question by definition is always a mistake." For those committed to a profession, though, "profession" is a term that carries a moral charge. What those philosophers are suggesting is, a professional might say, like defining "murder" as "an unlawful killing without legal justification or legal excuse" rather than as a violation of the moral rule against killing. The morally neutral definition may be adequate for some purposes (say, criminal prosecution), but it is fundamentally incomplete; it leaves out precisely what distinguishes murder from other sorts of killing (e.g., what explains the legal exceptions). Methodological neutrality would omit a crucial feature of profession. If we want to understand professions "from the inside," we must regard the demand for neutrality as involving a serious mistake in method (a refusal to consider professions from the inside). (For an extensive discussion of this mistake in another context, see Davis, 1983.)

The third way that professions connect with morality is that their special standards are *morally binding* on every member of the profession simply because of that membership. These binding standards are what constitute the profession's essential organization, not (as many sociologists suppose) its learned societies or regulatory agencies. But how is it possible for standards that are (according to the Socratic definition) morally permissible but not otherwise part of ordinary morality to be morally binding on members of a profession? That, I think, is a central question in the philosophy of professions. Here is my answer.

Professions must be "professed" (i.e., declared or claimed). Physicians must declare themselves to be physicians; lawyers must claim to be lawyers; engineers must say they are engineers; and so on. They need not advertise or otherwise *publicly* announce their profession. There is nothing conceptually impossible about a secret profession, for example, a profession of spies (assuming spying can be moral often enough for spies to constitute a morally permissible occupation). But even members of a profession of spies would have to declare their profession to potential clients or employers. Professionals must declare their profession to earn their living by it. (This is, I think, a conceptual truth—assuming certain general truths about us, such as that we cannot often, if ever, read one another's mind.) Members of a profession cannot be hired as members of that profession—say, as chemical engineers—unless potential employers know that they are "chemical engineers" (in the special-standards sense). They cannot, that is, be hired as a chemical engineer if they can only (truthfully) claim to know a lot about chemical plants, to have earned a living by designing, managing, or overseeing certain chemical plants for several years, and to be good at it. If chemical engineers have a good reputation for what they do, the (truthful) declaration of membership in that profession ("I am a chemical engineer") will aid them in earning a living by that profession. They will find appropriate employment. If, however, their profession has a bad reputation (or none), their declaration of membership will be a disadvantage (or, at least, no advantage). Compare, for example, our response to the declaration, "I am a chemical engineer," with our response to "I am an alchemist." Chemical engineers have proved themselves in a way alchemists—and, indeed, even chemists) have not.

Where members of a profession freely declare their membership, the profession's way of pursuing its moral ideal will be a voluntary practice. The members of the profession will be members because they were entitled to be, wished to be, and spoke up accordingly. They may cease to be members simply by ceasing to claim membership.

In general, members of an occupation free to declare membership in the corresponding profession will declare it if, but only if, the declaration seems likely to benefit them (i.e., serve at least one purpose of their own at what seems a reasonable cost). The purpose need not be self-interested, though it often is. There is nothing to prevent some, or even all, members of a profession entering it, for example, simply to help others in a certain way.

If hired (in part) because they declared their membership, members of a profession will be in position to have the benefits of the profession, employment as a member, because the employer sought such-and-such and they identified themselves as one. They will also be in position to take advantage of the practice by doing less than the standards of the practice require, even though the expectation was that they would at least do what the standards require (because they declared the appropriate profession).[10] In this respect at least, the practice will be cooperative. If cheating consists in violating the rules of a voluntary, morally permissible, cooperative practice, then every member of a profession is—because of that membership—in a position to cheat. Since, all else equal, cheating is morally wrong, every member of a profession has a moral obligation, all else equal, to do as the special standards of the profession require. The professional standards are morally binding much as a promise is.

I am not, I should add, describing the psychology of any individual member of the profession. A particular member of the profession may not know what moral ideal the profession serves. She may not understand the purpose of the discipline she has mastered (though it is hard to practice a complex discipline competently without understanding its purpose). She may have entered the profession because, and only because, "you make good money doing this." She may have no idea why the pay is good—or even care. She has nevertheless undertaken the obligations that go with the profession. If she fails to serve the ideal in question in the appropriate way, she should be disciplined (i.e., brought to understand what the profession requires of her). If she cannot be brought to understand what is expected of her (or, at least, to act as if she does), she may, and indeed should, be expelled from the profession. She is a threat to the profession's reputation—and to the long-term benefits of membership. She is a "free rider." The expulsion is not a punishment but an attempt to maintain discipline within the profession, the very discipline that generates the benefits she takes without doing her share to maintain the benefit-generating discipline.

In licensed professions, expulsion is simply a matter of withdrawing the license. But in unlicensed professions, such as journalism or university teaching, expulsion is more subtle. Members of the profession must cease to treat "expelled" members as members, for example, by refusing to work with them or to write letters of reference for them.

An occupation "professionalizes" by organizing as a profession, that is, by adopting special standards of the right sort; it "de-professionalizes" (ceases to be a profession) by abandoning those standards (without replacing them with something similar). "Professionalism" is (strictly speaking) simply acting as the standards of the (relevant) profession require. To be a "professional" (or "a real pro") is to be a member (in good standing) of the profession in question—or (by analogy) to act as if one were (i.e., to act in the way the relevant standards require or, perhaps, should require).

Professional standards are, of course, open to interpretation. Part of being a professional is interpreting the relevant standards in ways that the profession recognizes as competent, for example, interpreting a certain technical standard considering the moral ideal it was designed to serve. Conduct is "unprofessional" if it is inconsistent with the profession's standards (properly interpreted). Since only members of a profession are subject to the profession's standards, only they can violate them. Someone not a member of the profession can be a charlatan, mountebank, impostor, or fraud, but cannot engage in unprofessional conduct.

Another important difference between most conceptions of profession and this one is that, according to this one, *any* occupation can become a profession if (a) what it does is morally permissible and (b) it organizes to serve a moral ideal in a way beyond what law, market, morality, and public opinion require. Professions as such need not be "learned." (Learning, i.e., higher education, is necessary only for competence in some occupations, not all.) In practice, even some morally-permissible learned occupations are not professions. Perhaps some are not because they cannot agree on a moral ideal, but most are not because their members (or at least many of them) only want to do an honest day's work. They are willing to give up the special benefits of profession to avoid its special burdens. In the United States today, this seems to be the reason why, for example, the police, though now largely college educated, do not constitute a profession (or, at least, why police disagree among themselves about whether policing is or should be a profession).

Professional standards may, and generally do, vary from profession to profession. There is no reason why the professional standards of engineers should be the same as those of lawyers—or even of architects. A profession's standards depend, at least in part, on opinion within the profession and therefore change from time to time as opinion changes. A profession's standards are, within wide bounds, contingent. They generally appear in a range of documents, including admission requirements, rules of practice, disciplinary procedures, and the body of knowledge. A profession is organized (successfully) insofar as its special standards are realized in the practice of its members, in what they do and how they evaluate themselves and one another.

One of the documents stating professional standards may be (what is often called) "a code of ethics," that is, a formal statement of the most general rules of practice. Yet, while many definitions of profession require such a code as a condition of being a profession, the Socratic definition offered here does not. That omission is important for engineering's claim to be a global profession. While a formal code of ethics is a central feature of professions in the United States, Canada, Britain, and most other English-speaking countries and has been since early in the twentieth century, few such codes seem to have existed outside English-speaking countries until after World War II. I say "almost" because there certainly seem to have been some codes of professional ethics outside the English-speaking world well before World War II. For our purposes, the most interesting of them is the code of ethics that the engineers of Norway adopted in 1921, the code of ethics of the Chinese Institute of Engineers adopted in 1933 and the code of ethics of the Japanese Society of Civil Engineers adopted in 1938 ("Beliefs and Principles of Practice for Civil Engineers"). Perhaps, if we looked harder, we would find many more such examples (in their original language if not in translation).

4. IS THERE A PROFESSION THERE?

The use of the word "profession" in anything like the special-kind-of-honest-occupation sense discussed here seems to have begun in English-speaking countries only in the last hundred-forty years or so and to have spread elsewhere only in the last seventy. There is, I think, little reason to doubt that "profession" (in the sense discussed here) is an English invention much as the railroad engine and parliamentary democracy are—and, like the railroad engine and parliamentary democracy, has spread to much of the rest of the world. Every new thing must begin somewhere.

Yet, some non-English-speaking countries, though without their own word for profession or a formal code of professional ethics, seem to have entities otherwise like professions in the sense just described. So, requiring a *formal* code for a group to be a profession, or requiring the code to apply to something called a "profession," seems unnecessarily Anglo-centric—as well as prejudging what would otherwise be an interesting empirical question. It is therefore important that the Socratic definition offered here does not require a profession to be called "a profession" or to have a formal code of ethics but instead instructs us in how to determine by empirical research whether a particular occupation is organized in a certain way. What it tells us to look for is the triple connection between occupation and morality just described. It is this complex connection that, according to the Socratic definition (and no other), distinguishes profession from many otherwise similar forms of social organization, such as labor unions, learned societies, and licensed trades.

In many countries lacking formal codes of professional ethics, perhaps in all, technical standards incorporate the same standards a code of ethics would in England, Australia, or the United States, though implicit in details rather than explicit in the more general terms characteristic of a code of ethics. In those countries, the code of ethics *may*, in this sense, be both in writing (scattered across documents) and still "unwritten" (i.e., not formalized as a "code of ethics"). Whether the technical standards of engineers in a certain country in fact serve as an implicit code of ethics in this way will depend on the attitude that engineers there generally take toward those standards (assuming the standards to be morally permissible and designed to serve the same ideal that other engineers serve). If engineers in a country regard technical standards applying to them as (primarily) external impositions, the standards count as law, not as an (implicit) code of ethics (whatever the content). If, however, each engineer (or, at least, most of them after due consideration) regard the standards as what they want every other engineer there to follow even if that would mean having to do the same, that is, as part of a cooperative practice, then (all else equal) the standards do constitute a code of ethics (even if an unusually detailed one and even if enacted into law)—and a profession of engineering exists there.

I have informally carried on such empirical research for more than two decades, mostly by asking questions of engineers or professors of engineering I meet on my travels. I have been left with the impression that some countries, such as the Netherlands, clearly have an engineering profession even if their engineers do not have a formal code of ethics or a term for profession not borrowed from English.[11] I also have the impression that some countries *may* lack an engineering profession. For example, some French engineers I questioned seemed to understand themselves as government agents (even if working for a private employer). They serve "the state" (*l'état*), not some more familiar moral ideal (such as "the public welfare").[12] They understood themselves as bound by law and morality but not by a code of professional ethics (as I understand that term). Indeed, they initially understood "profession" to be a synonym for "occupation" (even when speaking English) and had great trouble understanding what I meant by "professional ethics." They thought I meant the application to engineering of moral theories (what philosophers teach in courses called "Ethics"). These conceptual difficulties notwithstanding, their resolution of practical problems of engineering ethics, including the reasons they gave, seemed to track what engineers in other countries would say. The state, they said, had a "civilizing mission"—and that mission included improving the material condition of humanity. I therefore regard France as an "interesting case" rather than as a clear example of a country with engineers (strictly so called) but without a profession of engineering.

My impression of Japanese engineers, that is, those reasonably fluent in English, is that they are more like the Dutch than the French, but most of them owe their fluency in English to having had part of their technical education in the United States. They cannot count as good examples of ordinary "Japanese-style" engineers. We will not know how far the profession of engineering extends in Japan or elsewhere until we (well, social scientists), go to all those places and ask engineers questions that bring out how they understand their work and their relationship to other engineers. My complaint about research so far done is that researchers asked the wrong questions (questions one or another sociological conception suggested) and therefore discovered much about certain occupations but almost nothing about professions as such. For an example of what I consider the right way to carry out such research, see Davis and Zhang (2016), Wei et al. (2018), and Wei and Davis (2020).

Understanding engineering as a profession has many consequences for both teaching and research. So, for example, if engineering is a profession everywhere, all engineers (and only engineers) belong to one community, engineering—whether they belong as well to other communities—a country, language group, religion, company, industry, or occupational category ("technologists"). To understand engineers as engineers, we must study their profession (as well as their function, discipline, and occupation). If we are to teach engineering ethics, we must take into account not only the substance of their code of ethics but also the special reason a professional has to obey it ("Don't cheat").

NOTES

Early work on this chapter was carried out in part under National Science Foundation grant SES-0117471. Early versions (under various titles and focusing on various professions) were presented to: the workshop, "Toward a Common Goal: Ethics Across the Professions," Sierra Health Foundation, Sacramento, California, August 26, 2006; the Research Group of Ethics, Faculty of Letters, Hokkaido University, Sapporo, Japan, February 14, 2007; the Second ASPCP International Conference on Philosophical Practice, Purdue University Calumet, Hammond, Indiana, May 19, 2007; Philosophy Section, Faculty of Technology, Policy and Management, University of Technology-Delft, The Netherlands, September 24, 2007; Center for Ethics and Technology, University of Technology-Twente, The Netherlands, September 27, 2007; and the Center for the Study of Ethics in Society, Western Michigan University, Kalamazoo, Michigan, October 4, 2007. I should like to thank those present at these lectures, as well as two reviewers for *Philosophia*, for many improvements in what became this chapter. Originally published as "Is Engineering a Profession Everywhere?" *Philosophia* 37 (June 2009): 211–225.

1. See, for example, Iseda, 2008; Downey et al, 2007. For a non-engineering example of how the definition of profession can shape practice, see Pinch et al., 2003.

2. I examine this history in some detail in the first two chapters of Davis, 1998. Of course, one *function* of engineering (building on a large scale) does go back to the beginnings of recorded history (and, indeed, earlier), but that function is something engineering shares with several other disciplines, including architecture and masonry.

Indeed, that function is something engineers share with ants, beavers, and coral. It cannot define "engineering" (in the sense we need).

3. For more on the enormous variety of sociological definitions, see Kultgen, 1988, especially, 60-62. See also the more recent exchange between: Sciulli, 2005; Torstendahl, 2005; and Evetts, 2006.

4. Durkheim, 1957, discusses the customs or standards that occupational communities typically subject themselves to (whether morally permissible or not). Durkheim seems to lack any sense of profession as I will be defining it. In contrast, Talcott Parsons, the most important representative of this tradition in the United States, really is a student of professions. See, for example, Parson, 1939 (revised 1954), p 34-49.

5. For an attempt to explain the attractions of the various sociological approaches, see Burrage and Torstendahl, 1990, especially the Introduction.

6. Often, those using the Cartesian approach avoid this conclusion only by adding morality as a side constraint. See, for example, Kasher, 2005. His conception, resting on action theory, would justify the same conclusion about the mafia, 70, did he not eventually connect profession with democracy, 83ff, democracy's morality providing a constraining "envelope." For another (more plausible) example of the Cartesian approach, one that builds in morality, see Koehn, 1994. Like Kultgen, Bayles, 1981, seems to offer a sociological definition. Even a philosopher may find sociology's promise of empiricism attractive.

7. Several occupations seem to have a profession as a subset. For example, financial analysts are divided into those who work as individuals (the non-professionals) and those who claim the status of Certified Financial Analyst (and satisfy the Socratic definition).

8. Of course, "never" may seem an overstatement. We certainly can imagine two radically different disciplines (law and medicine or engineering and journalism) changing over time until they become enough alike to become one occupation. The point, though, is that such changes would be so radical that at least one of the disciplines would be unrecognizable. We would have the usual problems of deciding when one individual ends and another begins.

9. There is no need for the moral ideal to be unique. Two professions may share the same moral ideal. So, for example, osteopaths (OD's) seem to have the same moral ideal as physicians (MD's). What distinguishes osteopaths from physicians are their special standards, especially their educational standards and standards of practice.

10. They are, of course, in position to take advantage of the professional practice, in large part at least, precisely because law, morality, market, and public opinion do not enforce those standards (or at least do not enforce them effectively enough to make following the standards prudent without the additional moral obligation arising from profession).

11. When the Netherlands' Royal Society of Engineers was working on its first code of ethics, the Dutch engineers I talked to (as late as 2011) seemed to think of the code as "documenting" what they already accepted rather than as setting a new standard.

12. I am assuming that "serving the state" is not a moral ideal. In political philosophy, there is a recent debate about (in effect) whether patriotism is a moral ideal—or, like political loyalty or nationalism, morally suspect (though not necessarily bad). I do not think we need worry about that debate here—because, as I shall soon explain, French engineers seem to understand the *French* state to serve a moral ideal (one we non-French can recognize).

REFERENCES

Abbott, Andrew. *The System of Professions* (University of Chicago Press: Chicago, 1988).
Bayles, Michael. *Professional Ethics* (Wadsworth: Belmont, California, 1981),
Burrage, Michael and Rolf Torstendahl, *Professions in Theory and History: Rethinking the Study of Professions* (Sage Publications: London, 1990).
Davis, Michael. "Liberalism and/or Democracy?" *Social Theory and Practice* 9 (Spring 1983): 51–72.
Davis, Michael. *Thinking like an Engineer* (Oxford University Press: New York, 1998).
Davis, Michael and Hengli Zhang, "Proving that China has a Profession of Engineering: A Case Study in Operationalizing a Concept across a Cultural Divide," *Science and Engineering Ethics* (December 2017): 1581–1596.
Downey, Gary Lee, Juan C. Lucena, and Carl Mitcham, "Engineering Ethics and Identity: Emerging Initiatives in Comparative Perspective," *Science and Engineering Ethics* 13 (December 2007): 463–487.
Durkheim, Emile. *Professional Ethics and Civic Morals*, Cornelia Brookfield, trans. (Routledge: London, 1957).
Evetts, Julia. "Short Note: The Sociology of Professional Groups," *Current Sociology* 54 (January 2006): 133–143.
Freidson, Eliot. *Professionalism: The Third Logic* (University of Chicago Press: Chicago, 2001).
Iseda, Tetsuji. "How Should We Foster the Professional Integrity of Engineers in Japan? A Pride-Based Approach," *Science and Engineering Ethics* 14: 165–176.
Kasher, Asa. "Professional Ethics and Collective Professional Autonomy: A Conceptual Analysis," *Ethical Perspectives* 11 (March 2005): 67–96.
Koehn, Daryl. *The Ground of Professional Ethics* (Routledge: London, 1994).
Kultgen, John. *Ethics and Professionalism* (University of Pennsylvania Press: Philadelphia, 1988).
Larson, Magali Sarfatti, *The Rise of Professionalism* (University of California Press: Berkeley, 1977).
Parsons, Talcott. *Essays in Sociological Theory* (Free Press of Glencoe: New York, 1954).
Pinch, Franklin C., L. William Bentley, and Phyllis Browne, *Research Program on the Military Profession: Background Considerations* (Canadian Forces Leadership Institute: Kingston, Ontario, 2003).

Sanders, John T. "Honor among Thieves: Some Reflections on Codes of Professional Ethics," *Professional Ethics* 2 (Fall/Winter 1993): 83–103.

Sciulli, David. "Continental Sociology of Professions Today: Conceptual Contributions," *Current Sociology* 53 (November 2005): 915–942.

Torstendahl, Rolf. "The Need for a Definition of 'Profession," *Current Sociology* 53 (November 2005): 947–951.

Wei, Lina, and Michael Davis, "China's unwritten code of engineering ethics," *Business & Professional Ethics Journal* (2018–2019): 169–206.

Wei, Lina, Michael Davis, and Hangqin Cong. "Professionalism Among Chinese Engineers: An Empirical Study." *Science and Engineering Ethics* (November 2019): 2121–2136.

Zussman, Robert. *Mechanics of the Middle Class* (University of California Press: Berkeley, 1985).

Chapter 2

Engineering—From Chicago to Shantou

Like "profession," the words "engineer" and "engineering" differ from most words, such as "dog" or "barking," in at least one important way. One can define "dog" or "barking" without upsetting any barking dog. Definitions of "engineer" or "engineering" will upset some engineers (or would-be engineers) if too narrow and upset both engineers and certain non-engineers (e.g., synthetic chemists) if too broad. Indeed, the definition may upset some of these people even if accurate. "Engineer" and "engineering" are not mere descriptors but "honorifics" conferring a desirable membership. They are "party terms" like "English," "democracy," and "profession." To offer a definition is to join a controversy about more than words.

The controversy is not easy to resolve for at least three reasons. First, there are engineers who (for reasons of history) are not called "engineers," such as naval architects. Second, there are non-engineers, such as the operators of railway trains, who (also for reasons of history) are called "engineers." Third, several disciplines are not so easily separated from engineering proper as driving a train is. Among these are not only software engineering, genetic engineering, and geo-engineering, but also architecture, industrial design, and surveying. Indeed, some languages have no distinct word for engineering (or engineer) but manage with a broader term, "technology" (or something similar), several narrower terms, or some combination of these. Yet, engineers (in the preferred sense yet to be explained) are generally clear about the distinction between themselves and "mere technicians and other technologists." Engineers engineer; other technologists (generally) do something else (architecture, biology, chemistry, or the like).

I therefore propose to reach engineering's definition by a long detour. My starting point will be a specific problem almost everyone will admit to concern engineering proper—a convenient "hypothetical case." I set it in

Shantou, a city of about five million in one of China's booming Economic Development Zones. I set it there both because "Shantou" sounds good next to "Chicago" and because the city is not only physically far from Chicago but also a place both North Americans and Europeans tend to consider culturally far (as they do not consider, say, Sydney, Australia, or Cape Town, South Africa, though they are physically farther). I might have picked Timbuktu or Shangri-La but neither of these famously faraway places has a local polytechnic. Engineers there would have had to come from elsewhere (or at least have trained elsewhere), making the problem less exotic, that is, less a test of our understanding of what engineering is.

The definition will emerge naturally in the course of solving the practical problem our hypothetical poses. But the definition will be institutional rather than linguistic, that is, will rely on a historically realized procedure for identifying engineers and engineering rather than on a verbal formula (as dictionary definitions typically do). Philosophers reflecting on this definition (and the rationale for it) may want to broaden their ideas of definition. This chapter will also illustrate two other points about method in the philosophy of engineering.

First, the problem discussed here concerns engineering ethics. We will use a problem of engineering ethics to understand engineering in much the way that Walter Vincente used more technical problems (Vincenti, 1990). Engineering ethics is not only a part of the philosophy of engineering (the attempt to understand engineering as a reasonable undertaking) but also, as we shall see, a useful tool in other parts of that field. This chapter is a contribution to the philosophy of engineering in part because it is a contribution to engineering ethics.

Second, the ethical problem discussed will concern "globalization." Problems of globalization often force us to be more explicit than we might otherwise be about what engineers share. Much that at home is too familiar to demand attention demands it in a strange place.

1. THE PROBLEM

You are an American-trained civil engineer working for a Chicago company that makes sophisticated industrial equipment. You are "degreed" (the graduate of an accredited engineering program) but not a PE (not, i.e., licensed as a Professional Engineer).[1] You are not unusual in this respect. Only about a third of degreed engineers in the United States are licensed. You are in Shantou to help install a chemical mixer, a hollow stainless-steel ball on stubby legs standing about thirty feet high and weighing several tons empty. Your job is quality control.

The specifications require that the legs be bolted to a concrete base. Bolting is important because the mixer vibrates when in operation and, without the bolting, might move about unpredictably or even fall over. The chemical mix is not dangerous and the mixing ball is designed to keep the chemicals inside even if the ball is on its side. But, once free of its bolting, the mixer would be a danger to anyone close to it (until it broke free of its power source and ceased operation). For a few minutes, it would be the modern equivalent of a "loose cannon."

The concrete, bolts, and installation are standard in the United States—and, indeed, around the world. So, to keep costs down, the concrete and bolts were to be procured in Shantou and the work done by local contractors. The concrete and bolts have been procured and you have tested them to assure quality. The bolts passed all tests "with flying colors," but the concrete proved marginal (some samples passing, some failing, but all close to the line). Even in the United States, you might reject the concrete shipment (i.e., refuse to sign the quality documents). Here, where there are (as you have been told) more problems of controlling quality during installation, you believe you need the full margin of safety that the original specifications give.

You reported your findings to your contact, the local engineer in charge of assembling the new production line. She responded (in tech-school English), "Ah, we must give this supplier benefit of doubt. People's Liberation Army owns company." You asked around and soon learned that dealing with the PLA is rather like dealing with some combination of the Pentagon, the mafia, and the old Mayor Daley. You know what you would do in Chicago. You would not risk workers being crushed by a loose mixer—not to mention the cost to the client in repairs and to your employer in lost reputation. As your supervisor back in Chicago likes to say, "Sometimes an engineer has to say no—and take the consequences." But this is Shantou. Should you act differently here?

The contract requires you, acting as your employer's representative, to sign off on the installation. If you do not sign off, the legal result is that your company is not responsible for the mixer's safety, reliability, or operation. The Chinese company cannot operate the mixer except "at its own risk." You inform the local engineer in charge. She responds, "That's good compromise. We don't have problem with worker liability here. No lawsuits. Don't sign. We can handle local inspectors. Everyone happy." Have you done all that you should? Has she? How are you to decide?

This case raises many questions, even ignoring the three just explicitly asked. We may begin to answer them by considering what our American engineer, "you," should do in the United States in a similar situation, then whether different standards apply to "you" outside the United States, and then whether different standards apply to a *Chinese* engineer working in China.

2. THE AMERICAN ENGINEER AT HOME

The first question we are to consider (what more "you" should do) seems the easiest to answer. For a civil engineer "back home," at least three codes of ethics are relevant: (a) that of the American Society of Civil Engineers (ASCE), (b) that of the National Society of Professional Engineers (NSPE), and (c) that of ABET (formerly the Accreditation Board of Engineering and Technology). In theory, the status of these codes is straightforward. Each applies to "engineers" as such, not to members of the enacting association (as one might expect of the ASCE or NSPE code) or to PEs (as one might expect of the NSPE code). All three codes state *professional* obligations rather than obligations arising from state license or membership in a technical, scientific, or social organization.[2]

In practice, however, the relationship between engineers and these codes is more complicated. While most engineers believe themselves to have a professional code, few engineers (as far as I can tell by asking engineers I meet) have ever consulted a code of engineering ethics to make a decision. Indeed, few can even recall seeing a code of engineering ethics. Yet, when I have gone through one of these codes line by line with American engineers, they generally responded in one or both of the following ways. First, they agreed to almost every line of the code, that is, they said something like "Yes, that's how I want other engineers to act and how I am willing to act if they do." Second, they reacted with some combination of surprise and relief: "I didn't know it was all documented like that—and that I agreed with other engineers on so much. I thought I was an outlier."

Engineering education seems to "hardwire" much of engineering ethics into engineers. The hardwiring is done so subtly that engineers often do not realize that they agree with other engineers concerning how engineers should act. Because they do not realize they agree, they are less likely to raise ethical issues than if they expected the engineers around them to agree. They are less likely to act ethically than if they expected their fellow engineers to agree. For that reason (among others), engineering education (and engineering practice) should routinely include explicit discussion of engineering ethics, especially the codes.

There are significant differences in language between the three codes. They are nonetheless consistent. One or another simply sets a higher (minimum) standard than the others on this or that point. For example, the ASCE and NSPE codes now have language about "sustainable development" that the ABET code lacks. More important, the provisions relevant to our case are, in substance, the same. The first "Fundamental Canon" in all three codes requires engineers to "hold paramount the health, safety, and welfare of the public." That brings us to our first question of interpretation: Are workers, such as those a loose mixer might crush, members of the public (in the relevant sense)? Are engineers responsible for their safety?

I have discussed this question before—first in an article on the Challenger (Davis, 1991). That article is often cited, and has even been reprinted several times, but no one has, as far as I know, ever objected to the answer I gave there. So, I think we may treat that answer as uncontroversial: "Public" refers to all those persons (even those working for an engineer's client or employer) who, owing to ignorance, powerlessness, or lack of competence, cannot protect themselves from what engineers do (alone or with the help of others). Sometimes a worker is a member of the public, for example, if the risk is concealed—as the unusual risk of the mixer breaking free because of its substandard bolting would be. Part of every American engineer's job is to protect workers from such hidden risks. A worker is excluded from the public only with respect to those risks she knows about, understands, and can reasonably avoid (e.g., the well-known risks of the job she can avoid by taking a different one).

How should an engineer protect workers from risks that workers do not know about, do not understand, or cannot reasonably avoid? Let us use the ASCE code to answer that question. Because its Fundamental Canons say nothing more relevant here, we must turn to the "Guidelines" that interpret the Fundamental Canons. According to one (1.2), "you" have already taken one action you may be required to take: "Engineers shall approve or seal only those design documents, reviewed or prepared by them, which are determined to be safe for public health and welfare in conformity with accepted engineering standards." You have declined to approve the quality control document that would say that the installation, which you consider to be substandard, meets the relevant standard. (We may, I think, interpret "design documents" to include quality control documents in support of carrying out an engineering design.)

That, however, is not all you are required to do. Your Chinese counterpart seems ready to overrule your judgment. The ASCE's Guidelines (1.3) also say: "Engineers whose professional judgment is overruled under circumstances where the safety, health and welfare of the public are endangered . . . shall inform their clients or employers of the possible consequences." Interestingly, the Guidelines do not consider this sort of informing to be solely a matter of protecting the public. There is a similar provision under the canon protecting clients and employers: Guideline 4.5 reads: "Engineers shall advise their employers or clients when, as a result of their studies, they believe a project will not be successful." If you consider the mixer's faulty installation a failure of the project you were sent to China to carry out—reasonable, I think, since you were sent to China to ensure the mixer's *proper* installation—you have an obligation to notify both your client (the Chinese company) and your employer back in Chicago.

Because you have already informed your Chinese counterpart of the risk, you must now decide whether informing her is enough to inform "the client." How much can you count on your counterpart to share with her superiors? How likely is what she tells her superiors to reach those managers who should decide such a question for the company? However you answer those questions, you should let your superiors back in Chicago know about the problem. You will need their backing. Their resources for constructing a solution are better than yours. They may, for example, be able to talk to the president of the Chinese company, something you probably cannot do, or augment your budget to pay for some "fix," something else you cannot do. And, of course, no manager wants to learn of a problem like this when it is too late to do anything about it (when, e.g., the mixer has broken loose and crushed a worker). Notifying your superiors is prudent as well as ethical.

Your responsibilities do not end with notifying client and employer. Should informing client and employer not resolve the problem to your satisfaction, the ASCE Guidelines (1.4) require you to do something more: "Engineers who have knowledge or reason to believe that another person or firm may be in violation of any of the provisions of Canon 1 shall present such information to the proper authority in writing and shall cooperate with the proper authority in furnishing such further information or assistance as may be required." "Proper authority" would certainly include the Chinese equivalent of the Occupational Safety and Health Administration, but might include international or American agencies as well. You would have to check with your company's legal department—and perhaps other experts—to know.

I may, for the sake of brevity, pass over the details of what the guidelines for the other two codes say about how an engineer should handle this situation. Though the language and arrangement of the relevant provisions differ somewhat, showing that they were not just uncritically copied from one source, they are in substance the same. The engineer must inform both client and employer of the risk to the public. If informing fails, he must report the problem to an appropriate authority. For an engineer, in the United States at least, the public safety, health, and welfare each take precedence over the interests of client and employer, including any interest in keeping business information confidential.

3. THE ENGINEER IN SOMEONE ELSE'S HOME

We have so far assumed that standards that apply to American engineers at home apply to their conduct when far from home—"in another culture" (as we sometimes say). If a culture is a distinctive way of doing certain things (together with the beliefs, commitments, and feelings supporting that way of doing them), then China certainly has a different culture. Among the differences may be what Chinese count as safe (or, at least, safe enough). The Chinese seem to be willing to take some risks Americans would not. The dark clouds that make many Chinese cities look like Pittsburgh a century ago certainly suggest that. Let us then agree that China has a different culture—and, in consequence, that the Chinese may not view safety as Americans do or even expect to be informed of risk as Americans would. Let us even agree that they prefer social harmony to individual autonomy. None of that matters now. While such differences may affect how an American engineer in China should act in many situations (e.g., when raising a sensitive topic with a superior), deference to local culture should not extend to any *ethical requirement*—or so I shall now argue.

The point of sending an American engineer to China is to have whatever advantages come from having an American engineer there. The Chinese have enough competent engineers. What then are the advantages of having an American engineer do the quality control when installing the mixer in China?

One advantage of having an American engineer do it, or at least having "you" do it, is having an engineer who knows the equipment and how it should be installed. This knowledge may seem merely technical. But almost nothing engineers know is *merely* technical. Engineering knowledge differs in at least two respects from "mere technical knowledge" (what one finds, say, in a report of research results in particle physics). First, engineering knowledge, to be of use, must be embedded in an engineer's judgment. And, unlike "pure knowledge," judgment is always an application of everything the judge knows, believes, or merely feels about the question before him.

Second, engineering knowledge has itself developed with certain practical ends in view. Often the end is obvious. For example, safety factors are developed to ensure a certain level of safety. They are not disinterested deductions from physics, chemistry, biology, or any other natural science or combination of them. They are expressions of practical sagacity (as we might say), for example, an intelligent response to what was learned from keeping good records of products in use. Thus, the safety factor for industrial bolts such as those to be used to anchor the mixer arose from experience with failure of earlier bolts, not only those failures that arose from normal use but also those that arose from common errors in manufacture, installation, or maintenance, misuse of machinery, and even effects of changing technology. The American engineer brings with him an understanding of what *American* engineers think safe—and judgments reflecting that understanding—an understanding only partially expressed in formal criteria. A Chinese engineer will have a similar but somewhat different understanding of engineering *in China*. That is why the judgment of one cannot substitute for the judgment of the other. They are both necessary for the proper installation of an American mixer in China.

The American engineer is in Shantou to exercise judgment, his American engineering judgment. If he does not do that, he might as well have stayed home, sending an installation manual instead. Since part of American engineering judgment is ethical, as the codes of ethics make clear, the American engineer must bring his ethical judgment with him. Indeed, it is part of his technical judgment. Of course, he should not impose his own ethics on his Chinese counterpart. Though he is more than an advisor, since he has the power to withhold his signature from the document certifying proper installation, he is something less than the superior of his Chinese counterpart. He can tell her what *he* should do, but *she* will have to decide what she should do in response. Can he do more? Can he, speaking engineer-to-engineer, explain to her why she—a Chinese engineer—should work with him to protect the safety of her Chinese workers rather than putting the (supposed) wishes or interests of her employer first?

4. THE CHINESE ENGINEER AT HOME

The answer to the question just posed must begin with a point I have already made. Engineering is itself a culture, that is, a distinctive way of doing certain things. Indeed, in some respects, engineering is a more powerful culture than, say, Chinese (or any other national) culture. Given her tech-school knowledge of English and a green card, our American engineer's Chinese counterpart could move to the United States tomorrow and work much as an American engineer would. She would find it about as hard to move to the north of China (where the local language, customs, and even food are quite different from Shantou's) and much harder to stay in Shantou but switch from engineering to law, medicine, or some other skilled occupation. In this respect at least, engineering is a global culture, one that can overrule a local culture.[3]

That is not the only respect in which engineering is a global culture. If we examine the curriculum at Shantou's polytechnic (the local engineering school from which the Chinese engineer graduated), we will find it differs in only small ways from that of any American engineering curriculum. If we seek for an explanation of this common curriculum, we may find that Americans brought that engineering curriculum to China more than a century ago. If not, we will still find that both the Chinese and American curricula ultimately originated in France about three centuries ago (Davis, 1998). Why would the Chinese, with several thousand years of technical innovation, large-scale manufacture, and impressive construction, adopt a European approach to such things?

The answer to that question is doubtless complicated, but any satisfactory answer will include at least two elements. First, the culture that Chinese engineers share with engineers elsewhere allows Chinese engineers to work well with engineers elsewhere, not only directly as "you" are doing with your Chinese counterpart, but also indirectly, for example, when writing a description of parts to include in a catalog for sale overseas or when anticipating the size of a bolt thread on a foreign machine or which way it will screw. Engineers form one world-wide technological network. The Chinese join that network (in part) by employing engineers (i.e., graduates of schools with a certain curriculum).

Second, the Chinese, like most other peoples, must have been impressed by what engineers achieve. There must, for example, be the Chinese counterpart of America's experiments with building bridges during the nineteenth century. For a time in the United States, anyone might design and oversee construction of a major bridge: architects, carpenters, inventors, gentlemen-amateurs, and so on. Eventually, though, experience with bridge failure taught Americans that engineers were better at building bridges than their competitors; governments began to require an engineer to approve the design of any bridge the public was to use.[4] Chinese engineers are as much beneficiaries of the world's experience with engineers as American engineers are. When a Chinese truthfully claims to be "an engineer" (someone trained, skilled, and intending to work in the appropriate way), she claims ("professes") membership in an international entity, one defined by its distinctive ways of doing certain things. Insofar as she benefits by that claim (e.g., by being hired as an engineer or by having other engineers treat her as one of them), she should do as engineers are supposed to do. To do otherwise would be not simply to "free ride" on engineering's reputation but to take unfair advantage of what other engineers have achieved in a way damaging their joint achievement.

There is, I think, a misunderstanding about codes of engineering ethics. When someone says "code," most people, including most engineers, think of a short document with the title "code of ethics," "ethical guidelines," "rules of practice," or the like. They do not think about the possibility that a code of ethics might be *implicit* in the technical standards all engineers share. Such an implicit code would (as explained in chapter 1) be "unwritten" (in the sense of *not* being written in a "code of ethics") and yet be in writing (i.e., implicit in formal technical standards). Though (mainland) Chinese engineers now have no formal code of ethics, they might still have an unwritten one (the one implicit in the technical standards they share with the rest of the world).[5] Indeed, even if they had a formal code of ethics (as American engineers do), much of their ethics might still be implicit in the technical standards (as much of the ethics of American engineers is).

Whether Chinese technical standards do constitute an implicit code of engineering ethics depends, in part at least, on how Chinese engineers understand those standards. We may define a code of ethics as any set of morally permissible standards of conduct (explicit, implicit, or a mix) that all members of a group (at their most reasonable) want all others in the group to follow even if their following the standards would mean having to do the same. (For an extended defense of this definition, see Davis, 2002.) If Chinese engineers view their technical standards in that way—at least when, in a cool and sober hour, they reflect on them—then the standards constitute (among other things) their code of ethics. If, however, even after due reflection, Chinese engineers regard those standards as mere external impositions that they have little interest in having other engineers follow, then the standards are not a code of ethics for them—and they are not, strictly speaking, engineers but some other sort of technician, technologist, or technical manager.

Whether Chinese engineers are engineers strictly so called or another sort of technician, technologist, or technical manager is, of course, an empirical question—one a philosopher cannot, *qua* philosopher, answer authoritatively. Yet, I can, I think, offer at least three reasons to think Chinese engineers are engineers properly speaking, reasons that should, all else equal, incline any reasonable person to accept my answer.

First, when I question Chinese engineers I meet, they generally understand engineering standards as other engineers typically do. That is, they regard those standards as helping to avoid waste, save lives, and do other good things, not as mere external impositions. They want other engineers to follow those standards; indeed, they want to do the same. I assume that the engineers I have met are, in this respect at least, a fair sample of Chinese engineers generally. This, of course, is an empirical claim, one others interested in engineering ethics may contest until the appropriate empirical research is done.

Second, I recently participated in three empirical studies of several hundred Chinese engineers (using questionnaires in Chinese). Responses to ethical questions did not (once translated) seem much different from what I would expect of American engineers (Davis and Zhang, 2017; Wei et al. (2018); and Wei and Davis (2020)).

Third, interpreting engineering standards properly requires understanding their purpose. The understanding in question cannot simply be an intellectual grasp of the sort that can generate plausible arguments; it must as well include the visceral commitment that typically expresses itself in good judgment. The quality of Chinese engineering is, then, itself evidence for the ethics of Chinese engineers. You cannot fake good judgment; good engineering requires good engineering judgment; and good engineering judgment includes good ethical judgment. An unethical engineer is not a good engineer (though she may pass for one for a time).

We may then imagine a conversation between our American engineer and his Chinese counterpart that takes into account the similarities in engineering culture as well as the differences in national culture. The American might begin by acknowledging the problem. "Well," he might say, "I've never had to face the Pentagon, the mafia, and Mayor Daley rolled into one. I concede the problem. But we are engineers. We still have a responsibility to protect workers from that mixer. Surely, you cannot believe risking their lives like that would be good engineering—and you want your engineering to be good, don't you?" Assuming that I am right about Chinese membership in the global profession of engineering, "your" Chinese counterpart should respond, "You're right. I don't think that would be good engineering." If she does answer in some such way as that, "you" (our American engineer) can continue: "So, what we need is a way to avoid conflict with the PLA while satisfying engineering requirements."

I see no reason why the Chinese engineer should not respond with a "yes" to this too. What remains, then, is an ordinary engineering problem and at least two possible solutions. First, the two engineers may contact the PLA's company (either directly or through some intermediary) to see whether the company knows about the problem with its concrete. Perhaps it does not and, being informed, would replace the concrete with a higher grade or provide some less expensive correction. (The PLA's company will have its own engineers and they too will not want to be responsible for harm to innocent workers; they may be able to sway the relevant managers.) Should that solution fail, the American engineer and his Chinese counterpart might then change the installation to make up for the marginal quality of the concrete, for example, by anchoring the bolts in steel plates beneath the concrete. They may, of course, have to go to their superiors should the cost of such a change be significant. But, as that supervisor in Chicago said, "Sometimes an engineer has to say no—and take the consequences." That is as true in Shantou as in Chicago.

This conclusion may explain what might otherwise seem quite odd. Codes of engineering ethics, even those adopted by national engineering societies, typically apply to "engineers" (without qualification), not to "American engineers," "German engineers," or "Japanese engineers." The codes are also quite similar (as we shall see). The explanation for that unexpected indifference to what most of us (especially social scientists) consider important, our nationality, should now be clear: engineers are engineers the world over—defined by a common culture.

5. DEFINITION?

The definition of engineering that arises from the analysis of the problem presented here clearly is not a classical "abstract" definition—genus and species, necessary and sufficient conditions, or anything of the sort. What I have done is point out an institution or practice, the profession of engineering. That profession was not identified by what engineers do (their function). Engineers do a great many things: design, discover, inspect, invent, manage, write manuals, teach, testify in court, test, and so on. Nor was the profession identified by a term, such as "engineer" or "engineering." Engineers were instead identified by a common curriculum imparting a common discipline (a culture, i.e., a shared way of doing certain things, the distinctive way of doing certain things speakers of English call "engineering"). The reason naval architecture is engineering (whatever it is called) while ordinary architecture is not (however similar it is to engineering in function), is that naval architecture shares a discipline with the rest of engineering while architecture does not. One has only to look at the curriculum of naval architecture to see that naval architecture is engineering, not architecture, and that the graduates of that curriculum will not be architects but engineers (or, at least, well on their way to becoming engineers). (For more on the differences between engineering's discipline and architecture's, see chapter 3.)

The reason disciplines similar to engineering in name—software engineering, genetic engineering, geo-engineering, climate-engineering, and the like—are not engineering is that they do not share engineering's curriculum, they have not been taught the distinctive ways of doing certain things that engineers consider part of their discipline (what an accreditation committee typically looks for when evaluating an academic program). That curriculum cannot be deduced from engineering's function, purpose, or moral ideal. It is a product of history, not abstract logic. Engineering might have had a somewhat different curriculum, one allowing industrial chemistry or software engineering to be part of engineering (as chemical engineering and computer engineering are). Indeed, nothing but practical considerations prevents that amalgamation from occurring tomorrow. What logic forbids is that engineering should absorb industrial chemistry or software engineering without some change in at least one of the disciplines. The disciplines are as real as nations, political parties, languages, or religions—and, like them, not something an abstract definition can capture. Philosophers must take that history into account if they are to define engineering.

There are, of course, many definitions of engineering useful in practice. Perhaps the best of these comes from ABET: "Engineering is the *profession* in which a *knowledge of the mathematical and natural sciences* gained by *study, experience, and practice* is applied with *judgment* to develop ways to utilize *economically* the *materials and forces of nature* for the *benefit of mankind*." Yet, even this best of definitions (italics added) applies to architecture, industrial chemistry, and so on, as well as it does to engineering. In the next few chapters, we must consider how those professions nonetheless differ substantially from engineering and, along the way, sharpen our understanding of engineering.

NOTES

Early versions of this chapter were presented as a Steelcase Corporation Endowed Fund for Excellence Leadership Lecture, College of Engineering and Applied Sciences, Western Michigan University, Kalamazoo, March 10, 2008; and as a Seminar of the Mechanical, Material, and Aerospace Engineering Department, Illinois Institute of Technology, Chicago, March 12, 2008. I should like to thank those present, as well as Kevin Cassell, Vivian Weil, and several Chinese students for helpful comments. Originally published as "Defining Engineering—From Chicago to Shantou," *Monist* 92 (July 2009): 325–339.

1. PEs have a few powers or rights that other engineers do not, such as the ability to "seal" documents or advertise their services to the public. For most purposes, however, thanks to what is known as the "industrial exemption," PEs and other degreed engineers are more or less interchangeable in the United States. However, this is not true in some other countries, such as Canada (which requires all engineers to be licensed) or the Netherlands (which does not license engineers at all).

2. Not all engineering codes of ethics are like this. The most important one that is not belongs to the IEEE (formerly the Institute of Electrical and Electronics Engineers). The IEEE code applies only to IEEE "members." There is a good explanation for this. Many IEEE members are not engineers but computer scientists, systems analysts, or technical managers. Their presence among IEEE's members may also explain why the IEEE code is so short (agreement between engineers and non-engineers being considerably less than among engineers alone). Recently, this difference between the IEEE and other engineering societies was made more explicit with the disappearance of "engineering" from the code so as not to "give the appearance of marginalizing those who are not engineers but nevertheless are dedicated technical professionals" (IEEE, 2018).

3. Engineers take this for granted, forgetting that many professions—including law and medicine—do not give their members the same freedom to move. When asked for an explanation for their freedom to move about the world, engineers are likely to point to scientific laws, saying that gravity works on a bridge in the same way anywhere in the world. They are, of course, right about gravity (which may explain why physicists, chemists, and other physical scientists can move about the world as freely as engineers). Engineers should, however, wonder why physicians cannot move about the world as freely as engineers even though the human body is much the same the world over and many diseases travel as freely as engineers. In any case, it is not bridges that are the same around the world but bridges built by engineers. Before engineers took over bridge building, building a bridge was a skill not easily transferred from one place to another. For example, the Spanish who first saw the Inca's suspension bridges had no idea what to make of them. (Waddell, 1925)

4. See, for example, Waddell, 1925, p 22–23: "The decade from 1850 to 1860 marks a very important advance in American bridge engineering, for the designing then came into the hands of educated engineers; and rational design really begins." Before this time, the failure rate for bridges seems to have been quite high, perhaps one in four. Watson, 1981.

5. Some engineering societies in (mainland) China now have a code of ethics. And Taiwanese engineers have a code of ethics with roots reaching back to the mainland in the 1930s. See Zhang and Davis, 2017.

REFERENCES

Davis, Michael, "Thinking like an Engineer: The Place of a Code of Ethics in the Practice of a Profession," *Philosophy and Public Affairs* 20 (Spring 1991): 150–167.

Davis, Michael, *Thinking like an Engineer: Essays in the Ethics of a Profession* (Oxford University Press: New York, 1998).

Davis, Michael, and Hengli Zhang, "Proving that China has a Profession of Engineering: A Case Study in Operationalizing a Concept across a Cultural Divide," *Science and Engineering Ethics* (December 2017): 1581–1596.

Davis, Michael, *Profession, Code, and Ethics* (Ashgate: Aldershot, England, 2002).

IEEE, "Board Approves Revision of IEEE Code of Ethics: Changes Reflect Technological Advances," January 17, 2018, http://theinstitute.ieee.org/resources/ieee-news/board-approves-revisions-to-the-ieee-code-of-ethics (accessed January 22, 2018).

Vincenti, Walter G. *What Engineers Know and How They Know It* (Johns Hopkins University Press: Baltimore, 1990).

Waddell, J.A.L., *Bridge Building*, vol. 1 (John Wiley & Sons: New York, 1925).

Watson, Sarah Ruth, "The Rate of Bridge Failure." *Technology and Culture* 22 (October 1981): 847–848.

Wei, Lina, and Michael Davis, "China's unwritten code of engineering ethics," *Business & Professional Ethics Journal* (Summer 2020): 169–206.

Wei, Lina, Michael Davis, and Hangqin Cong, "Professionalism Among Chinese Engineers: An Empirical Study." *Science and Engineering Ethics* (November, 2019): 1581–1596.

Zhang, Hengli, and Michael Davis, "A Brief History of Codes of Ethics in China," *Business & Professional Ethics* 37 (Spring 2018): 105–135.

Chapter 3

Why Architects Are Not Engineers

1. INTRODUCTION

The subject of this book is engineering. What then is engineering? What is this activity for which we are to do philosophy of? That is *not* a question about how the term "engineering" is used, a question of lexicography. In some of the languages some of us speak, there is not even a clear distinction between "engineer" and "technologist." As I have already observed in chapter 2, even in English, which has both words, there is some confusion. The linguistically proper application of the term "engineer" (or "engineering") is no guarantee that what is in question is an engineer (or engineering) in the sense that interests the philosophy of engineering. The custodian of my apartment building, with only a high school education, has an "engineer's license" that entitles him to oversee operation of the building's two boilers. The management refers to him as "the engineer." A few miles south of those boilers is the local office of the Brotherhood of Locomotive Engineers and Trainman. Neither my building's "licensed engineer" nor those "locomotive engineers" are engineers in the sense this book requires. They are engineers in an older sense indicating a connection with engines, that is, they are technicians much like mechanics, trainman, or bus drivers. The confusion is even greater when the term is "marine engineer." Some marine engineers are a kind of licensed mariner (technicians who operate and maintain the mechanical systems of modern ships)—while others practice a branch of engineering called "marine engineering" (the subject of which is the design of those same mechanical systems). Then there are the knockoffs of engineering proper already mentioned in chapter 2: "genetic engineering," "social engineering," and so on.

Equally confusing, I suppose, is that there is at least one field of engineering *not* called "engineering": *naval architecture*. In the United States, programs in naval architecture are accredited as engineering programs, describe themselves in those terms, and in fact look much like other engineering programs. Naval architects are called "architects" only because there was once a tradition of English shipbuilding in which gentlemen, not trained as engineers, used detailed drawings to instruct tradesman on how to build large sailing ships. Having a classical education, these gentlemen knew Greek and preferred to use the Greek term, "architect," rather than the English, "master shipwright" (a term analogous to "master carpenter" or "master builder," suggesting a tradesman with dirt on his hands). The name outlived the tradition.[1]

What, then, distinguishes engineers, in the sense appropriate here, from other technologists—assuming, for the moment, that architects, computer scientists, industrial chemists, industrial designers, and so on, are technologists but not engineers. The answer cannot be the *function* of engineers. Like other technologists, engineers design and "build," or otherwise contribute to the life (and death) of technological systems (i.e., relatively complex and useful artifacts embedded in a social network that designs, builds, distributes, maintains, uses, and disposes of them). *Equating* engineering with design or building involves at least two mistakes.

The first, already noted, is that equating designing, building, or the like with engineering makes distinguishing engineers from other technologists impossible. Architects, computer scientists, industrial chemists, and so on, also design, build, and so on. Equating engineering with designing, building, or the like thus cuts off any research on differences in the way different disciplines work (since they are, by definition, one discipline).

Second, and equally important, is that equating engineering with designing, building, or the like gives a misleading picture of what engineers in fact do. Not all engineers design, build, or the like. Some engineers simply inspect; some write regulations; some evaluate patents; some attempt to reconstruct equipment failures; some sell complex equipment; some teach engineering; and so on. Whether all of these activities are properly engineering, indeed, whether any of them are, is a question for the philosophy of engineering. My point now is that inspecting, writing regulations, evaluating patents, and so on are functions that engineers routinely perform not simply in the sense that they are functions some engineers happen to perform but in the philosophically more interesting sense that they are functions that some engineers are supposed to perform. Employers sometimes advertise for engineers rather than for other "technologists" to perform these functions. I agree that design (engineering design) is important to understanding engineering, but I do not see how designing can be *the* (defining) function of engineering—because, as I see it, there is no function that engineers, and only engineers, seem to perform (except, of course, engineering itself, which is what we are trying to define).

If not defined by its function, what can define engineering? Chapter 2 gave a double answer to that question. One answer is that if "define" means giving a classic verbal definition (e.g., by genus and species), there are only practical definitions, useful for a particular purpose. There can be no philosophical definition, that is, one that captures the "essence" of engineering—because engineering no more has an essence than you or I do. All attempts at philosophical definition will: (a) be circular (i.e., use "engineering," a synonym, or some equally troublesome term); (b) be open to serious counter-example (whether because they exclude from engineering activities clearly belonging or because they include disciplines clearly not engineering); (c) be too abstract or too detailed to be informative; or (d) suffer a combination of these faults. Consider again the ABET definition with which chapter 2 ended.

The other answer, the one that explains in part this first, is that engineering, like other professions, is self-defining (in something other than the classical sense of definition). There is a core, more or less fixed by history at any given time, which determines what is engineering and what is not. This historical core, a set of living practitioners who—by discipline, occupation, and profession—undoubtedly are engineers, constitutes the profession at a certain time. They decide what is within their joint competence and what is not. They also admit or reject candidates for membership in the profession, using criteria such as similarity in education, method of work, and product. Often these criteria function as algorithms. So, for example, the ordinary lawyer clearly is not an engineer (i.e., competent to do engineering), while the typical graduate of an ABET-accredited engineering program with a few years' experience successfully working as an engineer (as ABET and other organizations of engineers define working as an engineer) clearly is an engineer. But sometimes these criteria cannot be applied without exercise of judgment. Does someone with a degree in surveying who, say, has successfully managed large construction projects for five years, count as an engineer because what she has been doing is, in effect, "civil engineering?"

Chapter 2 has, I said, already argued for this double answer. I shall not repeat the arguments made there. Instead, I shall now provide indirect evidence for that double answer by comparing engineering to a closely related discipline, architecture. This chapter's thesis is that architects, though clearly technologists who design, are just as clearly not engineers. If architects are technologists who design but not engineers, not all technologists who design are engineers and engineering cannot be equated with technology. The philosophy of engineering is *not* simply the philosophy of technology (nor even the philosophy of technologists).

Architecture is a good choice for comparison with engineering because engineers have a tendency to claim some "architects" as engineers, for example, the chief builder of one of the great Egyptian pyramids or Renaissance artisans such as Michelangelo or Leonardo da Vinci (Kostof, 2000). I put scare quotes around "architect" here because these are only proto-architects—as well as only proto-engineers. Why? I shall soon explain.

There are at least three other reasons why architecture is a good choice for this comparison. First, engineers have often tried to integrate architecture into engineering. For example, the first attempt to found an American society of civil engineers (1852) was called the "American Society of Civil Engineers *and Architects*." The idea was that architects and civil engineers had enough in common to form one (technical) society.[2]

Second, the original *plan* for Paris' *École Polytechnique* (1794), the mother of engineering schools, included a program in architecture.[3] Several of the first American schools of architecture began as programs within engineering departments with curricula borrowing much from engineering. Even today, many engineering schools, such as IIT or Rensselaer Polytechnic Institute (RPI), have a school of architecture (though the *École Polytechnique* never did).

Third, architecture is also a good choice for comparison with engineering because architecture's history parallels engineering's. Architects and engineers appear in history at about the same time, in the same countries, and sometimes even doing the same work (e.g., designing bridges). Architecture and engineering are in fact siblings, sometimes competing and sometimes cooperating. Yet, as often happens with siblings, they are unmistakably different individuals (as I shall now show).

My argument proceeds in three stages. The first identifies several ways in which architecture (what architects typically do) differs from engineering (what engineers typically do). Though each of these ways they differ is significant in itself, what matters here is their accumulation. The *overall* difference between what the two disciplines do is surprisingly great—as great, say, as that between accounting and law, or medicine and dentistry. The second stage briefly contrasts the history of architecture with the history of engineering. While explaining how engineers and architects might claim the same ancestors, this history also sharpens our sense of how different the two disciplines are. The third stage draws conclusions about how to approach the philosophy of engineering. One conclusion, of course, is that any philosophy of engineering should be able to distinguish engineering from architecture. Technologists are not one but many. Each discipline is an historic individual, not a participant in the same Platonic form.

3. SOME DIFFERENCES BETWEEN ARCHITECTS AND ENGINEERS TODAY

How do today's architects differ from today's engineers? Let us begin with their education, the basis of their discipline. One important difference is that architects typically learn less mathematics and science than engineers do, much less, and what they learn is also different. In addition, engineers are generally required to take a year each of physics and chemistry. At some schools, such as IIT, architects will have to take one or two semesters of physics. At others, such as RPI, the only natural science architecture students must take is biology. There is not, I think, any architecture school that requires its students to take chemistry.

More important is the difference in the math that engineers and architects typically learn. Engineering students must take two years of calculus. Architects are nowhere required to take more than one semester of calculus and at some schools are not required to take any calculus or only to take a watered-down version of what the engineers take their first semester. While engineers in practice may seldom use calculus, attempts to eliminate it from the engineering curriculum have failed. Even substantially reducing the calculus requirement seems to turn an engineering program into a program training mere technical managers. Engineering seems to be a calculus-based discipline (in a way we do not understand); architecture is not.[4] That, I think, is a fundamental difference between the two disciplines (even though much of what architects do depends on someone else using calculus now and then).

What do architects learn instead of calculus? One thing is free-hand drawing. Free-hand drawing seems to be important. Many architecture schools have a two-semester sequence that students must take (in addition to computer-assisted design). While engineers do a lot of sketching, they are not taught to draw free hand. And, in fact, the average engineer seems to draw about as well as the average philosopher.

Another subject architects learn instead of calculus is the history of architecture. This is more important than it may seem. While calculus has had a central place in the education of engineers since well before the founding of the *École Polytechnique*, the history of engineering has never had such a place. The history of engineering is not a course required for engineers. Indeed, many engineering schools do not offer the course even as an elective. Apart from a few stories about classic engineering disasters, engineers know little about the history of their discipline (not even the names of "famous engineers"). History has little to do with what engineers typically do—or, at least, little *explicitly* to do with it. Much history is embedded in technical standards, methods, and attitudes, of course.

In contrast, the history of architecture (often a two-semester sequence) is everywhere a requirement for architects. The history of architecture is also often explicit in how architects are taught to work. Until the 1950s, the first thing an architecture student given a design problem was supposed to do was go to the architecture library to check for historical antecedents. In some schools, that is still the first thing a student should do. The names of "famous architects" appear regularly in the architect's classroom; and, indeed, even in the informal discussions of architecture students enjoy outside the classroom. Architecture students sometimes "quote" earlier buildings in their designs. They are *supposed* to use history as a source of inspiration—even if practicing architects use history no more (and perhaps no less) than practicing engineers use calculus. Architecture is a history-based discipline much as engineering is calculus-based.[5]

The rhythm of architectural study is also different from that of engineering study. Engineers typically divide their time between classes and labs. Though engineers generally work harder than the average liberal arts student, the engineer's semester is otherwise similar, building to a peak at midterm and again at finals. Architecture students not only have "design studios" in addition to class and the rare lab, those studios also dominate their lives for the full four or five years of the bachelor's program. It is the studios that require the large, well-lit spaces typical of architecture schools. Studios are organized around "projects," particular design problems. Being "best in your class" means being best at such projects. As the date that a project is due nears, architecture students tend to drop everything else—social life, classes, and even sleep. They disappear from campus life for a week or two and then reappear (different students disappearing at different times). Though there was always some logic to teaching architecture in an engineering school, the studio has, I think, tended to separate architecture from engineering even when, as at IIT or RPI, architecture was brought into the institution. A senior design course, itself relatively new to engineering education, demands much less time than an architect's studio, demands it only for one or two semesters, and tends to respect the demands of other classes in a way the studio often does not.

These differences in the way the two disciplines are taught deserve more attention than they have received. They do not seem to have an obvious explanation founded on differences in how architects and engineers practice (a purely "functionalist" explanation). In practice, architects and engineers work in ways that appear much more alike than their schooling would suggest. For example, both work on "projects." Both do a good deal of "designing" (though the engineers do engineering designs while the architects do architectural designs). Both are likely to adjust the hours they work to what the work demands. Even the physical spaces in which they work are often similar (much more alike than either is like the space in which lawyers or physicians typically work).

Of course, there are also differences between architects and engineers at work. There are the obvious but superficial differences, for example, that architects tend to dress better than engineers. Their dress is also more stylish. More interesting, I think, is the way they work. There are at least three differences worth mention here. First, when an engineer makes a presentation, she is likely to use flow charts, graphs, mathematical tables, diagrams, and blueprints. The architectural presentation is likely to keep such things to a minimum. Instead, the architect is likely to use physical models (like those architecture students carry around campus), color renderings (complex multicolor drawings in chalk, water color, or ink), and (recently) computer simulations of buildings. Architects emphasize appearance in a way engineers generally do not.

That is not surprising, of course. Architects are trained to pay attention to appearances in a way engineers are not. Indeed, architects understand themselves to be responsible for the appearance of the "built environment"—what the American Institute of Architecture (AIA) Code of Ethics calls "aesthetic excellence" (E.S. 1.2.). Architects regularly comment on the aesthetics of their own work and that of other architects. Aesthetic evaluation, so central to architecture, is largely missing from engineering. While many products of engineering, from bridges to circuit boards, are in fact beautiful, engineers seldom comment on that beauty. They certainly do not treat beauty as a routine design criterion. That is not because engineers as individuals do not care about beauty. Many certainly do (or, at least seem to). The reason engineers do not treat beauty as a design criterion is that beauty is not part of their discipline. What engineers tend to emphasize in place of beauty is cost, efficiency, and safety.[6] That is a second difference in the way engineers and architects work.

Perhaps that difference explains a third. Architects not only routinely design for crafts, they are also supposed to. So, for example, the AIA Code of Ethics (E.S. 1-5) explicitly asks architects to "promote allied arts." In contrast, engineering has a long history of trying to substitute unskilled labor or machines for allied arts. Engineers have no corresponding duty to any "allied art." Engineers tend to standardize work; architects to individualize it.

Closely related to these three differences in the way architects and engineers work is the way they present themselves. Engineers tend to present themselves as "scientists"—applied scientists. Though their training, the emphasis on math and natural science, may explain this, presenting themselves as applied scientists is nonetheless misleading. The point of saying that engineers are applied scientists is to claim the authority of the natural sciences for what they know. But most of what they know is not natural science, but the special knowledge engineering has itself developed ("engineering science"). The point here is not that engineers design and scientists do not. Scientists also design. Most scientists design experiments; and a few, such as synthetic chemists, design other things as well (e.g., molecules unknown to nature). My point is rather that much of what makes an engineer an engineer is not scientific knowledge, whether abstract or applied, but engineering knowledge, something scientists do not have.

Architects could claim to be applied scientists just as engineers do. Architecture does apply knowledge of materials, environment, and so on, much as engineering does. Instead, architects typically present themselves as artists.[7] They emphasize their creativity rather than their science. When searching for a job, they present a "portfolio" of their work rather than a "resume" (as engineers typically do). The portfolio will often contain their own free-hand renderings as well as computer-generated floor plans, sections, and "elevations" (i.e., the façade, sides, and back of the structure in question).

Another difference between engineers and architects is in what they work on. Engineers often work on weapons. Architects today do not work on weapons, not even fortifications, and have not for almost four centuries. This is odd because the most famous book about architecture, and one of the most influential, is Vitruvius' *Ten Books on Architecture*. Vitruvius took up (what he called "architecture") only upon retiring from a Roman legion after a long career as a "mechanic" (i.e., a siege master or "military engineer"). Architecture is an art of peace in a way engineering is not.[8]

The last difference between architects and engineers that I shall note here does not concern how they work or what they work on but how they have organized. Architects and engineers are organized more or less independently. They are taught in separate schools even when those schools are in the same university. They take different degrees. They belong to separate associations. They have separate accreditation systems, separate codes of ethics, and separate professional publications. Computer science is organizationally much closer to engineering than architecture is.

These are, I think, the most important differences between architecture and engineering today. I may, I admit, have overlooked some differences, and may have overstated some of the differences stated. But all I need claim, all I do claim here, is that this list is both sufficiently long and sufficiently close to accurate that we are entitled to conclude that architecture is not a kind of engineering (or engineering a kind of architecture). Not all technology is engineering.

4. HISTORY'S CONTRIBUTION TO THESE DIFFERENCES

Like engineers, architects like to trace their history back at least to the ancient Greeks. Architects have a somewhat better claim to that linage, however. Unlike the relatively new word "engineer," the word "architect" does come from ancient Greek (thanks to Vitruvius). And the architects (chief builders) of ancient Greece did build structures similar to those modern architects built until quite recently. Yet, the fact is that any history of architecture beginning even with the ancient Greeks would have several discontinuities. The most important of these is the European middle ages. That is a period of almost a thousand years in which great buildings—not only churches and castles, but also guild halls, hospitals, bridges, and so on—were built without anyone called "the architect" and, indeed, using methods, styles, decorations, and even proportions different from those the ancients used. The term "architecture" was raised from the dead as part of that general revival of ancient learning in Western Europe we call "the Renaissance." After about 1450, a Gothic building was just a building, but a classical building was "architecture."[9]

The revival of architecture took place in at least three related ways. The first was an act of scholarship. The general revival of classical learning seems to have generated a desire to build in the classical style. The builders of the day, stonemasons, did not know how to do that. Trained in guilds, they carried on (and developed) the local building traditions, usually some mix of Gothic, Moorish, and Romanesque. Those who wanted to build like the ancient Greeks (or their Roman successors) had to study Greek or Roman texts—largely in the original language—or travel to places where classical buildings survived in sufficient number, especially, Rome.[10]

Those who chose to study texts had to be "learnéd" (i.e., to have a reading knowledge of at least one ancient language). A good example of such a scholar is Leon Battista Alberti (1404–1472), a Florentine graduate of the University of Bologne, a canon lawyer by training (and so, someone with a working knowledge of Latin). He first wrote a book about architecture (*De re aedificatoria*) addressed to patrons (rather than to builders or architects) and only later began to receive commissions. (Ettlinger, 2000, 113–114)

To "build like the ancients" did not mean using classical methods of construction. Much of what the Greeks and Romans knew, their oral or tacit knowledge of construction, had been lost, and what little survived in Vitruvius was not easy to apply. Instead, building like the ancients meant achieving the same effects using contemporary methods. So, when Alberti used his classical knowledge to design a building (or, more often, to design new facades for existing building), he consulted a stonemason to determine what was in fact practical (Kostof, 2000, 114). He knew the elements and proportions that were necessary for a classical effect but not how to realize them in a durable form at a tolerable cost.

The second way in which the revival of architecture occurred was through the work of artists. Michelangelo (1475–1564) is fairly typical in this respect (though unusual in most others). Both engineers and architects like to claim him as their own, but Michelangelo was properly neither. He generally identified himself as a "sculptor"; his career looks like that of an artisan, that is, someone trained to use his own hands to make beautiful objects of stone, precious metals, paint, and so on. Michelangelo's formal education ended before his teens. He picked up knowledge of classical architecture from scholars in Florence, from observing recently constructed buildings in Florence, and from observing classical buildings old and new in Rome. Classical facades appeared in his paintings for many years before, at age fifty (1525), he was asked to design his first building, the Laurentian Library, for his patron, Lorenzino de'Medici, then the leading man of Florence. Michelangelo did as asked, and oversaw the construction—and that (along with a few unrealized designs) made him an "architect." Later commissions, including work on St. Peter's in Rome, gave him a portfolio of achievements sufficient for an enviable career in architecture. But, for Michelangelo, architecture was a sideline.

The claim that he was an engineer has even less foundation. Soon after Michelangelo finished the Laurentian Library, the Florentines revolted, drove the Medici out, and declared a republic. A supporter of the republic, Michelangelo was asked to help prepare the city's fortifications. He did his best for about a year, 1528–1529. In 1530, the city fell to Medici allies and Michelangelo fled, never to return.

These facts do not show that Michelangelo was an architect or an engineer. What they show instead is that Italy then had neither architects nor engineers. What Italy had were people who, at the whim of a patron or in a pinch, could function as an architect or engineer (or, rather, as an architect or engineer would later function). The Laurentian Library, though a wonderful building, is not the work of someone with a career in architecture. Would we be willing to call Michelangelo an architect if, all other facts being the same, the Library and all his later buildings had been disasters instead of successes? I assume the answer is no. It is, then, important that we have some way of identifying architects that is independent of their actual achievements. If a failed architect is still an architect, then a successful builder is not necessarily an architect.

The same point applies to engineers. The fortifications constructed for Florence under Michelangelo's supervision, though apparently satisfying the standards of the time, were his only "engineering." His qualifications for the job seem to have been that he was a famous artist, that he was a republican who had just overseen a large construction project (the Library), and that there was no one better qualified. That is not enough to make someone an engineer; it is, at best, a start.

One reason there was no one better qualified to oversee the fortification of Florence was that increasingly powerful artillery was then changing the principles of fortification. The high, thick walls in which stonemasons specialized, once almost certain to force a long siege, could not withstand cannon for more than a few days. Low earthworks were required instead. Their design was not well understood in 1528. There were in fact no experts in fortification anymore. Even an artist might have a good idea. Michelangelo was not alone among artists to offer ideas. Leonardo da Vinci did something similar—though, it seems, no one ever carried his out.[11]

Central Italy thus had a double tradition of classical building, one scholarly and the other artistic. About this time, a third tradition was developing in northern Italy. Andrea Palladio (1508–1580), a stonemason, undertook a series of large country houses, beginning with the Villa Godi (1537–1542). He seems to have learned the classical style in part from scholarly patrons, though he also studied briefly in Rome in 1541, a perceptive patron having suggested the trip to the thirty-three-year-old stonemason and paid his way. Thirty years later, after a successful career, Palladio gathered his designs into *The Four Books of Architecture* (1570)—a work in Italian, the language of stonemasons, not Latin, the language of scholars. Except perhaps for Vitruvius' book, Palladio's is the most influential in the history of architecture.

Palladio was an architect in a way neither Alberti nor Michelangelo was. Unlike Alberti, he understood the materials with which he worked. Unlike Michelangelo, he made a *career* of designing and constructing buildings. His book on architecture helped to define the new discipline: the design, construction, and retrofitting of buildings in the classical style. But there remained the problem of how to train architects. Palladio, a stonemason, was—like Alberti and Michelangelo—a self-taught architect—or, at least, not formally trained as an architect.

The three Italian traditions of architecture—the scholarly and artistic of central Italy, and the stonemason's in the north—remained more or less separate while influencing the other two. There was still no curriculum to define the discipline of architecture, only books, traditional guilds having related skills, buildings to study, public discussion of architectural design, and various educational experiments. We must wait a century for the next important step in the discipline's development—and go to another country.

Cardinal Mazarin, an Italian, was chief minister of Louis XIV during much of that king's minority. Mazarin thought France deserved modern art, something better than French artisans were providing, and conceived of a school to train the most talented students in France. They would be taught *drawing, painting, sculpture, engraving, and other media* necessary for the proper construction of the classical public building he was planning. In 1648, that school became a reality, the Royal Academy of Painting and Sculpture. The Academy itself was an association of prominent artists, much as the Royal Academy of Science was an association of prominent scientists. The art school operated under the auspices of the Academy, taking its name. Some of members of the Academy served as faculty. The members of the Academy were to provide the government with a convenient pool of mature talent; its school, to prepare the next generation. The French in this period were organizing schools for many purposes as a way to teach large numbers of students more quickly than apprenticeship did and to teach skills apprentices were unlikely to be taught. Often the school resembled an apprenticeship except that the students did not work in anyone's shop, had several masters rather than one, and had a set curriculum. These schools were entirely separate from France's universities—as "academy" in their name indicated: They did not offer doctorates or even baccalaureates; they did not teach Latin, much less Greek or Hebrew; they taught manual arts like drawing, painting, and sculpture, that no university would teach until the 19th century.

Though the Academy of Painting and Sculpture seems to have trained architects from the beginning, it was not until 1671 that the French established a distinct "Academy of Architecture" with its own school. The separate academy was, of course, a recognition that training architects differed significantly from training painters or sculptors. This recognition appeared in two ways. First, the later years of the student's time at the Academy included more design of buildings than before. Second, preparation for practice was now to include a term in an "atelier" of a royal architect, a formal apprenticeship under the supervision of the Academy.

Like many French institutions, the Academy of Architecture then remained more or less unchanged until the French Revolution. The only important improvements seem to have been the addition of formal lectures on design, mathematics, materials, and history during the mid-1700s. Graduates of the Academy of Architecture provided France with most of those who made a career of constructing large buildings of stone in a classical style. If the Italians invented architecture, the French invented architects.

The French Revolution initially abolished all the academies and then, after rediscovering the need for them, reestablished them in some form or other. After the restoration of the monarchy, the academies were reformed again. In 1816, the Academy of Architecture was combined with the Academy of Painting and Sculpture and the Academy of Music to form *l'Académie des Beaux-Arts* (the Academy of Fine Arts). Though merged for administrative convenience and some economies of scale, the three academies maintained distinct curricula. The architecture curriculum was again more or less what it had been before the Revolution.

The curriculum of the Academy of Architecture focused on classical arts and architecture. All architecture students were required to prove their skill in basic drawing before advancing to figure drawing and painting just as the art and sculpture students were. Only then did the architecture curriculum diverge from painting and sculpture, focusing on building facades and floor plans instead of the composition of paintings or sculpture. Architecture was understood—and therefore taught—as a "fine art."

One distinctive feature of the Academy's curriculum was a series of competitions around which all the work of the year was organized (the predecessor of today's studio). The competitions were judged by an outside panel of experts who did not know who had produced the drawing, plan, or design. The winner of the senior competition received the "Prize of Rome," an all-expenses-paid residence in Rome to study classical buildings, send back reports, and prepare a design to justify admission into the Academy as an architect.

The *Académie des Beaux-Arts* was an extraordinary success. Its graduates produced many fine buildings; its curriculum was copied by similar institutions in other European states and then in much of the rest of the world. So, for example, many of the "engineer-architects" who built the world's first skyscrapers in Chicago were civil engineers who, deciding they wanted to beautify exteriors and interiors of their buildings, went to Paris for a year or two to learn how—or to one of the new American schools of architecture offering something like the Beaux-Arts curriculum.[12]

Modern construction seems to have put pressure on architecture schools to include more about building than the Beaux-Arts curriculum did. That may explain why the United States began to have its own schools of architecture two decades before the first skyscrapers began to appear, why those schools were generally part of a university or engineering school (rather than independent entities like the French or like many of the contemporaneous schools of art or design), and why the curriculum typically had much more about construction than the Beaux-Arts curriculum did (though there was considerable variation until after World War II). Something similar seems to have been going on in Germany about the same time. The architecture curriculum of the *Académie des Beaux-Arts* slowly lost status during the twentieth century—until, in 1968, its own students went into the streets of Paris demanding an "American-style curriculum." Within a few months, *l'École Nationale Supérieure d'Architecture* was created as part of the University of Paris.

I shall not retell the history of engineering in the detail I have just told the history of architecture because it is enough for our purposes to note that the history of architecture laid out here is largely independent of the history of engineering. So, for example, the first engineers in France, officers in the *corps du genie*, were soldiers, not civilians, as the architects were. The engineers were designing fortifications, roads, bridges, warships, and other military works when the first architects were designing palaces, hospitals, theaters, and other public buildings. Only in the nineteenth century did engineering become useful enough for *non*-military purposes that large numbers of engineers could make careers as civilians. Even then, they tended to work on projects different from those architects worked on. For example, an engineer was more likely to design a factory or canal; an architect, a house or department store. Engineering schools (polytechniques rather than academies) were generally separate from architecture schools (even when, as became increasingly common, they shared a campus). The skills of engineers and architects are probably more alike today than in 1700, 1800, or 1900.

5. CONCLUSIONS

If we return to the long list of differences between architects and engineers in section 3, we can see that the origin of most, if not all the differences, is in the history of the two disciplines. The chief difference between early architects and early engineers is that the early architects were civilians working in or near large cities while the early engineers were military officers often working in hostile country or, at least, far from any city. The architects had access to all the skills of those cities: stonemasons, carpenters, painters, chemists, and other "allied arts." They were seldom in a rush. They worked like other artists. The engineers, in contrast, generally had to work in places, whether urban or rural, in which little skilled labor was available; where experts were far away; and where time was important. The main labor pool was the military itself, mostly soldiers whose only trade was war. Engineers therefore had to know much more about construction than architects did, had to plan much more carefully than architects did, and had to design for efficient use of unskilled labor. Beauty was not important to the military. Being able to put up something to the purpose quickly, using only unskilled labor, was. The engineer's knowledge of math, physics, and chemistry was necessary to an engineer's work in a way it was not to an architect's.

There is, I admit, nothing inevitable about these differences between architects and engineers. A different history might have given us different differences—or none at all—and might still do so in the next century or beyond. Indeed, I not only admit the lack of inevitability but stress it. The history of a discipline is not the unfolding of a plan even in the way gestation of a child from a fertilized egg is. Much of what a discipline is depends on historical contingencies. Past societies have arranged things differently; ours might have done the same.

The history of a discipline is not, let me emphasize, the history of the discipline's achievements—or, at least, not primarily that. We cannot work backward from artifacts to the social arrangements that produced them. For example, many societies have had canals not so different from the canals civil engineers now build. Yet, we know that some of those canals were the work of people whose training was radically different from what we expect today of civil engineers. Thus, the Erie Canal (1817–1825) seems to have been designed by, and built mainly under the supervision of, self-taught surveyors who, in those days, were more likely to be involved in selling real estate or resolving boundary disputes than in doing anything resembling engineering. No one with formal training as a civil engineer was involved in the Erie Canal, though the person in charge of the project, Benjamin Wright, held the title "Chief Engineer."[13]

A discipline is a distinctive set of practices taught to novices by adepts the learning of which is a condition of being accepted as one of the adepts. What constitutes the discipline is, at any time, largely a matter of history. Curriculum is important. A "discipline without a curriculum," a settled course of study, is not a discipline. In this respect, poetry, though an art, is not a discipline. The same is true of invention. Part of understanding engineering is understanding the differences between the way engineers approach certain problems and the way those in closely related disciplines, not only architects but also computer scientists, industrial designers, or public health officers, approach them. For example, in what ways do these disciplines approach safety differently? Before we accept someone as an engineer—whatever her title or achievements—we should ask how she was trained. If she lacks proper training, she is not an engineer (unless her experience and achievements are good enough to lead engineers to "adopt" her into the discipline).

Are the differences we may discover between engineering and another discipline merely historical or are they the product of some underlying logic? For example, does accident or necessity explain why we now have chemical engineers as well as chemists but not archeological engineers as well as archeologists? When does "science" need *engineers* to "apply" it? That is another question for the philosophy of engineering: in what ways, if any, might engineering have been different?

This chapter has focused on engineering as a discipline rather than on engineering as a profession. That was necessary because the history of engineering is much longer than the history of the profession. (Davis, 2003) But differences in profession are, where they exist, often important aspects of a discipline. I have hinted at how different by earlier quoting a few passages from the AIA Code of Ethics. Comparison of codes of ethics for engineers with codes of ethics for other professions would, I think, highlight other revealing differences (as we shall see in the next three chapters). The comparison would, of course, also highlight important similarities.

NOTES

My thanks to Kevin Harrington for catching many small errors in my description of architects—and for more than two decades helping me learn about them. Originally published as "Distinguishing Architects from Engineers: A Pilot Study in Differences between Engineers and other Technologists," in *Philosophy and Engineering: An Emerging Agenda*, ed. Ibo van de Poel and David Goldberg (Springer, 2010), pp. 15–30."

1. Applied physics (also sometimes called "engineering physics") is another of these misnamed fields; it is not physics but engineering (both by curriculum and accreditation). "Rocket science" would be another misnamed field—if the term ever appeared anywhere but in (something like) the humorous observation, "It's not rocket science."

2. This might have worked. Only five years before, the American Medical Association was founded with an equally inclusive ideal, admitting surgeons as well as physicians. (Surgeons were, by history and training, then at least as distinct from physicians as architects were from engineers.) The AMA's inclusiveness was not unlimited, however. The AMA declined to admit dentists (who, in those days, would have been oral surgeons and, like other surgeons, without a doctorate). The admission of surgeons to the AMA was successful (as the admission of architects to the ASCE was not). The exclusion of dentists was also successful. They remain a separate profession.

3. The claim that the *École Polytechnique* is the mother of modern engineering schools, especially the great number of nineteenth-century schools that put "polytechnic" into their name, is, I think, uncontroversial. But that claim is consistent with another, the claim that several schools of the *ancien régime* seem to have developed the curriculum that the *École Polytechnique* made famous. History has few bright lines. Most are a device of historians trying to put what they know into a form brief enough to be useful.

4. Non-engineers are unlikely to appreciate how central calculus is to the training of engineers. The four courses in calculus are only part of the story. The rest of the story includes three years of engineering courses in which students regularly use calculus to solve problems. Few architecture classes—and none of the core—use calculus at all.

5. Architecture is also a discipline that relies on "precedent," generally following modes of design and construction that have proved their reliability. In this respect, architecture resembles engineering. Engineers hate to "reinvent the wheel" (as they often put it). That, however, is a different point from the one I am making here. While architecture is history-based, architects are, in some respects at least, less attached to precedent than engineers are. For example, architects love to "reinvent the house."

6. I am, of course, talking about public presentations or the major presentation that "sells" the design. Architects are quite capable of discussing cost, efficiency, and safety—and regularly do it when dealing with technical issues of design. My point is simply that such considerations do not have as prominent a place in their discipline as they have in engineering.

7. In my experience, this is the dominant form of presentation. It is certainly not the only one. For example, some architects present themselves as craftsmen (rather

than artists), emphasizing durability and comfort over beauty; some, as "social engineers" (changing the way people live and think); and so on. This variety suggests the variety of things any architect must do while moving from beginning a design to completing a building. Engineers also have ways of presenting themselves other than as "applied scientists," but there are fewer of them—and they are different. The most common are, I think, inventor, manager, and businessperson.

8. Early "architects," for example, the proto-architects of the Renaissance, often did work on fortifications. But the point here is that, despite this early history, architecture has long (for several centuries) been "an art of peace." That it was not always so suggests the importance of history in defining architecture as a discipline, a point I shall come back to.

9. Oddly, Vitruvius' use of "architecture" (rather than a Latin synonym such as *aedificatoria*) may itself have been a revival, a word choice signaling that he wanted to bring back Greek styles at a time when the Romans were developing their own. The revival of Vitruvius' book 1500 years after he first published it was almost an accident. Only one copy survived the dark ages. One copy less and "architect" might not have been our word for the builder of classical buildings. We might instead have had the architectural equivalent of "engineer," for example, "edificer," the English equivalent of Alberti's word.

10. Compare Vasari, 1963, 151: "Rule in architecture is the measurement of antiques, following the plans of ancient buildings in making modern ones."

11. For two less well-know "architect-engineers" of about the same time, see Vasari, 1941, 211–221: Guilano and Antonio da S. Gallo, both trained in "woodcarving and perspective."

12. Cuff, 1991 and Draper, 2000. The first of these American schools was MIT (1865). By 1900, there were eight more: Cornell (1871), Illinois (1872); Syracuse (1874); Columbia (1881); Pennsylvania (1890); George Washington (1894); Armour Institute, later IIT (1895); Harvard (1895). The University of Michigan opened a school of architecture in 1876 but closed it two years later. (The date in parentheses after a school's name is the date when a distinct department or complete four-year program was established.) Weatherhead, 1941, 33–62 and 90–108.

13. See, for example, http://en.wikipedia.org/wiki/Erie_Canal (October 14, 2017): "The men who planned and oversaw construction were novices, both as surveyors and as engineers—there were no civil engineers in the US at the time. James Geddes and Benjamin Wright, who laid out the route, were both judges who had gained experience in surveying as part of settling boundary disputes." Both Geddes and Wright seem to have been frontiersmen with, at best, a primary-school education (though Geddes did briefly teach primary school in his native Kentucky before leaving for upper New York State). Those who worked under the two judges (such as Canvass White or Nathan Roberts) were at the time also at best "amateur engineers"—without formal training in engineering and, indeed, generally without any advanced education at all—and almost no experience of any sort building canals, much less building a canal as large as the Erie. Building the Erie Canal was a school for a whole generation of canal builders, the last before civil engineers took over U.S. canal building. (Calhoun, 1960, 16 and 26–27.)

REFERENCES

Calhoun, Daniel Hovey, *The American Civil Engineer: Origins and Conflict* (Cambridge, Massachusetts: The Technology Press, Massachusetts Institute of Technology, 1960).

Cuff, Dana, *Architecture: The Story of Practice* (Cambridge: Massachusetts Institute of Technology, 1991).

Davis, Michael, "Are 'Software Engineers' Engineers?" *Philosophy and the History of Science* 4 (1995): 1–24.

Davis, Michael, "Defining Engineering: How to Do it and Why it Matters," *Journal of Engineering Education* 85(1996): 97–101.

Davis, Michael, "Do the Professional Ethics of Chemists and Engineers Differ?" *HYLE* 8 (2002): 21–34.

Davis, Michael, "What Can We Learn by Looking for the First Code of Professional Ethics?" *Theoretical Medicine and Bioethics* 24 (2003): 433–454.

Davis, Michael, *Thinking like an Engineer: Essays in the Ethics of a Profession* (New York: Oxford University Press, 1998).

Draper, Joan, "The École des Beaux-Arts and the Architectural Profession in the United States," in *The Architect: Chapters in the History of the Profession*, ed. Spiro Kostof (Berkeley: University of California Press, 2000), pp. 208–235.

Ettlinger, Leopold D., "The Emergence of the Italian Architect during the Fifteenth Century," in *The Architect: Chapters in the History of the Profession*, ed. Spiro Kostof (Berkeley: University of California Press, 2000), pp. 96–123.

Kostof, Spiro, "The Practice of Architecture in the Ancient World: Egypt and Greece," in *The Architect: Chapters in the History of the Profession*, ed. Spiro Kostof (Berkeley: University of California Press, 2000), pp. 3–27.

Vasari, Georgio. *The Lives of the Painters, Sculptors, and Architects*, Vol 2, trans. by A.B. Hinds (London: Everyman's Library, 1963).

Vincenti, Walter, *What Engineers Know and How They Know It: Analytical Studies from Aeronautical History* (Baltimore, MD: Johns Hopkins University Press, 1990).

Weatherhead, Arthur Clason, *The History of Collegiate Education in Architecture in the United States* (Los Angeles, CA: Weatherhead, 1941).

Chapter 4

Distinguishing Chemists from Engineers

This chapter's subject is *not* the ethics of chemistry and engineering (abstract standards), but the ethics of chemists and engineers (the professions), not only those chemists and engineers at the forefront of science but the much larger number whose work for government, industry, or non-profits is, though useful—indeed, crucial to our health, prosperity, and comfort—unlikely to lead to publication. My subject is living practices, not timeless ideas. What I shall argue here is that chemists are not engineers in part because the professional ethics of engineers differs from that of chemists. This may come as a surprise to many chemists, especially those who work with engineers (as many do). The surprise is understandable for two reasons:

First, the differences I will point out are not necessary; they do not derive from the "nature" or "essence" of chemistry or engineering. The most striking of those differences only dates from the 1970s. Not only could engineers have standards more like the chemists'; not so long ago they did. Such merely empirical differences are hard to anticipate.

Second, the difference between the ethics of the two professions is not large—at least compared to the difference between the ethics of either and the ethics of, say, accountants, lawyers, or physicians. The difference between the ethics of chemists and the ethics of engineers is therefore easy to overlook—or, when noticed, to dismiss as an individual's idiosyncrasy.

The difference is nonetheless significant (as this chapter's last section will show). Awareness of the difference may not only help engineers understand chemists better; it may also help engineers understand their own profession better and suggest improvements.

1. SOME DIFFERENCES BETWEEN CHEMISTS AND ENGINEERS

On June 25, 2001, the Chicago *Tribune* reported that Phil Eaton, a professor of chemistry at the University of Chicago, had synthesized a new compound, octanitrocubane. Octanitrocubane was described in two ways, one emphasizing the practical, the other the esthetic. So, for example, the first paragraph of the *Tribune*'s report described octanitrocubane as "the world's most powerful non-nuclear explosive," while the second described it as "a cube-shaped molecule of flawless symmetry." The first time Eaton is quoted, it is to say, "I think it's gorgeous." Just before this, Bart Kahr, a chemistry professor from the University of Washington, gushed, "Eaton is to Chicago what Michelangelo was to Florence."

For those who think science is all about "learning nature's secrets," synthetic chemists like Eaton are anomalies. Eaton did, of course, learn how to make a new compound out of nature's elements. But the compound itself is not one of nature's secrets. As far as anyone knows, octanitrocubane has never occurred naturally. It is as much Eaton's invention as the light bulb is Edison's. Nature merely provided the raw material. In this respect, synthetic chemists resemble engineers more than they resemble analytic chemists; their object is (in part at least) to *make* something, not simply to know something new—not simply to know-that, know-why, or know-how.

In another respect, however, synthetic chemists resemble architects or industrial designers more than they resemble either engineers or other chemists. The beauty of Eaton's creation impressed both Eaton and Kahr. That beauty seems to be important to their assessment of Eaton's achievement. Octanitrocubane may turn out not to be as good an explosive as the *Tribune* claims (e.g., because it is too unstable to have any practical use), but Eaton's achievement in chemistry will remain. He has developed a method for synthesizing octanitrocubane, used the method successfully, and given chemistry this "molecule of flawless symmetry." For engineers, on the other hand, there is no achievement without practical use; neither knowledge nor beauty is a normal part of assessing an engineering achievement. That is not to say that engineers do not generate knowledge or beauty. They do (as I have had occasion to say before). But neither knowledge nor beauty is what engineers are likely to mention, much less boast of, when commenting on their achievements. Neither Eaton nor Kahr sounds like an engineer.[1]

What I have described so far is only a difference in *attitude* between chemists and engineers. Such a difference has only an indirect relation to ethics. A difference in attitude, while enough to alert us to the possibility of a difference in ethics, is not enough to prove it. To prove a difference in ethics we must look elsewhere; we must understand the professions in question.

Still, what I have already shown is significant; it is not at all what we might expect from the history of the two professions. Until almost the end of the nineteenth century, many chemists worked in chemical plants much as engineers do today, overseeing the plants' operation as well as checking the quality of their processes or inventing new compounds. Other chemists (metallurgists) did something similar in smelters, foundries, or steel plants before there was such a thing as a metallurgical or materials engineer. Then engineers, specially trained for the work, began to replace chemists in these jobs. Why?

That is another question for historians. But perhaps part of the answer is the enormous increase in the scale of the processes involved. Engineers seem to do better in large undertakings than chemists do. Perhaps another part of the explanation is a change in the way chemists were trained or saw themselves. During the last part of the nineteenth century, chemical education, like chemistry itself, became more "scientific," more theoretical. And perhaps yet another part of the explanation is that engineering educators at last found a way to train engineers for certain jobs that chemists had previously done. Whatever the explanation, what is clear is that chemists were once enough like engineers for engineers to take over whole categories of work that chemists had been doing. Chemists remain enough like engineers that many chemists still work beside engineers in jobs of similar description.

To argue that chemists and engineers belong to different professions, each with its own ethics, will therefore require us to be clear enough about what we mean both by "ethics" and by "profession" to tell when we have a difference in ethics between professions. I must say a bit more about those two terms, "ethics" and "profession," than I have in the preceding chapters.

2. ETHICS IN GENERAL

"Ethics" has at least five senses in ordinary English. In one (as noted in chapter 1), it is a mere synonym for ordinary morality, those universal standards of conduct that apply to moral agents simply because they are moral agents. Etymology justifies this first sense. The root of "ethics" (*ethos*) is the Greek word for custom just as the root of "morality" (*mores*) is the Latin word for it. Etymologically, "ethics" and "morality" are twins (as are "ethic" and "morale," "ethical," and "moral," and "etiquette" and "petty morals"). In this first sense of "ethics," chemists and engineers must have a common ethics (since we all do). This sense of ethics would make our present question trivial. Since the question does not seem trivial, this is probably not the sense of "ethics" that concerns us.

In four other senses of "ethics," "ethics" differs from "morality." In one, ethics consists of those standards of conduct that moral agents *should* follow (what is sometimes also called "critical morality"); morality, in contrast, is said to consist of those standards that moral agents actually follow (what is also sometimes called "positive morality"). "Morality" in this sense is very close to its root *mores*; it can be unethical (in our first sense of "ethics"). "Morality" (in this sense) has a plural. Each society or group can have its own "moral code," indeed, even everyone can have a code of her own. There can be as many moralities as there are moral agents. But even so, ethics remains a standard common to everyone (or, at least, *may* be such a standard, depending on how "critical morality" is understood).

"Ethics" is sometimes contrasted with "morality" in another way. *Morality* then consists of those standards every moral agent should follow. A universal minimum, morality (in this sense) is our standard of moral right and wrong. Ethics, in contrast, is concerned with moral good, with whatever is beyond the moral minimum. Ethics (in this sense) is whatever is left over of morality (in our first—universal—sense, which includes both the morally right and the morally good) once we subtract morality (in this third—minimum right-only—sense). Since (as earlier chapters suggested) professional ethics consists (in large part at least) of moral *requirements*, this cannot be the sense of "ethics" with which we are concerned.

The fourth sense of ethics, a field of philosophy, is closely related to the second (or "should") sense. When philosophers offer a course in "ethics," its subject is various attempts to understand morality (all or part of morality in our first sense) as a reasonable undertaking.[2] Philosophers do not teach morality (in our first, second, or third sense)—except perhaps by inadvertence or *hubris*. They also generally do not teach critical morality, though the attempt to understand morality as a reasonable undertaking should lead students to dismiss some parts of morality (in its second, descriptive, sense) as unreasonable or to feel more committed to morality (in its first or third sense) because they can now see the point of it.

"Ethics" can be used in yet another sense, the special-standards sense we focused on in preceding chapters. In this sense, Hopi ethics is for the Hopi and no one else; business ethics, for people in business and no one else; and professional ethics, for members of a profession and no one else; and so on. Ethics—in this sense—is relative even though morality is not. But ethics (in this sense) is not therefore mere *mores* or custom. Ethics (in this sense) must at least be morally permissible. As noted in chapter 1, there can be no thieves' ethics or Nazi ethics, except with quotes around "ethics" to signal irony or analogy.

This fifth sense of "ethics" is, I think, the one implied in the claim that one profession's ethics differs from another's (or, at least, the one likely to yield the most interesting interpretation of that claim). So, for example, while a philosopher's course in Chemical Ethics might differ in many ways from her course in Engineering Ethics (in codes of ethics, problems discussed, and facts considered relevant to resolving a problem), such differences would not answer our question. We could still ask whether the professional ethics of chemists (in our fifth sense of "ethics") differs from the professional ethics of engineers and reasonably expect a nontrivial answer.

3. PROFESSIONAL ETHICS

What then is "professional ethics" (given our fifth sense of "ethics")? That, of course, depends on what we mean by "profession." Unfortunately, "profession" resembles "ethics" in having several senses. Chapter 1 explained why we should adopt the following definition of profession when studying professions: *a number of individuals in the same occupation voluntarily organized to earn a living by openly serving a moral ideal in a morally permissible way beyond what law, market, morality, and public opinion would otherwise require.*

Like a promise, professional ethics (the special standards of the profession) imposes moral obligations. These standards may, and generally do, vary from profession to profession. Indeed, it is possible to have several professions sharing a single occupation, one profession being distinguished from another only by its distinctive professional standards. So, for example, professional standards, perhaps including somewhat different moral ideals, seem to be all that make physicians (MDs) one profession of medical healer and osteopaths (ODs) another.

The special standards of a profession generally appear in a range of documents, including standards of admission, practice, and discipline. The appearance of such standard generally signals the transition from occupation to profession. But, in the United States at least, it is the publication of a formal code of ethics, not just any special standards, that is taken to be the unambiguous signal that an occupation has organized itself as a profession. An occupation's status as a profession is (more or less) independent of license, state-imposed monopoly, or other special legal intervention. Even dogs are licensed and the state grants monopolies for many activities that are not professions, such as the copying of books or the cutting of hair.

While professions often commit themselves to obey the law, they need not. Indeed, insofar as the laws of a particular country are unjust (or otherwise fall below the moral minimum), any provision of a professional code purporting to bind members of the profession to obey the law would be void (just as a promise to do what morality forbids is void).[3]

4. CHEMISTRY AND ENGINEERING: TWO PROFESSIONS

Chemistry is, I think, clearly a profession in the sense just explained—or, at least, clearly is in those countries where chemists earn a living by chemistry and have adopted a formal code of ethics. Consider, for example, the "The Chemical Professional's Code of Conduct" (2016) of the American Chemical Society (ACS). The code apparently applies to all chemists, or at least to all "chemical professionals" (the occupation), not merely to ACS members. (The term "chemical professionals" seems to have been coined to distinguish those chemists who share the occupation of chemistry from those "chemists" who are merely skilled amateurs, for example, those school children belonging to a special group of volunteers who test water for pollutants on weekends—chemists in function but not members of the profession.) The code's preamble states a recognizable moral ideal, "the improvement of the qualifications and usefulness of chemists through high standards of professional ethics, education and attainments." Every reasonable person at his most reasonable—that is, if well informed about what (professional) chemists do, not drugged, drunk, tired, or otherwise mentally infirmed, and with time to deliberate under good conditions—would want (professional) chemists to be qualified and useful.

The body of the code states "responsibilities" to the public, the science of chemistry, the profession, the employer, employees, students, associates, clients, and the environment. Some of these responsibilities clearly go beyond what law, market, ordinary morality, and public opinion demand. For example, chemists are supposed to "ensure that their scientific contributions, and those of their collaborators, are thorough, accurate, and unbiased in design, implementation, and presentation." Chemists do not allow themselves the hit-or-miss approach to chemistry that would pass without comment in a plumber, pilot, or politician.

"The Rules of Conduct" of the American Institute of Chemists (April 29, 1983), though different in detail, is equally a code of professional ethics for American chemists. The moral ideal stated in its preamble is "To protect the public and maintain the honor of the profession." Among its special "duties" are "To avoid associating or being identified with any enterprise of questionable character." Does the existence of this second code of ethics mean that the US has two professions of chemistry? Not necessarily. The ACS is the more general association of chemists, including academics as well as chemists working in industry; the membership of the American Institute (AIC) is primarily for chemists working in industry. For that reason, I think, the AIC code is much more specific about employment practices and altogether silent on other subjects about which the ACS code has something to say, for example, the treatment of students. In addition, the AIC code has a somewhat different function. Its final numbered paragraph imposes a duty on chemists, "To report any infractions of these principles of professional conduct to the authorities responsible for enforcement of applicable laws or regulations, or to the Ethics Committee of The American Institute of Chemists, as appropriate." The ACS code has no equivalent provision for enforcement. Because it can be used in disciplinary proceedings, it seems reasonable for the AIC code to be both more specific and less demanding than a code designed merely to guide conscience. While designed as a legal or quasi-legal document, the AIC code does not seem to contain any provision inconsistent with the ACS code. I therefore think it reasonable to view the AIC code as offering specifications of the ACS code for a certain purpose, not as defining a second profession of chemistry.

For engineers, perhaps the nearest equivalent of the ACS code is still ABET's "Code of Ethics of Engineers" (1977). Like the ACS code, the ABET code apparently applies to all members of the profession, not just to members of ABET (which are engineering societies, not individuals). Also, like the ACS code, the ABET code (1977) begins with a statement of moral ideals ("Fundamental Principles"):

Engineers uphold and advance the integrity, honor, and dignity of the engineering profession by:

using their knowledge and skill for the enhancement of human welfare; being honest and impartial, and serving with fidelity the public, their employers and clients.

Even with this partial statement of the preamble, we begin to see differences between the ABET code and the ACS code. The ABET code (like other engineering codes) commits engineers to using their knowledge and skill for "the *enhancement* of human welfare." While chemists aspire to make themselves "useful," engineers aspire to *improve* the overall welfare of human beings. An engineer is committed to human progress in a way that chemists are not. For engineers, human progress is a *professional* commitment.

Now, it might be argued that I am making too much of this difference in preamble. After all (it might be said), the first principle of most codes of engineering seems much like the "responsibility" of a chemist to the public:

Chemical professionals should actively be concerned with the health and safety of co-workers, consumers and the community. They have a responsibility to serve the public interest and to further advance the knowledge of science. Public comments on scientific matters should be made with care and accuracy, without unsubstantiated, exaggerated, or premature statements.

While this responsibility of a chemist *sounds* much like the engineer's, it also differs substantially. There is, first, the absence of any mention of *improvement* in human welfare. For all the ACS code says, serving the public interest may consist of no more than being "actively ... concerned with [*maintaining*] the health and safety of co-workers, consumers and the community." Chemists need not be actively "concerned" to *improve* human health or welfare. The only improvement in the human condition that the ACS code seems to recognize is in "knowledge of science." That, and that alone, is to be "further[ed]."

If we now compare this first "responsibility" of chemists with the corresponding first "fundamental canon" of the ABET code, we shall see a second difference between the ACS code and the ABET code:

> Engineers shall hold paramount the safety, health, and welfare of the public in the performance of their professional duties.

Here again, the engineers seem to have set a higher standard than the chemists. The chemists need only "serve the public interest." When the public interest comes into conflict with another responsibility (e.g., a responsibility to the employer or the environment), the ACS code provides no guidance. Presumably, chemists are to try to satisfy both responsibilities as much as sensible, perhaps trading off some satisfaction of one to obtain more satisfaction of the other. For engineers, however, there is no question of trade-off. The engineers must hold the public safety, health, and welfare "paramount"; the public safety, health, and welfare are to take precedence over all other considerations. Engineers are not allowed to look after the interests of employer or environment until they have taken care of the public.

This difference between the ethics of chemists and engineers is not a characteristic of the engineering profession as such, or at least it was not always, even though something like the paramountcy provision is now present in most codes of engineering ethics. The first codes of engineering ethics we know of, those adopted in Canada, Britain, and the United States late in the nineteenth century or early in the twentieth, did not contain such a provision, though there were proposals for it a few years later (AAE, 1924).[4] The first major U.S. code to include such a provision did not appear until 1974. Thereafter, the provision spread quickly to other engineering codes.

The paramountcy provision has, as far as I know, not reached the code of ethics of any scientific society. The only non-engineering societies to adopt such a provision have been technical societies the membership of which includes both engineers and scientists. For example, the Chemical Institute of Canada (CIC) is open to "professional chemists, chemical engineers, or chemical technologists." Its code of ethics (March 9, 1996) specifically requires that members (among other things) "accept and defend the primacy of public well-being." The Australian Institute of Mining and Metallurgy (AIMM)—"an assemblage of scientists, engineers, and technologists"—takes a somewhat different approach. The first rule of its code of ethics (1997) reads: "The responsibility of members for the welfare, health, and safety of the community shall at all times come before their responsibility to the profession, to sectional or private interests, or to other members." AIMM's code seems to allow its members to tradeoff the public's welfare against, say, the environment (since the environment is not a merely private, sectional, or professional interest). In both the CIC code and the AIMM code, the engineers seem to have moved the chemists in their direction but not won them over entirely. Engineers are not released from their higher standard, but chemists who join the CIC or AIMM are bound to a higher standard than other chemists.

I should perhaps point out that neither the CIC code nor the AIMM code is a professional code. There are at least two (related) reasons why they are not. First, both the CIC and AIMM codes apply only to members of the society in question, not to all members of any "profession." Any engineer or chemist can avoid application of the code to her simply by avoiding membership in the society; there is no need to change profession. Second, neither the CIC nor AIMM can have a code of ethics that at once is a professional code and applies to all its members. The members of CIC belong to three different occupations: chemical technologists, chemical engineers, and professional chemists. Much the same is true of AIMM's membership. Its members include technologists, engineers, and scientists. A professional code—by definition—applies to members of one profession simply because they are members of that profession. Because there are at least three different occupations represented in each society (and perhaps at least three different professions as well), there cannot be one professional code for all. And, however similar they may be in content, the codes, to be professional codes, will have to be written so that they apply only to one profession and to be subject to change by that one.

I have, I believe, now established that the professional ethics of chemists differs from the professional ethics of engineers and that they (chemists and engineers) therefore belong to different professions. We must now consider why the differences might matter.

During this discussion, we have had to quote codes of ethics that may seem to use certain terms as if they are interchangeable, especially "duty," "obligation," and "responsibility." Chapter 9 will try to clarify the relation of those terms.

5. BACK TO EATON

Both the *Tribune* report I quoted, and an earlier article in *Nature* (January 19, 2000), raised a question about the morality of Eaton's work on explosives.

Both compared Eaton's work with that of another chemist, Alfred Nobel, who, having invented dynamite and many ways to use it effectively in weapons, came to regret what he had done, devoting much of the considerable fortune derived from his inventions to avoiding their use in war. Much of the money for Eaton's research, several million dollars, came from the U.S. military. Unlike Nobel, Eaton must have known from the beginning of his research that he was working on a weapon. He is, in any case, now clear about his reason for working on that weapon:

> I don't consider the military an enemy. I'm damn glad we're safe here. I do believe it's important that the country be able to defend itself. The Army people really deserve a lot of credit for sticking it out and providing a lot of funding [for search on octanitrocubane].

Kahr, on the other hand, is not so sure: "Would it trouble me to know that one of my projects might be used for ill? Absolutely. I wouldn't accept a grant from a military agency under any circumstances." But—Kahr goes on to make clear—this is only an individual judgment. Chemists have no *professional* obligation to refuse work that "might be used for ill": "I am not trying to seem superior. We all draw our line in the sand at whichever place is comfortable."

Like chemists, engineers sometimes work on weapons; and, like chemists, engineers sometimes wonder whether it is appropriate for them to engage in such work. But the engineering version of the question has a somewhat different structure. For chemists, the chief questions are: a) will what I do be useful? and b) will it serve the public interest? For chemists, public safety is doubtless an element of the public interest, but it is not a matter of *special* interest. Chemists are free to trade off safety against other interests. For example, a chemist may, without professional impropriety, treat the risk that octanitrocubane will eventually fall into the hands of terrorists, making it easier for them to conceal explosives of a certain power, as just one consideration among others (including the interest in furthering knowledge of chemistry). For an engineer, however, the public safety, along with public health and welfare, is paramount. Engineers working on the equivalent of octanitrocubane should be building public safety into the compound (or, at least, into the manual that describes the appropriate conditions under which it should be stored, used, and disposed of). That is part of their professional responsibility. Where chemists and engineers work together, the engineer's special concern with safety may strike chemists as a bit much.

Engineers do, of course, balance safety against some other considerations, that is, those other "paramount" considerations, the public welfare and public health. Their profession does not require them to make anything "perfectly safe"—where "perfectly safe" means something like "has a zero probability of causing unintended harm to anyone." For engineers, safety is a relatively complicated notion. Generally, safety is defined for a specific product, activity, or system. Among considerations relevant to defining safety for a specific product is what the public knows or expects and what the public thinks is "safe enough" for that product. The standard for a safe Ford Fiesta need not be the same as for a safe BMW. The public has a right knowingly to trade safety for comfort, convenience, affordability, or even pleasure (within certain limits).

Public knowledge or expectation is, however, not the only relevant consideration for engineers. For many products (activities or systems), there is a governmental agency that acts on behalf of the public. A city or state may adopt a building code in part to assure the public that buildings will meet certain minimum standards of safety (relatively low levels of asbestos or lead, for example). The Department of Energy may adopt safety standards for fast breeder reactors. And so on. These governmental regulations also help to define safety for engineers (more or less independently of what the public knows or expects).

For engineers, safety also depends in part on the "state of the art" at the time the product is made. If, for example, an engineer finds a way to make some product safer without additional cost (say, by substituting one alloy for another), she has an obligation to make the product safer even though the public does not expect it and no government agency requires it. The engineer's discovery has changed the state of the art. If, however, the improvement would add significantly to the cost of the product, the engineer will have to balance the public safety against the public welfare. Sometimes the balance will be clearly enough in favor of making the improvement; then the engineer should make it. When, however, the balance in favor does not seem clear enough (e.g., when even well-informed engineers disagree about whether the benefit justifies the cost that the public would pay), the engineer should seek some way of letting the public decide where its overall interest lies, for example, by offering both the new product (at a higher price) and the old product (at its old price), by asking the government to decide whether to require the improvement, or by initiating a public debate.

Should chemists change their code of ethics to give safety the same priority engineers give it? That is a question for chemists (so long, of course, as what they choose does not fall below the moral minimum). The answer is not obvious. While it is probably true that chemists could win back some jobs lost to engineers if they treated safety the way engineers do, it may also be that chemists who pay that sort of attention to safety will not be as adventurous (and therefore as useful in their own way) as chemists now are. We have different professions in part at least because serving some moral ideal well may (in practice at least) be inconsistent with serving another well.

NOTES

This chapter was first drafted during the six weeks of July and early August 2001 that I was a Visiting Fellow at the Center for Applied Philosophy and Public Ethics (CAPPE)-Canberra, Australia. I read that draft at a CAPPE seminar, Australian National University, Canberra, July 13, 2001; a later draft at CAPPE, University of Melbourne, August 1, 2001; and yet another draft at CAPPE, Charles Sturt University, Wagga Wagga, August 8, 2001. I should like to thank all those present, as well as Jeffrey Kovac, for many helpful comments. Originally published as "'Do the Professional Ethics of Chemists and Engineers Differ'? *HYLE* 8 (Spring 2002): 21–34."

1. For more on the anomalous status of synthetic chemistry within the natural sciences, see Rosenfeld and Bhushan, 2000, pp 187–207.
2. Note that I have not defined philosophical ethics (moral theory) as concerned merely with "right and wrong." Much right and wrong is outside philosophical ethics, e.g. what is the right move in a chess game or the wrong answer to a math problem. For those who want a definition of "right" or "wrong," here is one I have found useful: the right is whatever satisfies the appropriate standard; the wrong is whatever does not satisfy the appropriate standard (or, more strictly, whatever *violates* the appropriate standard). For example, the right answer to a math problem is the one that satisfies the standards for solving the problem; a wrong answer, any answer inconsistent with those standards.
3. I would like to thank Seumas Miller for questioning me until I saw the need to make this point.
4. Never very large, the American Association of Engineers seems to have disappeared a few years after publication of this code.

REFERENCES

American Association of Engineers (AAE), "Specific Principles of Good Professional Conduct (1924)," http://ethics.iit.edu/ecodes/node/6182 (accessed December 20, 2017).
ABET, "Code of Ethics of Engineers" (1977), http://ethics.iit.edu/ecodes/node/3225 (accessed December 20, 2017).
American Chemical Society, "The Chemical Professional's Code of Conduct" (2016), http://ethics.iit.edu/codes/CPCC%20(2016).pdf
American Institute of Chemists, "The Rules of Conduct" (April 29, 1983), http://ethics.iit.edu/ecodes/node/3887 (accessed December 20, 2017).
N. Bhusham and S. Rosenfeld, *Of Minds and Molecules* (Oxford University Press: New York, 2000).

Chapter 5

Will Software Engineering Ever Be Engineering?

"Even with another name, social engineering would not be mistaken for real engineering."—Anonymous

In 1996, I published an article in the *Journal of Engineering Education* arguing—among other things—that software engineering was not then engineering (Davis, 1996). This was a short version of Davis (1995), which became Chapter 3 of my first book on engineering, *Thinking like an Engineer* (Davis, 1998). In 2010, the editors of *Communications of the ACM* asked whether enough had changed to make me think software engineering had become engineering. My answer was (and remains): much has changed—with some changes weakening the separation between engineering and software engineering and some re-enforcing it—but, overall, the separation still stands. This answer will surprise those who, unaware of my earlier articles, think software engineering's status as engineering is settled (and obvious). I therefore think it wise to precede any explanation of why software engineering is not engineering by disposing of a few unexamined presumptions the examination of which might make software engineering's status as engineering seem less obvious. This examination should help us understand engineering better.

As I observed in chapter 2, "engineering" has at least four senses in English. One, the oldest, understands engineering as minding engines (originally, "engines of war"). Casey Jones was an engineer in this sense; and so too, is the sailor rated "marine engineer." Neither are engineers strictly speaking nor are software engineers (in this sense).

Almost the opposite of this first sense is what we might call the *functional* sense, engineering-as-invention-of-useful-artifacts. In this sense, the first engineer may have been the proto-human who invented the club, cutting stone, or fire pit. Though this sense would certainly make software engineers engineers, there are at least two reasons to reject it here. First, the functional sense is too broad. Architects, industrial designers, and even weekend inventors are all engineers in this sense, making software engineering's claim to be engineering uninteresting. Second, the functional sense is misleading. It takes a sense of "engineering" that did not exist much before 1700 and applies it to proto-humans, carpenters, tinkerers, and the like humans before 1700, who would have understood themselves quite differently, and perhaps even to beavers (and other animals that "build" and "mind" things with uses similar to what engineers also build or mind, such as dams).

The functional sense of engineering nonetheless may seem relevant here. Software engineering's official Body of Knowledge offers this definition of software engineering: "the application of a systematic, disciplined, quantifiable approach to the development, operation, and maintenance of software, and the study of these approaches; that is, the application of engineering to software" (Bourque and Dupuis, 2004, p 1–1). The Body of Knowledge assumes, without argument (a mere "that is"), that engineering is a certain function, namely, any "systematic, disciplined, quantifiable approach to the development, operation, and maintenance [of something]." That assumption must be false. If it were not, many activities we do not count as engineering would have to count as engineering. For example, accounting—a field no one supposes to be engineering—would have to count "financial-records engineering" (since accounting is a systematic, disciplined, quantifiable approach to the development, operation, and maintenance of financial records).

Closer to our subject is a third sense of "engineering," engineering-as-discipline. A discipline is a distinctive way of carrying on an activity, some combination of knowledge, skill, and judgment that must be learned. Any craft or trade has its discipline—as do many activities that are neither craft nor trade, such as poetry or calisthenics. In this sense, neither architects, nor industrial designers, nor weekend inventors are engineers. Architecture and industrial design each have a discipline easily distinguished from engineering's (as chapter 3 showed for architecture). Weekend inventors have no discipline at all; they may invent any way they like.

Software engineering is not engineering in this third sense. The Body of Knowledge that engineers are supposed to learn differs in important ways from software engineering's Body of Knowledge. So, for example, engineers have to take courses concerned with the material world, such as chemistry and statics; software engineers do not. Software engineering's official Body of Knowledge was in fact an important step in clarifying the distinction between engineering proper and software engineering. The Body of Knowledge requires software engineers to know things engineers typically do not (e.g., how to organize the design of complex software) and allows

Will Software Engineering Ever Be Engineering? 69

them not to know some things engineers are supposed to know (such as thermodynamics).

The last sense of engineering we need to distinguish here is engineering-as-profession. A profession is (as we have said before) a number of individuals in the same occupation voluntarily organized to earn a living by openly serving a moral ideal in a morally permissible way beyond what law, market, ordinary morality, and public opinion would otherwise require. An occupation is a discipline by which one may, and some do, earn a living. Both engineering and software engineering are now occupations but, having (as just noted) different disciplines, must be different occupations. That is one reason why they cannot share a profession. There is another.

The Software Engineering Code of Ethics and Professional Practice differs in significant ways from all the most important engineering codes of ethics discussed in earlier chapters. Software engineers are, for example, supposed to "[m]oderate the interests of the software engineer, the employer, the client and the users with the public good" (1.02).[1] Engineers do not now have that duty to "moderate"; indeed, engineers have a duty to put the public good first.

Software engineering has, indeed, become a profession. What it has not become is part of the engineering profession. Anyone who claims otherwise must find a sense of engineering different from those distinguished here, one that makes software engineering a part of engineering without including as well disciplines, occupations, or professions, such as architecture or accounting, that clearly are not part of engineering.

Professions are voluntary associations. You cannot become a member simply by claiming to be one. You must be admitted (by the profession, not just by a technical society like the IEEE or ACM). Engineering has a long history of other occupations claiming to be engineering: Among recent examples are: genetic engineering (a kind of tinkering with genes), re-engineering (a fad in management), and financial engineering (gambling on Wall Street). Software engineering actually began with an attempt to copy engineering practices, making its claim to be engineering more respectable than most. But the enormous complexity of software has forced software engineering to develop in ways engineering has not—and may never.[2] Many of the very methods that make software engineering useful distinguish it from engineering. Engineers have good reason to continue to treat software engineers as belonging to another profession.[3]

I have, I hope, just explained why it is reasonable to think software engineering is not engineering (in a sense that engineers should recognize). I now want to point out four reasons to think that engineering *might* someday merge with software engineering. All four are, oddly, changes in engineering, not in software engineering. Having pointed these out, I shall conclude with three reasons to think that, though possible, the merger is unlikely.

First, electrical and computer engineering (ECE) is often thought to be the field of engineering closest to software engineering. Over the last decade, ECE has become less committed to traditional engineering courses concerned with the material world. So, for example, a number of ECE departments, including the highly respected department at the University of Illinois at Urbana-Champaign, have stopped requiring statics, dynamics, and thermodynamics. If that trend continues, either ECE will split off from the main body of engineering or engineering's core of required *engineering* courses will increasingly resemble software engineering's core.

Second, since the 1700s, engineers have had to know just two natural sciences: physics and chemistry. Recently, some programs in environmental engineering, biomedical engineering, and agricultural engineering have begun to allow students to substitute biology for physics or chemistry. For engineers, this makes sense, since several of the new frontiers of engineering rely on biology rather than (as until recently) physics and chemistry. But, if this trend continues, engineering's required *science* courses will increasingly resemble the science courses that software engineers take to satisfy general distribution requirements.

Third, engineers are increasingly replacing mechanical systems with software. Not only do most engineers now use software regularly, but many also write specifications for software, modify existing programs themselves, or even write (simple) programs. Whether or not software engineers do any engineering, engineers increasingly engage in activities that look like software engineering (even if these engineers do not call themselves "software engineers," do not work the way that software engineers would, and indeed typically work on a smaller scale). Whether some fields of engineering will dissolve into software engineering seems an open question.

Fourth, computer science used to have an accreditation body separate from engineering. That is no longer true. All computer science programs, including software engineering, are now under engineering's accreditation body, ABET. Of course, ABET's accreditation process and standards distinguish between engineering programs and computer-science programs (and mere "technology programs"). But that distinction does not preclude eventual merger. ABET has always distinguished between various fields (or subdisciplines) of engineering. So, for example, it typically sends mechanical engineers to review a mechanical engineering program; electrical engineers, to review an electrical engineering program; and so on. The expansion of ABET's accreditation powers makes it easier than before for software engineering to merge into engineering, indeed, for all of computer science to do that.

Having pointed out four reasons that seem to point to software engineering's eventual merger with engineering, I shall now point out three reasons to believe that the merger will not happen soon, if at all:

First, all engineering is still fundamentally about physical systems; software engineering is not. Even a field so closely allied to software engineering as computer engineering must take physical factors into account in design (to a degree software engineers do not), for example, heat produced in a microchip or speed of electrical current limit what computer engineers can make their machines do in a way the constraints of "pure software" do not.

Second, software engineering is today a large profession, indeed, one of the largest—about the same size as medicine or law, though still half the size of engineering. With so many practitioners, software engineering is more likely to divide than to join up with another large profession.

Third, if computer science ever ceased to be the home of software engineering, the most likely new home might well be management information systems or information technology management. These business disciplines resemble software engineering at least as much as engineering does. In practice (as far as I can tell), most software engineers work more with managers of information systems than with engineers.

Whether knowledge of the future is possible is a perennial question in philosophy. What is certain is that prophets are seldom right on any important question about which many others disagree. So, I make no claim to know whether software engineering will ever merge with engineering. I only claim to know that—despite the common term "engineering"—software engineering is not now engineering. It is "merely" another important profession.

NOTES

Thanks to Keith Miller and Rachelle Hollander for helping me think through this chapter after reading an early draft. Originally published as "Will Software Engineering Ever Be Engineering? *Communications of the ACM* 54 (November 2011): 32–34."

1. Software Engineering Code, 1999. For history of this document, see Davis, 2007.
2. Young and Faulk, 2010, esp. pp. 439–440, argue that engineering has much to learn from software engineering—inadvertently making clear how much engineering's discipline differs from software engineering.
3. For a darker route to this conclusion, see Parnas, 2010. Note that Parnas, though a star of software engineering, is an electrical engineer—both by discipline and declaration—looking at software engineering the way knowledgeable engineers typically do.

REFERENCES

Bourque, Pierre, and Robert Dupuis. ed. *Guide to the Software Engineering Body of Knowledge—2004 Version* (Washington, DC: IEEE Computer Society, 2004).

Davis, Michael, "Are 'Software Engineers' Engineers?" *Philosophy and the History of Science* 4 (October 1995): 1–24.

Davis, Michael, "Defining Engineering: How to Do it and Why it Matters," *Journal of Engineering Education* 85 (April 1996): 97–101.

Davis, Michael, *Thinking like an Engineer* (Oxford: Oxford University Press, 1998).

Davis, Michael, *Code Writing: How Software Engineering Became a Profession*, Center for the Study of Ethics in the Professions: Chicago, 2007 (http://hum.iit.edu:8080/aire/sea/1/book/index.html).

Parnas, David L. "Risks of undisciplined development," *Communications of the ACM* 53, 10 (October 2010): 25–27.

Software Engineering Code of Ethics and Professional Practice (1999), https://www.acm.org/about-acm/code-of-ethics (accessed January 22, 2018).

Young, Michal and Stuart Faulk, "Sharing What We Know About Software Engineering," *Proceedings of the FSE/SDP workshop on Future of software engineering research (Foser '10*, ACM, 439–442.

Chapter 6

Engineering and Business Management
The Odd Couple

"Managers think in quarters; engineers, in decades."

—Anonymous

"What is in opposition is in concert, and from what differs comes the most beautiful harmony."

—Heraclitus

1. THE CHANGING RELATION

The Odd Couple is a play (and movie) about a cohabitation that seemed in prospect certain to fail. When fussy Felix became suicidal over his impending divorce, his best friend, disorderly Oscar, took him in. Within days, Felix and Oscar were finding each other hard to live with. *The Odd Couple* is a serious comedy about the benefits and costs of that "marriage of convenience."

There are at least three reasons *The Odd Couple* seems to me a useful metaphor for the long cohabitation between engineering and business management. The first reason, and least important, is that Felix seems to have the engineer's typical urge toward order and material improvement; Oscar, the manager's typical tolerance of imperfection and changes of plan. Felix is shy and socially awkward; Oscar, talkative and socially adept. The metaphor has a profound appeal. Second, their cohabitation depended on mutual interest. Oscar lived alone in a large apartment that divorce had emptied of wife and child; his housekeeping had turned the apartment into a health hazard. The cohabitation would not have lasted so long as it did had Felix not needed Oscar's company and housekeeping as much as Oscar

73

needed a place to live and someone to listen to him try to understand why his marriage had fallen apart. Third, and most important, both Oscar and Felix changed over time because of living together. Both were better people when they ended their cohabitation than when they began it. Each benefitted from the compromises, experiments, and revelations that their cohabitation forced on them.

There is, of course, also at least one important difference between *The Odd Couple*'s cohabitation and that of engineering and business management. Felix and Oscar ended their cohabitation after a few weeks; engineering and business management show no sign of ending theirs after many years.

The last reason I gave for taking *The Odd Couple* as a useful metaphor for the relationship between engineering and business management was that the odd couple's relationship changed over time, benefiting both. I counted that reason as the most important because scholars tend to overlook how much the relationship between engineering and business management has changed in the two centuries since engineers first entered business in significant numbers—and that change tells us something important about both engineers and business managers, especially about the ways in which they benefit from the relationship, especially the disagreements.

Two centuries ago, engineers were more likely to be independent consultants hired for a job than long-term employees. Like the Roeblings, those early engineers would have had a post-secondary degree in engineering. Business managers, in contrast, were then typically proprietors ("capitalists") educated in the "school of hard knocks." So, for example, Cornelius Vanderbilt (1794–1877), a railway magnate and one of the nineteenth century's richest men, ended his formal education at age 11. Most of what he knew of business he learned from running his own, starting with a ferry service he began at age 16. Such too were the managers that Thorstein Veblen seems to have had in mind in *The Engineers and the Price System* when he described the "business man" of the nineteenth century as one who "came more and more obtrusively to the front and came in for a more and more generous portion of the country's yearly income which was taken to argue that he also contributed increasingly to the yearly production of goods" (Veblen, 1921, 28). Veblen contrasted these businesspeople with the new breed of "financial manager" who "under the limitations to which all human capacity is subject"—because of the "increasingly exacting discipline of business administration"—were "increasingly out of touch with that manner of thinking and those elements of knowledge that go to make up the logic and relevant facts of mechanical technology" (Veblen, 1921, 39–40). The "entrepreneur" of old was evolving into a mere "chief of bureau," an employee knowledgeable about finance but ignorant of technology in a way the older entrepreneurs were not (Veblen, 1921, 41). The new business managers were bureaucrats much like their counterparts in the civil service.

A close reading of *The Engineers and the Price System* will, I think, reveal that Veblen knew little about engineers as such. Indeed, what he sometimes calls "production engineers" (Veblen, 1921, 53), he also calls "technologists" (Veblen, 1921, 61). The list of "technologists" can vary a good deal within a few pages. For example, on page 44, the list is "industrial experts, engineers, chemists, mineralogists, technicians of all kinds"; but on pages 60–61, the list is "inventors, designers, chemists, mineralogists, soil experts, crop specialists, production managers and engineers of many kinds and denominations." For Veblen, the important contrast was between "financial managers" whose focus is on taking a profit every quarter for the *owners* and "technologists," including technically trained managers, whose focus is on increasing the quantity and quality of goods, reducing waste, and otherwise adding to *society's* wealth.

Nonetheless, Veblen did identify an important problem in the relationship between engineers, by then already mostly employees, and business management, by then also mostly employees, an increasing difference between their respective skills, knowledge, and aspirations. The financial manager's focus on profit might often "sabotage" (Veblen's word) the efficient production of useful goods that engineers typically seek. No doubt, it was at least in part this difference between financial managers and engineers, even engineers ranking high in a large corporation, that contributed to what Edwin Layton called "the revolt of the engineers" (Layton, 1971).

The story of the business-engineering nexus does not end with that revolt, however. In the century since 1921, the number of engineers working in business has grown into the millions while the other "technologists" Veblen mentioned now number only in the tens of thousands. Engineers (along with computer scientists) are now central to most large businesses to a degree other technologists are not. What gave engineers this preeminence? The answer is obvious: the ways in which engineers differ from both business managers and other technologists.

Over the past century, business management became a popular field of study in universities. Indeed, many managers today have an advanced degree, typically a Masters of Business Administration, while their engineers typically have only a bachelor's. Business management has itself become a science-based technology, though one resting on economics rather than (as engineering does) physics and chemistry.

Yet, the division that Veblen remarked has not gone away, merely changed. In the 1920s, management ("business administration") seemed destined to join architecture, engineering, law, medicine, nursing, social work, teaching, and the like occupations as a profession. Schools of business management taught students that business should seek to serve society, not simply make a profit (Abend, 2013). But, by the 1960s, it was already clear that business management was *not* going to be a profession (in our preferred sense). Business managers were happy to declare that their primary loyalty was to their employer; their primary goal, to "maximize" their employer's profit. Indeed, some scandals of the 1950s, such as price-fixing in the electrical industry, suggested that managers might believe that loyalty to employer overrode even legal and moral obligations. Senior managers not only broke anti-trust laws for their employers but also lied about it to the press, Congress, or the courts (Herling, 1962).

The introduction of "business ethics" into the curriculum of business schools a decade later was in fact a re-introduction. Courses under that name (or near synonyms) had existed in many elite business schools as early as the second decade of the twentieth century, though most seem to have vanished by 1950 (Abend, 2013).

This new business ethics differed from the old in at least two notable ways. First, the new business ethics developed as a field of research as well as a course of study. There were soon several academic journals (as well as several textbooks and monographs; DeGeorge, 1987). Second, almost from the beginning, philosophers seem to have had an important part in both the research and teaching of the new business ethics.[1] These philosophers seem to have drawn on philosophy's recent experience with medical ethics, especially the emphasis of medical ethics on resolving ethical problems case by case rather than restating old reasons to accept a predetermined answer. The new business ethics was analytic rather than homiletic. But, like the old business ethics, the new did not seem to be a "revolt of the managers" so much as a revolt of their employers, the public, and the government, a response to scandals in which educated managers thought they had done all they should, and only what they should, when they single-mindedly sought (more or less successfully) to maximize short-term return on investment (as they had been taught).

According to some common sociological definitions of "profession" (advanced education, high income, and so on), business management was a profession well before 1960. Yet, by the definition that the professions themselves implicitly accept (and chapter 1 formulated), business management had long since ceased even to aspire to be a profession (Khurana, 2007). Management was definitely not a number of individuals serving a moral ideal. Maximizing return on the capital of one's employer is not a moral ideal (an objective all reasonable persons recognize as good); indeed, maximizing return on investment may not even be the objective of the manager's actual employer. If we take corporate "vision statements" seriously, many employers seek only a reasonable return on their investment so that they can continue to provide a useful product or service.

Rather than becoming a profession, business management had devolved into a mere "money-making calling" in at least two respects. First, of course, managers understood themselves as competing with each other to make as much money as (legally) possible for their respective employers. Profit was the chief measure of their success. The good of society was no longer understood as even among their objectives (though they might point to the social good they happened to do as a reason to be allowed to go on seeking profit). Second, each manager typically understood herself as a mercenary rather than a professional, that is, as a mere individual seeking to make as much money as possible for herself, not as a member of a group seeking to improve the skills, conditions of work, reputation, or the like in the group's common discipline. To have the loyalty of such a manager, an employer had to offer the proper "incentives," especially a high salary, bonuses for achievement, and opportunities to do work leading to "advancement," that is, to a position with an even higher salary and bonuses. We can measure business's increasing awareness of management as a mercenary calling not only by the increasing size of the individual managers' income relative to that of other employees but also by the increasing share of that income coming from bonuses (and other incentives) rather than from base salary.

Unlike the old business ethics, the new was to be not so much an alternative to the money-making conception of management as a constraint on it. Money-making management was to be bridled in certain ways (e.g., by the employer's code of ethics); its energies redirected in other ways (e.g., by replacing the "single bottom line" of profit with the "triple bottom line" of profit, social responsibility, and environmental responsibility).

2. BUSINESS ETHICS VERSUS ENGINEERING ETHICS

In principle, business ethics could be: (a) about how individual employees, including managers, should fulfill their moral obligations as employees, citizens, and human beings in the marketplace; (b) about how businesses should conduct their affairs within the bounds of morality, managers understood as mere agents of their employers (the ethics of profit-seeking organizations smaller than society at large); (c) about what society should expect of business and how it might go about getting it; or (d) some combination of these. In practice (judging from the textbooks), courses in business ethics are today primarily about how businesses, especially large corporations, should conduct themselves (essentially, b).

A typical course in business ethics today will have four divisions. First, there will be an introduction to the central concepts of business ethics, such as moral theories, "stakeholder analysis," commercial law, the market, and the moral status of a corporation (and the people it employs). Second, there will be discussion of moral issues that arise *within* the business, such as affirmative action, conflict of interest, confidentiality, employment at will, drug testing, fair wages, insider trading, occupational health and safety, sexual harassment, and whistleblowing. The emphasis in this second division will be not on how individual managers, much less individual employees, should deal with particular situations involving such issues, but on how the business as a whole should respond (the managers acting as faithful agents of the business). Third, there will be discussion of moral issues that arise between a business and its community, competitors, customers, regulators, suppliers, or others outside. Among these issues will be truth in advertising, influencing government ("lobbying," facilitation payments, and bribery), intellectual property, spying on competitors, legally permitted pollution, mergers and acquisitions, product safety, and social responsibility (especially, treatment of neighbors, suppliers, and society at large). The fourth division will reconsider the first three divisions in the context of "globalization," especially the variety of local customs, cultural differences, and different legal systems that a business is likely to meet when it establishes sales offices, factories, subsidiaries, or even suppliers in another country, especially a relatively poor country. Should a business take its ethics (its special standards) with it wherever it goes, change its ethics to suit the customs, culture, or laws of each country in which it operates, or respond in some other way? (Compare DeGeorge, 1987).

Increasingly, a course in business ethics will discuss "ethics infrastructure": ethics audits, ethics officers, ethics "hot lines," ombudsmen, and so on. This discussion may include corporate codes of ethics, codes of ethics adopted by trade associations, or the like. But I have yet to see a text in business ethics with anything to say about *professional* ethics, much less one noting that many employees in any large business (actuaries, computer scientists, chemists, lawyers, and so on) will belong to a profession and therefore have moral obligations in addition to those of ordinary employees. A few social scientists specializing in business have, it is true, noted the presence of large numbers of professionals in business (see, e.g., Shapero, 1985; or Raelin, 1986). But, to this day, courses in business ethics seem to divide the inside of a business into "management" (a collection of the employer's agents) and "employees" (the rest inside), with management answering to "the stockholders" (or "stakeholders") and controlling "the employees."

I speak here only of texts in (general) business ethics, texts designed to train (generic) "managers." Many business schools have programs in accounting, finance, human resources, or the like that have their own course in ethics (the ethics of the profession in question). These courses have their own texts, ones much more like texts in engineering ethics than the typical texts in business ethics.

Like much of the business school curriculum, the course in business ethics will typically be organized around in-depth study of "cases," some fictional but most actual (whether or not names are changed to avoid lawsuits). Some are law cases, but most are a summary of facts or a collection of documents. Among cases often included are some that are quite old, such as The Ford Pinto (from the 1970s) or The Space Shuttle Challenger (from the 1980s). Others are relatively new, such as the tardy 2014 recall by GM of 800,000 small cars to have their ignition fixed to resolve a safety problem, or the 2015 scandal concerning VW's modification of its diesels' software so that pollution controls worked during tests but not on the road.² Like these four cases, many standard business ethics cases also appear (or at least could appear) in texts in engineering ethics. Such shared cases are, in fact, evidence for a close connection between business ethics and engineering ethics.

Nonetheless, in the United States at least, the course in engineering ethics arose (or, more accurately, re-arose) more or less independently of business ethics, though at about the same time. The same seems to be true of engineering ethics as a field of academic research (Davis, 1990). There are doubtless many reasons for that independence. Among the most obvious are these four: first, engineering schools and business schools, even when located on the same campus, have historically had little to do with each other. Second (and perhaps explaining the first), the atmosphere of business schools is quite different from that of engineering schools (as Veblen would have expected). For example, engineering students are typically much more interested in making *things* work than business students are; business students, much more interested in how to get *people* to make things. Third, though philosophers were as involved in early work in the new engineering ethics as in the new business ethics, they were rarely the same philosophers. Both business ethics and engineering ethics are (what philosophers call) "applied philosophy." Applying philosophy to a practice outside philosophy means learning a good deal about the practice. Learning enough about business to be useful to businesspeople probably left little time to learn enough about engineering to be useful to engineers—and *vice versa*. The economics of applied philosophy made it likely that there would be little overlap among philosophers in fields developing at about the same time but focused on quite different practices. Fourth, the two fields tended to attract different kinds of philosophers. So, for example, philosophers interested in social justice seem more likely to have become involved in business ethics; those interested in technology or professions, to have become involved in engineering ethics.

Not surprisingly, then, a course in engineering ethics typically differs in fundamental ways from a course in business ethics. Perhaps the most important of these differences is that engineering ethics typically is a course in *professional* ethics. There is an attempt to define "profession" and explain how engineering fits that definition. There is a discussion of engineering's code of ethics and practice applying the code to particular practical decisions ("problems" rather than "issues"). Engineering ethics texts typically reprint at least one code of engineering ethics. There may even be an introduction to engineering's professional associations, technical standards, and licensing bodies. The overall message is that engineers have a moral obligation to their profession at least as weighty as their obligation to their employer, whether or not the employer is a business. Engineers are *not* "mere employees."

Teaching engineering ethics is, however, not limited to a course in that subject. Such teaching goes on both explicitly and, more often, implicitly, in engineering's "technical" courses. Though I have written a good deal about explicitly integrating professional ethics into engineering's technical courses, I believe explicit integration is still relatively uncommon. So, I shall say no more about it here (for more, see, e.g., Davis, 2006; and Davis et al., 2016.) What does seem to be a common practice is the *implicit* integration of engineering ethics in at least some of engineering's technical courses. The integration goes on using such terms as "accuracy," "documentation," "efficiency," "reliability," "safety," and "sustainability." Such terms denote technical standards in engineering—standards government, engineering associations, or independent standard-setting bodies have elaborated in considerable detail. In general, engineering's technical standards are ethical insofar as they are morally binding guides to conduct that each engineer (at her most reasonable) wants every other engineer to follow even if the others following them would mean having to do the same. For engineers, their profession's ethics is (or, at least, should be) not so much a supplement or constraint on their main pursuit as a component of what they seek to accomplish. To be a good engineer is to help improve the material condition of human beings in the way engineers typically do, not to make a lot of money for self or employer (though, of course, money is always welcome). Accuracy, documentation, efficiency, reliability, safety, sustainability, and the like are part of good engineering, not mere constraints on what engineers as such do.

Engineering is sometimes described as a "captive profession." The description is supposed to recall a time when engineers were as "free" as most other professions were, but now they (today's engineers) only survive in "cages," the large organizations in which engineers now typically work, especially modern business corporations (Noble, 1977; Goldman, 1991). This description of engineering seems to be mistaken for at least five reasons.

First, much of the plausibility of claims for engineering's captivity seems to arise from confusing the *function* of engineers (building, designing, and so on) with the *discipline* of engineers (the special knowledge, skill, and judgment, largely taught in engineering school, that engineers bring to building, designing, inspection, and other work engineers typically do). While the function, or rather functions, of engineers have been carried on in many societies, including some quite ancient, and under many names (architect, builder, inventor, machinator, mechanic, technician, and so on), the discipline (as I have said before) seems to be much newer, originating in the French army in the late 1600s. Engineering became a civilian career only in the early 1800s when civilian technology, beginning with railroads, became demanding enough to benefit from engineering's special discipline (Davis, 1995)—and a profession only late in the 1800s when engineers began to work for a living. While some of those who have functioned as engineers in earlier times may have done so free from any large organization, those sharing the discipline of engineering have not.

Second, because professions are, by definition, ways to earn a living, no profession can long survive without employers, people to pay the cost of carrying on the profession. Even the freest profession must generally do what its employers want or cease to exist. Engineering has never been free of employers—nor could it be without becoming an (expensive) avocation rather than a profession. That is as true of other professions—including law and medicine—as of engineering.

Third, engineers have never been able to do much on their own. Even in the days when a lone engineer might oversee a siege, he could do little without the large organization that determined where he employed his siege craft and provided the labor, supplies, and protection necessary to carry out his plans. Today, good engineering generally requires the resources of a large organization, including the cooperation of other engineers. One engineer alone is, and always has been, of little use, an engineer only in the sense of having the potential to do engineering if others provide the resources.

Fourth, all this is as true of engineers working for government, a socialist enterprise, or a non-profit as of engineers working for a business. The word "captive" in "captive profession" sounds bad but in fact tells us nothing about engineering. While profit is a constraint on engineers working for a business, it corresponds to the constraint of budget characteristic of government, socialist enterprise, or non-profit. Business has not captured engineering—in any interesting sense of "capture." Engineering is, instead, a profession having a symbiotic relation with large organizations, whether for-profit or not.

Fifth, the idea that projects that are "intrinsically technically challenging and interesting but without a market" (Holt, 2001, 498) would have precedence in engineering but for the profit-motive of business seems to involve at least two mistakes. One mistake is the assumption that only business constrains engineers in some such way as this. In fact, every organization for which engineers are likely to work must direct the effort of engineers away from the merely technically challenging toward the useful, however, prosaic. Few engineers are free to do what they want even in a government laboratory. Few engineers are hired to do "pure science." The other mistake is to assume that the intrinsically technically challenging project should be the aim of engineers once freed of practical constraints. The moral ideal engineers seek to serve is (more or less) improving the material condition of human beings, not high-tech at any cost. A project without a market is unlikely to improve the material condition of human beings. It is therefore unlikely to count as good engineering. Hence, it is hard to know what the (seldom-used) term "pure engineering" might mean.

3. IMPORTANCE OF DISAGREEMENT BETWEEN ENGINEERS AND MANAGERS

The line between engineers and business managers is not as sharp as the discussion so far may suggest. The manager overseeing the work of engineers is likely to be an engineer as well (whether or not holding a business degree in addition to an engineering degree). Indeed, even the senior management of many large businesses will include a significant number of engineers. For example, of Lockheed Martin's eight vice presidents, three are engineers[3]; of GM's twenty-four senior officers, seven are engineers.[4] Many disagreements between engineers and business management are (in part at least) disagreements between engineers.

But besides, below, or above such "engineer-managers" may be managers trained only in accounting, computer science, industrial design, law, marketing, or another non-engineering discipline. Many of the ethical problems that engineers face in practice arise (as they did in Veblen's day) as a disagreement between engineers and such non-engineers. Some of these disagreements set engineering against finance (the constraints of budget), but some may set engineering against aesthetics (what designers think looks good), "culture" (what marketing thinks customers expect), or law (what lawyers think necessary to protect the employer against legal liability). Products of a modern business (like products of government) typically involve complex negotiation between many "stakeholders," some of them inside the business.

It is easy to assume (as Veblen did) that when there is disagreement between engineers and "financial managers," the financial managers must be wrong. They are wrong sometimes, of course, but certainly not always. Some engineering solutions may be both beyond an organization's resources and, while morally desirable, not morally required. Much of the time, the right answer, or even the least bad answer, about what to produce or how to produce, sell, maintain, or dispose of it may be unclear, especially at first. The work of business is increasingly carried on by interdisciplinary teams because no discipline has a monopoly on answers to the complex problems that modern businesses face.

What has been called "the revolt of the engineers" may be understood as part of a larger and longer negotiation both within engineering and between engineering, its fellow professions, managers, and their common employers concerning what engineering is, what it should do, and why it should do it (Sinclair, 1980). The "revolt" focused primarily on two issues: one about management (the power that engineers should exercise in corporate decisions); the other, about the welfare of "bench engineers" (their salary, conditions of work, opportunities for advancement, and other reasons ordinary engineers should have for doing their job). Meanwhile, engineers were making themselves increasingly necessary, especially for businesses making or operating complex artifacts, such as airplanes or electric power plants. Engineers made themselves increasingly necessary by developing technical standards, publishing them through professional organizations, and then trying to follow them. The standards were developed to reduce waste, increase safety, protect health, and so on. Insofar as the standards did what they set out to do, they served long-term business interests, tying business to engineering even as engineering seemed ever more subordinate to business. Even as the "revolt" collapsed during the 1920s, a revolution in the relationship between engineers and business management continued: The "master" increasingly became dependent on the "slave."

Consider, for example, the sealed-beam headlight. It was developed by engineers concerned to improve safety on night-time roads. It was adopted as the industry standard in 1939, a time (the Great Depression) when engineers are supposed to have been most subservient to business. The new headlight, though a technological leap, was a natural extension of standards that two engineering associations, the Illuminating Engineering Society and the Society of Automotive Engineers, had jointly been working on since 1918. The headlight was developed by engineers at General Electric (GE), especially Val Roper, the leader of an applied research team at GE's Automotive Lighting Laboratory in Cleveland, Ohio. Technical feasibility was established in 1937.

From the perspective of the typical "financial manager," the decisive barrier to adopting the new headlight was, however, not technical feasibility but financial:

> [In] 1937, General Electric, as a diversified company, had no compelling motive to overhaul a segment of their lamp business which was already profitable, growing, and arguably producing state-of-the-art products. In fact, some in the company argued that it would be wrong to require depression-beleaguered Americans to buy and install expensive new headlights. The market would buckle to popular resistance, and G.E. would be left with sizable losses from the venture. (Meese, 1982, 12)

Roper argued in response that failing to bring the new headlight to market was to continue tolerating the horribly high rate of nighttime auto accidents. More importantly, Roper was soon drawing on a network of engineers—in GE itself, in American auto manufacturers (such as GM), in state bodies regulating auto safety, and in headlight manufacturers to whom GE sold light bulbs but with whom GE might soon be competing with its new headlight—to work out a plan to overcome the legitimate worries of the financial managers while simultaneously stressing the importance that the safety of the public should have in the final decision:

> [Roper credited] the rapid introduction of the Sealed Beam headlight to the responsiveness and flexibility of General Electric management [primarily senior engineer-managers], the industry-wide cooperation regarding the exchange of technical information at the engineer-to-engineer level, the restraint of A.A.M.V.A. [American Association of Motor Vehicle Administrators] to withhold preemptive new regulation, and the persistent efforts of the S.A.E. Lighting Committee and the I.E.S. Headlighting Committee. (Meese 1982, p 16–17)

There is, I suggest, nothing unusual in this story of engineers leading the way in making a business decision except for the scale of the achievement. This story nonetheless has at least three lessons to teach concerning the relationship between engineering and business management (and, indeed, between engineers and managers generally).

The first lesson concerns breadth of vision. It is often said that engineers are narrowly technical while managers, being generalists, see the big picture. While some engineers may be narrowly technical, many are not. As in this story, the difference in vision may not be breadth so much as direction, with engineers looking one way and (financial) managers looking another. The safety of the public is certainly at least as broad a concern as GE's financial welfare. In another respect, however, it is the financial managers who plainly have the narrower vision. Not being professionals, their chief commitment (beyond morality's minimum) must be to their employer. They are expected to look beyond that commitment only if their employer instructs them to do something seriously morally wrong. Engineers, in contrast, have commitments to do good extending well beyond their employer's instructions, commitments arising from their profession.

The second lesson concerns political skills. Engineers are often thought of as politically helpless while managers are politically astute. The story of the sealed-beam headlight is, however, the story of engineers who were politically astute—at least while working within a network of engineers. The truth is probably that financial managers are generally good at working with other financial managers but not with "technical people." For dealing with senior management, especially senior managers who are not engineers, the financial managers may be better able to speak the common language—which, after all, is money. But for dealing with outside regulators, or engineers at suppliers, customers, or competitors, engineers may be better able to speak the language—which is more likely to be engineering than money.

The third lesson concerns the relative sterility of financial management. Like the older term "administration," "management" as such is typically about overseeing, reporting, or making arrangements, not inventing. Engineering, in contrast, is typically (in part at least) about inventing, improving old artifacts or creating new ones. From the perspective of engineers (and the rest of us), financial managers (whether in business, government, or non-profit) will either go along with the engineers, helping with their projects, or be impediments—"saboteurs," as Veblen would say. Of course, labeling financial managers as saboteurs is unfair, even in the story of the sealed-beam headlight. The sealed-beam headlight would have saved few lives had it quickly bankrupted GE (or simply not been accepted by auto manufacturers or the public). If a business is to do good in the long term, it must survive in the short term. One important function of business management, especially financial management, is to think about the short term when no one else is.

4. A PROPOSAL

The forgoing analysis seems to suggest a major change in the curriculum of business schools: Business schools should systematically teach about professions. What they should teach is, however, not best described as "managing professionals" but as "managing *with* professionals." "Managing professionals" suggests that professionals are passive, and managers are in control. The addition of "with" suggests instead not only that some managers will be members of this or that profession but also that managers must work with professionals, even if the professionals are not themselves managers, rather than merely control them.

Among the topics that should be stressed when teaching managing with professionals is the importance of disagreement between professionals and their managers. Professionals, though experts, are not mere experts. In addition to their special knowledge, skill, and judgment, professionals have commitments different from those of the ordinary manager. Professionals, such as engineers, are in fact hired in part because of those commitments. So, for example, one reason to hire an engineer, rather than an ordinary manager, to supervise safety testing is that engineers are committed to safety to a degree ordinary managers are not—whether the business makes the hire because it values safety as such, because the law requires an engineer to supervise certain safety tests, or because the legal department urged the hire to reduce liability should some accident occur. The engineer will serve the employer by carrying out those safety tests according to engineering standards even if the results hurt the employer in the short term. Out of a disagreement between a manager worried about that short-term harm and an engineer concerned to maintain engineering standards may come an agreement satisfying both and better than either original alternative ("the beautiful harmony" of which Heraclitus spoke).

Of course, such an agreement is more likely to come out of initial disagreement if the manager has learned how to carry on the discussion necessary to reach such an agreement. A course in business ethics should, therefore, include role play in which some students play engineers and some play managers engaged in trying to reach agreement that respects the concerns of the engineers as well as of the managers. Both business ethics and other management courses should pay more attention to the discussions out of which important decisions, as well as unimportant ones, come. Indeed, I think today's emphasis on "leadership" in business is a mistake. Leaders are typically people who know where they should go and how to get others to follow. In many situations involving engineers, especially the most important, neither managers nor engineers are (initially at least) in a position to lead (in this sense). Like the odd couple, they must work their way to solutions they cannot anticipate, helping each other along. Better than leadership are the compromises, partial solutions, and inventions of cohabitation.

NOTES

Originally published as: "Chapter 2. Engineers and Business Managers: The Odd Couple," in *The Engineering-Business Nexus – Symbiosis, Tension and Co-Evolution*, edited by Steen Hyldgaard Christensen et al. *Philosophy of Engineering and Technology*, v. 32 (Springer, 2019): 25–38.

1. The only philosopher I have come across in the old business (and professional) ethics is Carl F. Taeusch (1926).
2. For the details of this case, see chapter 16.
3. See biographies of: Patrick M. Dewar, Executive VP; Dale Bennett, VP for Mission Systems and Training; Richard F. Ambrose, VP for Space Systems, http://www.lockheedmartin.com/us/who-we-are/leadership.html (accessed October 17, 2015).
4. See biographies of: Mary T. Barra, Chief Executive Officer; Alan Bately, Executive Vice President and President, North America; Alicia Boler-Davis, Vice President of Global Connected Customer Experience; James B. DeLuca, Executive Vice President, Global Manufacturing; Grace Lieblein, Vice President, Global Quality; Karl-Thomas Neumann, Executive Vice President & President, Europe; Mark Reuss, Executive Vice President, Global Product Development, Purchasing and Supply Chain; Matt Tsien, Executive Vice President and President, GM China. http:

//www.gm.com/company/aboutGM/GM_Corporate_Officers.html (accessed October 17, 2015).

REFERENCES

Abend, Gabriel, "The Origins of Business Ethics in American Universities, 1902–1936," *Business Ethics Quarterly* 23 (2013): 171–205.
Davis, Michael, "The Ethics Boom: What and Why," *Centennial Review* 34 (1990): 163–186.
Davis, Michael, "An Historical Preface to Engineering Ethics," *Science and Engineering* Ethics 1 (1995): 33–48.
Davis, Michael, "Integrating Ethics into Technical Courses: Micro-Insertion." *Science and Engineering Ethics*, 12 (2006): 717–730.
Davis, Michael, "Is Engineering a Profession Everywhere?" *Philosophia* 37 (2009): 211–225.
Davis, Michael, Laas, Kelly, Hildt, Elizabeth, "Twenty-Five Years of Ethics Across the Curriculum: An Assessment," *Teaching Ethics* 16 (Spring 2016): 55–74.
DeGeorge, Richaed T. "The Status of Business Ethics Past and Future," *Journal of Business Ethics* 6 (1987): 201–211.
Goldman, Steven, "The Social Captivity of Engineering." in Paul Durbin (Ed.), *Critical Perspectives on Nonacademic Science and Engineering* (Bethlehem, PA: Leheigh University Press, 1991), 121–146.
Herling, John, *The Great Price Conspiracy* (Washington: Robert B. Luce, Inc., 1962).
Holt, J. E. "The Status of Engineering in the Age of Technology: Part I. Politics of Practice," *International Journal of Engineering Education* 17 (2001): 496–501.
Khurana, Rakesh, *From Higher Aims to Hired Hands: The Social Transformation of American Business Schools and the Unfulfilled Promise of Management as a Profession* (Princeton: Princeton University Press, 2007).
Layton, Edwin, *The Revolt of the Engineers* (Cleveland: Case Western Reserve University Press, 1971).
Meese, George E. "The Sealed Beam Case: Engineering in the Public and Private Interest," *Business & Professional Ethics Journal*, 1 (1982): 1–20.
Noble, David E. *America by Design* (New York: Alfred A. Knopf, 1977).
Raelin, Joseph A. *The Clash of Cultures: Managers and Professionals* (Cambridge, MA: Harvard Business School Press, 1986).
Shapero, Albert, *Managing Professional People* (New York: Free Press, 1985).
Sinclair, Bruce, *A Centennial History of the American Society of Mechanical Engineers, 1880–1980* (Toronto: American Society of Mechanical Engineers, 1980).
Taeusch, Carl F. *Professional and Business Ethics* (New York: Henry Holt and Company, 1926).
Veblen, Thorstein. *The Engineers and the Price System.* (New York: R. W. Huebsch, 1921).

Part II

THE STUDY OF ENGINEERING AS A PROFESSION

Chapter 7

Methodological Problems in the Study of Engineering

Gather round me, boys, and you will hear
The story of a brave engineer:
Casey Jones was that roller's name—
On a 68-wheeler, he won his fame.
—folk song

1. FUNCTION

Once, while unsuccessfully seeking a position at a certain large technological university, I briefly met with its president. To make conversation, I remarked how unusual it was for a university to have, as his did, both a School of Engineering and a School of Applied Science and Technology. How, I asked, did he decide which programs went into which school? His answer was: "The School of Applied Science and Technology consists of all those programs which look like engineering to me but not to the Dean of Engineering." I thought that answer showed considerable theoretical insight, especially the emphasis on procedure rather than abstract knowledge). I hope my reason for so thinking will be clear by the end of this chapter. In any case, I propose to use the term "technology" (and "technologist") in the same spirit as he did— as a catchall that includes not only what engineers think is engineering (or an engineer) but also what others think is engineering that engineers do not.

By "technology," I shall continue to mean any useful artifact embedded in a social network that designs, builds, distributes, maintains, uses, and disposes of such things. So, for example, while a hammer lost in space is only an artifact, a hammer at work in a factory is technology (part of a technological system). A technologist is anyone with a *significant* role in technology. A young child who lifts a hammer is not a technologist, but a carpenter doing the same is.

Like other technologists, engineers design, "build," or otherwise contribute to the life (and death) of certain technologies. Indeed, designing, building, or the like is (some might say) "*the* function" of engineers, what engineers, and only engineers, exist to do. It is what defines engineering.

This way of defining engineering is mistaken for the reasons I gave in Chapter 2. Engineers have many functions besides designing, building, or the like: some engineers simply inspect; some write regulations; some attempt to reconstruct equipment failures; some sell complex equipment; and so on. What defines engineering is a number of individuals sharing a common history.

We can now begin to understand what was going on at that unnamed technological university. The president, not himself an engineer, was applying a functional definition of engineering, one that could not distinguish between engineering and closely related technologies. The Dean of Engineering then applied engineering standards (especially, ABET's list of engineering subdisciplines), which did distinguish between engineering and closely related technologies. These standards recognized naval architecture and applied physics as engineering but excluded "engineering technology," "packaging science," software engineering, architecture, computer science, industrial design, engineering management, synthetic chemistry, and so on.

But, it will be objected, surely the theoretical question of what is and what is not engineering cannot be settled in such a practical way. There are good reasons beside curriculum for, say, excluding architecture and synthetic chemistry from engineering while including naval architecture or applied physics.

I agree. But those reasons are themselves a consequence of history, that is, a consequence of decisions that, over several centuries, made the discipline of engineering what it is today. The discipline might have been different, indeed, so different that it would not count as engineering at all. A theory that draws the opposite conclusion has to engage with that history.

2. ORIGIN OF THIS APPROACH

Those who know the history of philosophy may have recognized that my approach to defining engineering is (more or less) "Hegelian." Those who know me will find this Hegelianism surprising. I am—by training and in most other respects—plainly an analytic philosopher. I am, in short, not someone to abandon the ideal of abstract definition without a crisis, nor someone to turn to Hegel without first trying all the obvious—and some not so obvious—alternatives. I will now briefly describe that crisis. I will do that not because I suppose many readers want more of the autobiography I began in the Preface but because the story should clarify both the usefulness of following the method I propose and its origins not in the abstractions of philosophy but in the practicalities of empirical research.

When I began teaching engineering ethics in the 1980s, the field was in difficulty. Unlike philosophy of technology, engineering ethics presupposed that engineers could choose whether to act as they should or do something else. Most of the engineers I worked with at IIT accepted that presupposition. Yet, the way the social scientists then understood the hierarchical structures in which most engineers worked—whether government or large businesses— there was little or no room for autonomous choice. Engineers (it was thought) were told what to do and, if they did not do exactly that, were fired; they were, as well, allowed to see only a small part of the project they were working on, making foresight of its consequences rare. They were the captives of those who employed them.

This understanding turned out, upon examination, to rest on theory or on extrapolation from studies of ordinary workers, not from empirical studies of engineers in particular. The social scientists might be wrong. How to know? Thus began the thinking that, a few years later, became an interdisciplinary research project funded by the Hitachi Foundation of America. (Davis, 1998, Chapter Nine)

We (the research group with which I worked) eventually conducted sixty on-site interviews at ten companies. Our research protocol initially called for interviewing an equal number of "engineers" and "managers of engineers" (three of each in each company). We relied on the companies to give us the names and telephone extensions of engineers and managers willing to be interviewed. The first time I called a manager to make an appointment, I ran into a problem: When I explained the project and informed him that he was one of the managers we wanted to interview, he responded that he would be happy to be interviewed but thought I should know that he was an engineer who, for the time being, held the title "Manager." When I asked why he thought he was still an engineer, even though his employer did not count him as one, he told me about his education, about his early work for the company, and about how important it was for those who managed engineers to be engineers themselves.

When that first response turned out to be a common response, I drew a conclusion not then in the literature: engineers were like academics. Just as academics are commonly managed by other academics, engineers are commonly managed by other engineers. Technical management is a stage in an engineering career, not a distinct occupation entered by leaving engineering. That conclusion led to another. I needed a definition of "engineer" distinct from the job title or description so that I could pick out managers who were engineers from managers who were not. Self-identification was one way, but I thought it better to have a more objective way to do it. I soon decided that that first manager had provided such a way, one that was also practical: ask about education and work history.

When I applied that method, I learned not only that most managers of engineers were themselves engineers, but also that the managers' self-identification tended to track the objective criterion. Engineers did not cease (objectively or subjectively) to be engineers when they became managers. I also found that non-managing engineers tended to draw the line the same way, for example, saying that a certain manager was a good technical manager even though not an engineer (having only a degree in, say, physics or technology management). I also discovered that a few people that the company counted as engineers were actually scientists (chemists, computer scientist, or the like). They could be counted as competent in a certain engineering role without counting as engineers—in their own eyes or the eyes of engineers strictly speaking. By treating engineering as a discipline, not a function, I learned a lot more about engineering than I would have had I insisted on treating engineering as a function.

Most studies of engineers I see, even today, begin with the assumption that the researchers can tell whether their subjects are engineers without asking them or checking their education and work history. They rely on title or function. That, I think, is a serious mistake—analogous to an anthropologist undertaking a study of a tribe without determining who is related to whom and in what ways.

That insight (that education and work history were relevant to identifying engineers) led me, eventually, to rethink the history of engineering. If engineering is defined by its disciple (education and experience), not its function, how far back can we go before that discipline disappears into precursors, prototypes, or analogs?

3. DISCIPLINE

By "discipline," I mean any set of standardized ways of carrying on a specific activity, developed over time and taught in some structured way. Breathing is not a discipline but the special breathing required for meditation is. Building is not a discipline but building according to the standards of the Guild of Masons was. Inventing is not a discipline, but engineering is.

The history of engineering is in large part the history of its discipline. The way I now tell that story, the discipline began to take shape after the French created the *corps du génie* in 1676. Had the French given a different name to that organization (say, *corps de l'artifice* or *corps du mécanisme*) as French then allowed, we today might well have a different word for engineering (say, "artifice" or "machining"). Before 1676, the term "engineering" (or its equivalent) referred to a function (primarily, the management of sieges, whether defense or assault, and whatever skills were necessary for that function). An engineer was simply someone who managed sieges (catapults, artillery, trenching, sapping, and so on). Within a few decades after 1676, the term "engineering" (or, rather, *le génie*) referred to the French way of doing such things. By then, engineering was a discipline.

To say that engineering (in the sense relevant here) did not exist before 1676 is not to say that there were no technological achievements before then that might now count as engineering. There were, of course, for example: the invention of the ax, sling, and spear; the Passage Tomb at Newgrange (3200 BC); the Egyptian pyramids (2575–2150 BC); and the Beijing-Hangzhou Grand Canal (581–618 AD). To say that engineering did not exist before 1676 is to say instead both that no one called "engineer" (in today's sense) did any of these things and that those who did do them did not work as engineers typically do but according to another discipline—or no discipline at all. The history of engineering is only a small part of the history of technology.

During the 1700s, the French slowly developed a curriculum—a sequence of formal courses—to teach the new discipline to those who were to be engineers in (roughly) our sense, *officieurs du génie* (not enlisted men with shovel or saw, an older sense of "engineer," but officers, men who did not use tools but oversaw their use). There was much curricular experimentation, some of which—from our perspective—may seem ridiculous, such as (for a time) including riding, dancing, and fencing in the curriculum. But, by the late 1700s, the curriculum was recognizably what it is today: calculus, physics, chemistry, mechanical drawing, statics, dynamics, and so on. There was a common core lasting three years; then, in the fourth year, the engineers specialized, choosing artillery, military engineering (fortification and sieges), mining, bridges and roads (for military use), cartography, or shipbuilding. Though the engineering curriculum has changed much since then (e.g., adding electricity and computing), today's engineering curriculum resembles that of 1800 more than it resembles any other discipline's curriculum then or now.

Generally, it is this curriculum, or rather the distinctive ways of doing certain things (the discipline) resulting from it, that distinguishes engineers at any time from the non-engineers around them, whether they have "engineer" in their job title or in the name of their discipline. So, for example, Joseph Paxton designed and oversaw construction of the Crystal Palace to house Britain's Great Exposition (1851). It was only later in the nineteenth century that engineers (strictly so called) came to dominate large building projects. Today, the Crystal Palace could not be built without engineers involved at every stage after the initial sketch. But that fact does not mean Paxton was an engineer, only that he functioned as an engineer at the time. Paxton was a gardener without any formal education at all.

Have I not (it might be objected) put too much emphasis on the curriculum as the means of distinguishing the engineering discipline from other technological disciplines? The first year or two of the engineering curriculum today differs little from the corresponding curriculum in math, physics, or chemistry. In the last two years, it does differ from these, but the curricula of the major fields of engineering differ from each other considerably too. Does that not make it hard to see engineering as a single discipline—without falling back on generalities that make it hard to distinguish engineering from the physical sciences? Indeed, the problem of line-drawing may be getting worse. As noted in chapter 5, at least one field of engineering, electrical and computer engineering (ECE), seems to be abandoning courses that have helped to define the engineering curriculum since the eighteenth century, especially statics, dynamics, and thermodynamics. Such changes may not individually, or even collectively, mean that ECE has ceased to be an engineering discipline but, if they do not, what does make engineering a single discipline—if it is?

This question points to two unusual features of the way I have understood engineering. The first is that I have described the engineering profession as making decisions concerning the similarity (or difference) between the candidates for admission to the profession and those disciplines already in. Similarity is always a matter of degree. Matters of degree are often matters of judgment. Matters of judgment are subject to reasoned disagreement even among those competent to decide. There may be no simple "fact of the matter." I therefore have no reason to be concerned if some people, even some whose judgment on matters of engineering I respect, have doubts about the status of some subdiscipline of engineering when I do not. What I have said about engineering stands as long as there is a historical core about which there is no dispute. That core can then make decisions about the others— decisions anyone, even a philosopher, can approve or criticize (just as judges make legal decisions which anyone may approve or criticize). But, just as with judges, so with the engineering profession: their judgments concerning membership matter in a way mine do not.

The other unusual feature of the way I have understood engineering is that inclusion of a subdiscipline in engineering is (in part) a matter of history. That means, among other things, that what has already been included matters to what will be included later. Consider ECE again. It might be that, if ECE were invented today, it would—like software engineering—not be recognized as engineering (strictly speaking). On the other hand, because of past decisions, ECE, already is an engineering discipline and, therefore, is likely to remain so in part at least because it is itself part of the comparison group. The departure of what was formerly an engineering discipline is unlikely to occur until the difference between that discipline and the rest of engineering has become so great that working as a single discipline seems too inconvenient. The inconvenience will consist in part of differences in curriculum (what engineers are supposed to know) but in part too in what happens after members of the discipline enter practice. Right now, the various disciplines of engineering do not seem to have trouble working together as engineers. Indeed, engineers not infrequently migrate from one subdiscipline of engineering to another during their career. If, as a result of changes in curriculum, a certain engineering subdiscipline can no longer work with other engineering subdisciplines without standing out as alien, then either the changes in curriculum will be abandoned or the former engineering subdiscipline will eventually be accounted something other than engineering (strictly so called)—as happened, for example, with "scientific management" (which began within mechanical engineering and ended up as operations management and research, a business discipline).

So, I agree that recent changes in the ECE curriculum do not "mean" that ECE is no longer an engineering subdiscipline. Of course, "mean" suggests that there is a sharp line between engineering and everything else, one making judgment unnecessary. Either recent reforms in the ECE curriculum have obviously crossed that line or they have not. As I have defined engineering, though, not only is there no sharp line between engineering and everything else (except what engineers choose to draw) but also that there is a process of deciding a subdiscipline's status as engineering that may take years to reach a conclusion. The objection thus seems to miss the point of my historical definition. Ultimately, it is history that decides—with abstract reasons (similarities and differences) constituting only some of the relevant considerations.

4. OCCUPATION AND PROFESSION

Though the curriculum of engineering is recognizable by the late 1700s, engineering did not become an occupation until many decades later. This will seem a strange claim to anyone who does not appreciate how much is built into the term "occupation." By "occupation," I mean (as before) any fulltime activity defined (in part at least) by a discipline by which one can (and a significant number of people do) *earn a living*. Not all disciplines are occupations. So, for example, fencing, though certainly a discipline, is (in the United States at least) not a way to earn a living (though teaching fencing may be).

Engineering could not become an occupation until it ceased to be an exclusively military activity and became something more or less independent. Until then, engineers were a certain kind of military officer. They did not have a "calling" of their own. Engineering (strictly speaking) did not separate from military engineering much before the 1830s when railroads became the first important civilian employer of engineers (strictly so called). It was about then that the earlier distinctions between kinds of military engineering, including "civil engineering" (roads and bridges), became a distinction between military engineers (of all kinds) and civil engineering (in the modern sense—the building of great works for *civilian* purposes). Mechanical engineering developed a decade or so later as it became clear that civil engineers were not prepared to deal with powerful mobile boilers (such as powered railway engines and then ships).

But even after civilian engineering separated from military, engineering still could not be an occupation. Engineers were still gentlemen. And, until well after 1830, a gentleman could not *earn* a living. To earn one's living meant "going into trade" or becoming a "hired man" (or, worse, a servant). For a gentleman to go into trade or become a hired man was to cease to be a gentleman. Gentlemen were supposed to have enough inherited wealth to live decently (or, at least, were supposed to act as if they did). Any money a gentleman received for what he did when following a "calling" was not *earned* (the way wages, pay, or salary is earned) but given as an *honor* (much like the modern "tip" but without its demeaning suggestion of subordination). What to us would clearly be payment for services rendered was a "pecuniary acknowledgement" (as physicians called it), a way of showing gratitude. Even today, professionals tend to refer to the price of their services as "my fee"—a word recalling "the knight's fee," that is, the land given to a knight so that he could afford weapons, armor, horse, and the time to fight for his lord. Gentlemen did not work to live but, if they worked at all, lived for their work, whether reimbursed or not. Engineering could not become an occupation until that conception of "gentleman" lost its force (or until engineers became tradesmen or hired men).

The term "gentleman" did not die—as, for example, its opposites, "villain," "varlet," and "churl" (more or less), did. Instead, gentlemanliness was reconceived as one or more of its former implications, especially, good manners, good character, and good education (college or its equivalent). In the rough markets of the late nineteenth and early twentieth century, being a gentleman in this sense (polite, decent, articulate, and well-educated) was not necessarily an advantage. Eventually certain occupations, those that tended to attract gentlemen, began to organize to help gentlemen earn their living as gentleman (in something like the new sense of "gentlemen"). Each of these occupations was, or at least was intended to be, "a profession, not a mere trade or money-making calling."[1]

As noted in Chapter 1, the term "profession" has several senses today. In one, it is just a synonym for "occupation." A professional in this sense is the

Methodological Problems in the Study of Engineering 99

opposite of an amateur. In another sense, a profession is an honest occupation (one it is safe to profess, i.e., to declare openly). Its opposite is a clandestine enterprise ("professional thief"). In a third sense, a profession is a "learnéd art" (one requiring a knowledge of Latin and, hence, a college education). The opposite of a professional in this sense is a "mere" artisan, mechanic, tradesman, or the like. All three of these senses are quite old. But, during the late nineteenth and early twentieth century, "profession" came to have a new sense (in English), one that provides an interpretation of the slogan, "a profession, not a mere trade or money-making calling." Chapter 1 already argued for that new definition.

While each profession (in this sense) is a historical individual, profession as such is an ordinary concept (or conception), one developed by considering what the individuals apparently collected under the term share (and do not share). The professions do not have the final say on which groups count as professions in the same way each profession has the final say on who is (or is not) a member of that profession.

Formal statements of the concept, that is, attempted definitions of profession, might change over time both because the concept itself is changing or because our understanding of the concept has changed (or for both reasons). So, for example, the definition of "water" is different now than what it was, say, three hundred years ago. That is in part because the concept no longer includes all clear, colorless, odorless, and tasteless liquids, but also in part because we have learned that water is H_2O (i.e., that most of what was once called water consists of this chemical compound while some liquids once counted as water, such as *aqua vitae*, do not). Those who seek the meaning of "profession" in the origin of the term misunderstand how language works. Though the origin of the term can be suggestive, it can never be more than that. The concept that a term names stands at the other end of its history—as does any conception.

Professions have been mocked as "gentlemen's clubs." Those so mocking them generally do not explain what is wrong with gentlemen's clubs. They should. After all, there is much to be said for a gentlemen's club if the alternative is, say, a criminal gang, illiterate clique, or collection of charlatans. My guess is that what is supposed to be wrong is the criteria of membership. If, as with an ordinary gentlemen's club, membership in a profession were determined by sex, race, family, religion, wealth, or the like, then there would be something objectionable about professions. A gentlemen's club in which the membership is determined merely by sex, race, family, or the like marks of companionability would still be a gentlemen's club. Indeed, it might even be a good one. The purpose of a gentlemen's club is, after all, to please its members in a certain way (providing a home away from home, good company, and so on). A gentlemen's club makes no pretense of doing anything more exalted. Gentlemen's clubs differ in this respect from other similar voluntary associations, such as the Kiwanis or Lions Club, which have a higher purpose (charity). Professions also differ from a gentlemen's club in this respect. To be a profession, a voluntary association must—as the definition above says—seek to serve a moral ideal (in a morally permissible way beyond what law, market, morality, and public opinion would otherwise require).

A moral ideal is a state of affairs every reasonable person (at her most reasonable) recognizes as a significant public good, that is, as something desirable enough that she wants everyone else to aid in achieving it, whether by positive support or merely by not interfering, even if the others' doing so would mean having to do the same. Among moral ideals are: justice, public health, knowledge, and beauty. The moral ideal of engineering is (roughly) improving the material condition of humanity. To serve that ideal as engineers, engineers must be competent in their discipline, honest in its practice, and so on. The sex, race, family, religion, class, or the like of an engineer is (more or less) irrelevant. Indeed, taking those factors into account in the selection of engineers is likely to exclude some candidates who would be good engineers or to include some candidates who would not be (depending on which criteria are used and whether they are used to include or exclude). Hence, insofar as engineering seeks to serve its moral ideal, it should not select its members in the way a gentlemen's club properly selects its members. Selecting members by sex, race, family, and so on would tend to impede serving engineering's moral ideal.

5. PROFESSION AND CODES OF ENGINEERING ETHICS

Like other professions, engineering seeks to serve its moral ideal by setting (morally permissible) standards that require more of engineers than law, market, morality, and public opinion otherwise would. These are the "higher standards" that are supposed to distinguish a profession from a mere trade or money-making calling. They are "higher" in the sense that they require (morally permissible) conduct that law, market, morality, and public opinion do not require (or at least, do not require until the profession has established the standards in question). These standards are "special" insofar as they apply to the profession in particular, not to all moral agents as such or even to all professions as such.

A profession's special standards are correctly identified as the profession's "ethics" and incorrectly identified with the profession's "code of ethics." I have already argued that professional ethics is best understood as those morally permissible standards of conduct that every member of a group (the profession in question) wants (at her most reasonable) the other members of that group to follow even if their doing so would mean having to do the same. Given this special-standards definition of professional ethics, it is, I think, obvious that the ethics of engineers includes a good deal more than what is called "the code of engineering ethics." Among the standards that are ethics in this sense are (reasonable) technical standards for safety, efficiency, quality, and documentation. Or, to put the point another way, the entire discipline of engineering—apart from those few standards in dispute at any time—constitute the ethics of engineering. What engineers call "a code of ethics" is simply the most general statement of the discipline.

To say of some statement (or command) that it is an (actual) "standard of conduct" is to make two implicit claims. The first is that the statement *generally* guides conduct, that is, that its instructions are generally followed, that those it governs *generally* use it to evaluate their own conduct or that of others in the relevant group, and that members of the group *generally* use it to criticize publicly their own conduct or that of others in that group. If the standard does not at least generally guide conduct, it is an ideal (or model) standard, but not an actual standard—that is, not "really" a standard at all. An actual standard resembles a scientific law insofar as it allows us to predict (with reasonable success) what those it supposedly governs will do.

The other claim implicit in saying that some statement is a standard of conduct is that, though it generally guides conduct, the standard does not always. Statements that always "guide" conduct are not standards but scientific laws (strictly speaking). So, pointing to a few violations of a code of ethics does not refute the claim that it is an actual standard of conduct. A few violations may be explained away as the result of factual mistakes, differences of opinion (rather than as indifference to ethics), or simply as anomalies. To refute the claim that a code of ethics is a living practice requires showing that there are so many violations that the code tells us little, if anything, about what those whom the code supposedly governs will do or even think about what they do.

I am therefore inclined to dismiss those critics of ethics codes who move from a few obvious violations of a code to the conclusion that the code in question is "mere window dressing." Certainly, codes are (or, at least, may be) "window dressing," that is, something put on display to potential customers to attract them into the store behind the window. There is nothing wrong with window dressing as long as the store provides what it displays in the window. The problem is with *mere* window dressing, that is, with displays that mislead potential customers concerning the stock inside (a kind of "bait and switch"). On the evidence I have, codes of ethics in general, and codes of engineering ethics in particular, are not mere window dressing. I have myself interviewed several dozen engineers and found them to be serious about engineering ethics. I have also been assigning students in Engineering Ethics a paper requiring them each to interview one engineer of their own choosing. Generally, they have found those they interviewed not only serious about engineering ethics but knowledgeable enough to give a reasonably good answer to a problem of engineering ethics that the interviewer posed. We definitely need more empirical work on the question of how much engineers actually follow their ethics, including their technical standards, but absent such a study showing the opposite, I think the evidence points to the conclusion that engineering ethics is a living practice.

Indeed, it could hardly be otherwise—or, at least, otherwise for long. Engineering's employers, mostly sophisticated businesses and governments, employ engineers for certain jobs when they could employ other technologists—and, in the past, did. Apparently, they employ engineers now because they suppose engineers to have certain ways of doing certain tasks different from their technological competitors. They suppose some difference because engineers have routinely done a better job at those tasks (installing boilers, maintaining large bridges, managing chemical plants, designing computer chips, and so on) than their technological competitors. Like a trademark, the term "engineer" is valuable only so long as individual engineers generally confirm the expectation that the term invites. Once engineering's special standards become *mere* window dressing, not much time would pass before only a fool would employ an engineer. Engineering would go the way of alchemy, dowsing, and phrenology.

I have not claimed, please note, that most engineers have ever read their code of ethics, much less that they regularly consult it. The interviews that led me to the conclusion that engineers generally act as their code of ethics requires have taught me that most engineers could not even recall seeing a code of engineering ethics. The engineering code seems to be "hardwired" into engineers. Of course, "hardwired" is a metaphor for a process we do not understand very well. Yet, we can be pretty sure that the process is not the self-selection by which students choose engineering (or, at least, not primarily that). Those of us who teach engineering students in their first year as well as in advanced courses can see that many of the attitudes we take for granted in fourth-year engineering students are not present in first-years. The hardwiring seems to occur during the four years of engineering school. Since few engineering courses (at least until recently) explicitly discussed engineering ethics, my best guess is that most engineers learn ethics through instruction in technical standards (which goes on almost everywhere in the engineering curriculum). The students learn engineering ethics much as native speakers learn their own language, that is, while doing something else. We nevertheless teach English to native speakers through the grades (K through 12) so that they can communicate better than the unschooled. Much of education is making explicit what we all know—sort of.

Like many other professions, engineering seems confused about the moral status of its code of ethics (but not, I think, its technical standards). There are at least four reasons for that confusion. First, there is the question of how many codes there are. On the one hand, there seem to be dozens because so many engineering associations have their own code. The American Society of Civil Engineers has one; the National Society of Professional Engineers (NSPE) has another; ABET has another; and so on. Yet, these codes differ in language more than substance and even many differences that seem substantial at first disappear upon inquiry. (For example, engineers whose code of ethics does not yet include a provision concerning sustainable development seem to interpret the environmental or public welfare provision their code already has as including sustainable development.) I have therefore come think of the many formal codes as much like the many dictionaries of (American) English. Though they differ, they are reporting the same underlying reality. One code simply omits what another includes because of a different purpose, style, editor, or the like. One includes an interpretation that might be helpful in a certain context or fails to take account of recent change (because of the date of publication). And so on. This variety in formal statement is consistent with (more or less total) agreement on the "unwritten code" ("the substance").

The second reason engineers may be confused about the moral status of their code of ethics is the supposed source of a code's moral authority. There are in fact *at least* two possible sources.[2] Some codes of ethics are supposed to be morally binding because those governed have taken an oath, made a promise (or commitment), or otherwise given the code an "external sanction." The IEEE's code of ethics is a good example of this sort: the Preamble states that IEEE members "commit" themselves to it when they join the IEEE.

Another source of a code's moral authority (the one defended in Chapter 1) is "internal" to the practice, much as the moral obligation to follow the rules of a morally permissible game arises from one's voluntary participation in the game. (A good sign that we have such a code before us is that it applies to "engineers" as such, rather than to members of some formal association.) The idea is that, when a person voluntarily claims the benefits of a code of ethics—for example, the special trust others place in those whom the code binds—by claiming to be a member of the relevant group ("I am an engineer"), that person thereby takes on a moral obligation, an obligation of fairness, to do what the code says. Because a code of ethics applies only to voluntary participants in a special practice, not to everyone, a code, if it is generally followed, can create trust beyond what ordinary moral conduct can. It can create a special moral environment. So, for example, if engineers generally "issue public statements only in an objective and truthful manner [including] all relevant and pertinent information" (as the NSPE Code of Ethics, like most other codes of engineering ethics, requires), public statements of engineers will generally (and justifiably) be trusted in a way that those of politicians, lobbyists, and even ordinary private citizens would not be. Engineers will therefore have a moral obligation to do as required to preserve that trust. They will have a special moral obligation to provide all relevant and pertinent information even when others do not have such an obligation.

The third reason engineers have to be confused about the moral status of their code of ethics is controversy concerning whether—to be more than "mere window dressing"—the code must be *enforced* in the way laws are enforced, that is, by formal penalties (such as reprimand, fines, suspension, or expulsion). The legal (or "compliance") model of ethics often leads to calls for mandatory licensing of engineers, enactment of the code as "professional regulation," and an official body with the power to bar an engineer from practice for serious violation of the code of ethics. While there may well be good reason for legal enforcement of some aspects of a code of ethics, understanding ethics as primarily about law-like enforcement, that is, formal means of holding engineers accountable (such as expulsion from a professional association), simply confuses ethics with law. Law, custom, and other external guides to conduct do not claim to be standards everyone in a group (even at their most reasonable) wants everyone else to follow. Law, custom, and the like regulation must, then, depend heavily on external enforcement. They are "rules with teeth." Ethics, on the other hand, need not depend heavily on external enforcement. Insofar as individual engineers can see how everyone following the standards in question serves their (common or particular) interest, they have reason to do their share to maintain the trademark's value, that is, they have reason to act as engineers should. If they are dishonest, or simply indifferent to long-term consequences, they may (even at their most reasonable) find that reason unconvincing. They will then be incapable in principle of joining the profession (whatever their education and work history). In practice, they are likely to be driven from engineering by peer-pressure, employer avoidance, civil damages, or even criminal punishment. Most engineers, however, may be counted on to do their fair share (insofar as they understand it) because they are relatively reasonable and morally decent and understand that doing anything else would, all else equal, be morally wrong (cheating the ethical engineers).

The fourth reason engineers have to be confused about the moral status of their profession's code of ethics is that different codes formally apply to different engineers. Some codes apply only to members of an association, some apply only to a class of engineers not defined by organization, and some apply to "engineers" generally. The IEEE's Code of Ethics is a good example of the first; the (Asian) Declaration on Engineering Ethics, of the second; and the code of ethics of the NSPE, of the third. The first sentence of the IEEE code says that IEEE *members* "do hereby commit ourselves to the highest ethical and professional conduct and agree [to the ten rules constituting the body of the code]" (IEEE, 2013). The suggestion is that, but for IEEE membership, the engineers in question would not have those obligations. The Declaration (adopted by the national academies of engineering of China, Korea, and Japan in 2004) speaks instead of "Asian engineers." Interestingly, the only significant difference between the standards of the Declaration and the IEEE or NSPE code seems to be the last: "Asian engineers shall . . . Promote mutual understanding and solidarity among Asian engineers and contribute to the amicable relationships among Asian countries" (Asian Code, 2004). The NSPE Code (2007), in contrast, speaks only of "engineers." There is no distinction between ordinary engineers and (licensed) Professional Engineers, American engineers, and others, or NSPE members and non-members. The suggestion is that the obligations arise from being an engineer, that is, from membership in the profession of engineering, not from membership in any technical, scientific, or occupational association.[3] Only codes of ethics that apply to members of the profession as such are properly codes of professional ethics; the others are "organizational codes" (such as the IEEE's) or "sub-professional codes" (such as the Asian Declaration).

6. CONCLUSION

I have, I hope, now explained the importance of the distinction between function, discipline, occupation, and profession for the study of engineering ethics. While doing that, I tried to dispose of several objections commonly raised to this way of understanding engineering. Some of the objections seem to make the error of trying to refute a general claim with a few counter examples, forgetting that *general* claims (those claiming to be true "for the most part") cannot be refuted with a counter-example or two in the way that universal claims can be. The other objections seem to rely on empirical claims that, if true at all, remain unproved. The error of these objections is putting the burden of proof on the wrong party and then leaving the room, a good strategy for a lawyer or rhetorician, perhaps, but not for a philosopher.

NOTES

Originally published as: "Engineering as Profession: Some Methodological Problems in its Study," *Engineering, Development and Philosophy: American, Chinese,* and *European Perspectives*, Steen Hyldgaard Christensen, et al., editors (Springer Science + Business Media B.V, 2015): 65–79.

1. This discussion of the concept of gentleman relies not on any scholar's work but on my own observations, including observations of the literature of the eighteenth and nineteenth century, various biographies, codes of ethics, dictionaries, and similar documents. As far as I know, the concept of gentleman still awaits its scholar.

2. I ignore a third possibility here, that the code has moral authority because the code's content consists of rules derived (either by deduction or determination) from general moral rules (a kind of natural law approach rather than the two variations of social contract offered there). See, for example, the American Medical Association code of 1847. It presented itself as a work of "deontology" (Davis 2003). I ignore that possibility because it does not seem to have anything to do with the present confusion among American engineers about the moral status of their profession's code of ethics.

3. Some professional societies, such as the American Medical Association, have gone back and forth between the first and third kind of code. For details, see Davis (2003).

REFERENCES

Chinese Academy of Engineering (2004), "[Asian] Declaration on Engineering Ethics," http://ethics.iit.edu/ecodes/node/5076 (accessed January 14, 2013).

Davis, Michael, *Thinking like an Engineer: Essays in the Ethics of a Profession* (Oxford University Press. New York, 1998).

Davis, Michael, *Profession, Code, and Ethics* (Ashgate: Aldershot, England, 2002).

Davis, Michael, "What Can We Learn by Looking for the First Code of Professional Ethics?" *Theoretical Medicine and Bioethics* 24 (2003): 433–454.

IEEE Code of Ethics, http://www.ieee.org/about/corporate/governance/p7-8.html (accessed January 22, 2018).

National Society of Professional Engineers (NSPE), "Code of Ethics for Engineers," http://www.nspe. org/Ethics/CodeofEthics/index.html (accessed January 14, 2013).

Weingardt, Richard, *Engineering Legends: Great American Civil Engineers* (Reston, VA: ASCE, 2005).

Chapter 8

Profession as a Lens for Studying Technology

Engineering ethics is a subfield of both the ethics of technology and the ethics of professions, namely that subfield focusing on certain people (engineers) who help to shape technology in certain ways (by engineering) rather than focusing on the processes, products, or systems they help to shape ("technology" in one everyday sense). As we have seen, engineering ethics provides a means of distinguishing engineers from other technologists, such as architects or chemists. It also suggests questions the answers to which should help researchers, whether social scientists or philosophers, to understand better not only the place of engineering in technology but also the place of other professions there, especially their distinctive contributions to the ethics of technology.

This chapter has four parts: (1) a brief recap of the conception of profession we have been using; (2) some comments on the relation of profession so conceived to the ethics of technology; (3) a sketch of some advantages that studying technology through the lens of profession so conceived can have; and (4) advice concerning how a researcher should set about using that lens to study technology, including its engineering ethics.

Though there are many technological professions, some of which have already been discussed in some detail in preceding chapters, I shall continue to concentrate on engineering for at least three reasons. First, engineering is the technological profession I know best. Second, the subject of this book is engineering, not another candidate for global profession. Third, and most important, engineering seems to be the technological profession *par excellence*. Not only is it among the oldest and largest of technological professions; it is also the one that seems to come first to mind when someone says "technologist." Those who know nothing of engineering can know little of technology.

1. PROFESSION AND SOME RELATED CONCEPTS

By "profession," I continue to mean a number of individuals in the same occupation voluntarily organized to earn a living by openly serving a moral ideal in a morally permissible way beyond what law, market, morality, and public opinion would otherwise require. This definition differs in an important way from definitions of "profession" that social scientists typically use. Unlike those, this one is not a mere list of facts that *often* go together (high income, social prestige, advanced education, licensure, and so on). It is not the result of mere observation of people called "professionals" (everything from athletes to zoologists). Instead, this definition is the result of discussion with self-described professionals in which I tried to state, in a way satisfactory to them, what they meant when they described themselves as members of a profession (especially when they were claiming more than that they earned a living by some trade, art, or calling). The definition is meant to be true of *all* professions (strictly so called)—and immune to clear counter example—not true simply of "most professions," "the most developed professions," or "the ideal type of profession." Since I have defended this definition at length in Chapter 1 (and elsewhere), I shall now simply point out seven of its features especially relevant to the study of technology's ethics.

First, the definition distinguishes the professional from the mere expert. A professional is a member of a profession (or, by analogy, someone resembling a member of a profession in good standing, for example, when we describe someone without a profession as "a true professional" or "real pro"). While there can be just one expert in a field (though, generally, there are more), there can no more be a profession having only one member than there can be an army having only one soldier. A profession always has several members, generally thousands.

Second, the definition distinguishes profession from mere occupation. An occupation is a number of individuals who earn a living by the same (typically full-time) activity. An occupation is just an aggregate of individuals, such as porters or sales clerks. A profession, in contrast, is *organized*. Its members seek to earn their living by maintaining *shared* standards of competence and conduct, standards beyond what law, market, morality, and public opinion would otherwise require. To claim to be a member of a certain profession is to claim to adhere to those standards. Being a member of a profession is, therefore, always more than having a certain social function (such as designing or building) or practicing a certain discipline (such as calculus or carpentry). A profession is a *shared* discipline, shared not simply in the sense in which independent activities can share features because they are similar but in the sense in which participants in a cooperative practice share the practice. Members of a profession rely on one another to work in ways of which they approve and to avoid working in ways of which they disapprove. A profession is a cooperative practice.

One important question for those studying a profession is how this cooperation is achieved. Plainly, much of what is necessary for such cooperation must go on in the appropriate professional school, learning the basic discipline, but almost as plainly much seems to go on after the student leaves school and "enters the profession." Many professions, including engineering, seem to think that members are unlikely to know enough of the profession to count as full members until they have practiced for several years after graduation from an appropriate professional school. Until then, they are still "in training." We can therefore learn much about a profession from its curriculum (both formal and informal), but much too from what typically goes on in the early years of a professional's career.

Third, a profession is inherently value bearing. Whether technology as such is value bearing may be an open question, but a technological *profession* cannot view the technology it works on as a matter of indifference. For example, an engineer who doubts that the product she is working on will improve the material condition of humanity has a reason, as an engineer, to consider ceasing to work on that product. Like other professions, engineering has a moral ideal it seeks to serve. A moral ideal is an outcome that all reasonable persons, at their most reasonable, recognize as an objective the pursuit of which deserves, all else equal, assistance or at least non-interference. Public health is such an ideal; so are justice, beauty, and knowledge. Engineering's moral ideal is (roughly) improving the material condition of humanity. What does not promise such improvement is, strictly speaking, not engineering (or, what comes to the same thing, not "good engineering").

The reverse, however, is not true. That an activity serves a profession's moral ideal does not guarantee that the activity belongs to the profession. For example, that a certain invention will improve the material condition of humanity does not mean that the invention is engineering, much less good engineering. The invention must also be designed, tested, and so on in the way engineers are supposed to do such things (i.e., according to engineering's standards). A profession is a discipline designed to achieve a certain objective *in a certain way*, not simply any activity that achieves that objective.

Fourth, the definition makes ethics an essential feature of profession. Of course, by "ethics" I do not merely mean ordinary morality (general standards of conduct such as "Don't cheat," "Keep your promises," or "Help the needy"), much less philosophical ethics (the attempt to understand morality as a reasonable undertaking). Instead, by "ethics" I mean those morally permissible standards of conduct that apply to members of a group simply because they are members of that group. A profession's ethics (in this sense) consists of morally permissible standards (rules, principles, or ideals) that everyone in the profession (each at her most reasonable) wants everyone else in the profession to follow even if their following the standards would mean having to do the same.

How these special standards become morally binding, if they do, is an important question in the philosophy of professions. My answer to that question is (as I said in earlier chapters) that the standards are morally binding because a professional is, as such, a participant in a voluntary, morally permissible cooperative practice. To violate the standards of such a practice is to violate the moral rule "Don't cheat" (Davis, 1991). Answers others have given include: (a) that professional standards are morally binding on members of a profession because the profession has a contract with society (one in which moral obligation is exchanged for licensure, high social status, and so on); (b) that the standards are morally binding because they are mere specifications of ordinary morality (an application of an ordinary moral rule such as "Don't harm" to the special powers engineers wield); and (c) that the standards are morally binding because society has defined the professional's role to include a moral obligation to act according to the profession's standards (and society has the moral right to define morally binding standards).

Given the special-standards sense of "ethics," a profession's *technical* standards are part of its ethics—at least those technical standards (typically, the vast majority) that members of the profession at their most reasonable want all other members to follow even if their following them means having to do the same. What those shared standards actually are is, of course, an empirical question (but one the answer to which depends on both an analysis of "most reasonable" and an operationalization of that analysis).

Fifth, while a professional may be in business, no profession is a business and no professional is, as such, simply in business. By "business," I mean any effort, whether individual, collective, or corporate, to make a "profit" (i.e., to take wealth from an activity beyond what one puts into it). The profits of business come mainly from buying, selling, or exchanging goods or services (rather than from gift, taxation, theft, or other noncommercial activity).[1] Though professionals typically practice their profession to make a living, their membership in their profession means (or, at least, should mean) that they always have the purpose of serving the profession's moral ideal (in addition or instead of making a living). Professionals may seek to serve that ideal even when such service seems unlikely to be profitable; hence, the explicit endorsement in many professions of some unpaid work *pro bono publico*.

Sixth, the definition does not require that a profession be called "a profession" to be a profession (in this sense). It only requires that the occupation in question (or some part of it) be organized in a certain way. This sixth point is important because this sense of "profession" seems to be both relatively new (probably not much over a century old) and still largely confined to

English-speaking countries. In many languages, "profession" (or its nearest equivalent) is still little more than a synonym for "white-color occupation."

The literature on professions outside the English-speaking world and, indeed, even in it, seems to suffer from confusion concerning the difference between having a word for a certain concept (or conception) and having the concept (or conception). But, just as it is possible to recognize the difference between blue and cyan even if one does not have a word for cyan, so it is possible to organize an occupation in a certain way without having a word for that sort of organization. So, for example, whether engineering is a profession in China (in the relevant sense) cannot be answered simply by finding no word for "profession" in a Chinese dictionary or even by asking English-speaking engineers in China, "Is engineering a profession in China?" Instead, a researcher must ask a series of questions such as the following (all avoiding the term "profession"):

Do you claim to be an engineer?
Why do you claim to be an engineer?
Do you follow the technical standards that engineers share rather than others? Why?
Do you care whether others who claim to be engineers follow those technical standards? Why?
Do you think others who claim to be engineers would care whether you follow those standards? Why?
What, if anything, do you expect of those who claim to be engineers that you do not expect of non-engineers?

The point of these questions, and of others that might have been asked instead or in addition, is to bring out the interviewee's sense (or lack of a sense) of belonging to a group organized around a moral ideal in the way a profession is (for an example of such research, see Davis and Zhang, 2017).

Seventh, since this definition of profession does not require any of the institutional arrangements often associated with profession, such as licensure, accreditation of educational programs, technical societies, or a formal code of ethics, the argument that Harris et al. (2014, p. 189–190) offers for the claim that "the concept of 'professionalism' is a Western idea which is not universally applicable" actually misses the point. "Professionalism" simply means acting as members of the profession in question are supposed to act (or, by analogy, acting as members of a would-be profession should act). Though "professionalism" probably is a Western idea, since "profession" is, its Western origin does *not* mean that it cannot apply everywhere (just as paper money, though originally a Chinese invention, is now used everywhere). Indeed, perhaps aware of the weakness of their argument from

Western origin, Harris et al. seek evidence for their claim in the argument that Iseda (2008) made for a pride-based approach to teaching engineering ethics in Japan. Harris et al. fail to notice two difficulties in Iseda's argument. First, Iseda's claim that Japan needs a pride-based approach because Japan still lacks a profession of engineering relies on a misunderstanding of profession. Iseda assumes that profession requires licensure, high pay, accreditation, prestige, and so on. Yet, several chapters earlier Harris et al. (2014, 13–14) accepted the definition of profession presented here, one that does not make that assumption. Second, Iseda's proposal for improving the teaching of engineering ethics, an appeal to student pride *in engineering*, seems to presuppose that engineering students understand themselves as belonging to a group, engineers, who have undertaken to serve a certain moral ideal in a certain way. They are to take pride in working as engineers are supposed to work. Iseda is, in effect, assuming that Japanese engineering student understand engineering as a profession (one more like nursing or teaching than law or medicine in its social rewards). Though including what seems good advice about how to teach engineering ethics in Japan (and perhaps everywhere else as well), his argument is in fact evidence for the concept of "professionalism" (as I have defined it) being universally applicable, not evidence against. What it is evidence against is the universal applicability of the contract-with-society explanation of professional obligation. I think Iseda and I agree on that.

2. RELATION OF PROFESSION TO ETHICS OF TECHNOLOGY

What then is the relation between profession so defined and the ethics of technology? By "technology," I mean any (more or less) systematic arrangement by which people conceive, develop, manufacture, maintain, use, or dispose of artifacts (where "artifact" refers to any physical thing made for use, anything from artillery shells to biscuits, from circuits to software, from cattle to carrots). I take this to be the common understanding of "technology" among philosophers and social scientists now studying technology (though perhaps not an "everyday understanding").

Many of the people who conceive, develop, manufacture, maintain, use, or dispose of artifacts belong to one profession or another. They are actuaries, computer scientists, dentists, or the like. Working according to their respective disciplines, they typically give the technology they work on a character that it would not otherwise have. So, for example, engineers will typically insist on a specific "safety factor" for any artifact that they design, a safety factor different from that architects, geneticists, physicists, or the like would use (if they would use any). The reason engineers have for a safety factor will typically come neither directly from biology, chemistry, or any other natural science nor from law, market, morality, or public opinion. It will instead come from reflection on the experience of other engineers (or their own) with the artifact in question (or artifacts that seem similar enough). (For examples of how engineering standards actually come into being, see Vincenti, 1990, or Wells, Jones, and Davis, 1986.)

Since professional standards are largely a product of history, not natural or social science, the history of a profession is largely the history of its discipline, that is, a description of how its standards changed, including the failures of particular standards and the refinements such failures provoked. Professions are no more timeless abstractions than an individual human is. Every profession has its own biography. The standardization of a profession's curriculum is an important part of that biography, indeed, often a defining part.

Given these definitions, we can see that the term "ethics of technology" is ambiguous. While it can (among other things) simply mean the application of ordinary moral standards to what people involved with the artifacts of technology are doing, the term can also mean—in addition or instead—the special standards members of a group, such as a profession, bring to what they do with the artifacts of technology or the attempt to understand how technology can be a morally good or, at least, a morally permissible undertaking (when it is). Insofar as technology is an activity involving people (rather than a mere arrangement or collection of artifacts), the study of technology is always, at least implicitly, the study of ethics in at least one of these senses and, often, in two or even all three of them.

3. ADVANTAGES FOR STUDY OF TECHNOLOGY

We may distinguish at least three advantages that studying technology through the lens of profession can have. The first is opening a range of questions about what a particular profession might bring to technology that others, nonprofessional occupations as well as other professions, do not, not only what values they might bring but also what special knowledge, methods, or skills. For example, recall Benjamin Wright, "Chief Engineer of the Erie Canal" from 1817 to 1828: He was a self-taught surveyor with only a primary-school education. He learned surveying to aid his speculation in undeveloped land. Though he certainly *functioned* as an engineer when directing the building of the canal, he is in fact proof that one could then still be a great builder without being an engineer (Weingardt, 2005, 5–9).

Why do I deny that Wright is an engineer (strictly speaking)? First, his education. Not only did he not have a technical education much like that an engineer would have today, he did not even have the post-secondary technical education engineers in his time would have received at, say, the *École Polytechique* or the military academy at West Point. Second, we can explain how he could have functioned as Chief Engineer of the Erie Canal without supposing him to be an engineer, indeed, without supposing even that he worked much as engineers would then have worked. Building canals, even canals as large as the Erie, required only relatively simple technology, simple enough that similar canals had been built thousands of years ago (and in many different civilizations). Anyone who can survey can lay out a canal. Digging the canal, stabilizing its walls, waterproofing them and the floor, and so on are acts most builders can work out. If the canal needs locks (as the Erie did), their design and placement is still relatively simple, something an intelligent person can puzzle out from an old book and a little experience. The hard part of building the Erie Canal was political (maintaining funding over more than a decade) and logistical (feeding, housing, and productively employing thousands of workers in the north of New York State, what was then still largely a forested wilderness).

Once we see that not all builders, not even all successful builders of great works, are engineers, we can ask what engineers—their distinctive ways of planning, testing, documenting, and so on—could have brought to building a canal that other technologists (such as surveyors) did not. We might then want to study what engineers did for railroads, since railroads seem to be the first civilian technology that absorbed many American engineers trained much as engineers are trained today. If, instead of focusing on the profession of engineering (especially, its discipline), we were to focus on the function of engineers (designing, overseeing construction, and so on), we could not even distinguish between the history of technology (which includes the building of the Erie Canal) and the history of engineering strictly speaking (which does not). Every technologist, or at least every successful technologist, would be an engineer (because he *functioned* as engineers now function) whatever his discipline and however far back in time we go. Even the builders of the Ring of Brodgar or the village of Skara Brae would be engineers. Many interesting research questions, such as those about differences among technological professions, would be defined away, impoverishing the study of technology.

A second, but related, advantage of studying technology through the lens of profession is that studying technology that way invites inquiry into the naming of certain technological activities, for example, the many activities now called "architecture" (computer architecture, virtual architecture, landscape architecture, and so on) or "surgery" (tree surgery, surgical strikes, pet surgery, and so on). Which of these activities actually belong to the profession its name implies? Which are similar enough for a strong analogy? And which are merely connected metaphorically? What purpose, if any, is an analogical or metaphorical naming intended to serve? Is the purpose legitimate? Such questions are open only if a researcher can distinguish between the profession strictly speaking (even when called something else, for example, "naval architecture" instead of "naval engineering") and activities called by the profession's name though not belonging to that profession strictly speaking ("building engineer" for a custodian or janitor).

I note the possibility of such inquiry because engineering seems especially subject to having its name taken for activities that are not engineering strictly speaking. Some of the taking is historical accident. For example, as noted before, the drivers of railroad trains today take the name "engineer" simply because long ago, before there were any engineers strictly speaking, anyone in charge of an "engine" (such as a catapult or winch) might be called an "engineer" or "engineer." (Perhaps Benjamin Wright's title, "Chief Engineer," carried that old sense when he assumed it in 1817, since he might as informatively have been called "Master of Works" or "Chief Builder.") Some of these takings, though deliberate, may nonetheless be justified. For example, "software engineers," though not engineers strictly speaking, intended their new profession to resemble engineering closely. The taking was a commitment to that resemblance. But, often, the taking seems to be only an attempt to appropriate the reputation of engineering for an activity not deserving it: "social engineering" (for large-scale social experimentation), "re-engineering" (for a certain fad in organizational restructuring), "financial engineering" (for risky investment strategies using computers and statistics), "genetic engineering" (for biologists' tinkering with genes), "climate engineering" (for scientists' tinkering with climate), and so on (Compare Hansson, 2006). These are all activities from which both engineers and their typical methods are (largely or altogether) absent. Much might be learned about the ethics of technology from inquiring into such rhetorical misappropriation. But such inquiry is open only if researchers have a way to distinguish legitimate uses of a profession's name from illegitimate. The lens of profession can provide that.

A third advantage of studying technology through the lens of profession is that we can see that most large technological organizations—and, indeed, many smaller ones—include several professions (not only accountants, engineers, and lawyers, but also botanists, chemists, computer scientists, librarians, mathematicians, technical writers, and so on). We can ask what each of these professions contributes to the overall work of the organization. The answer is not always obvious—and often quite informative. For example, when I visited Argonne National Laboratory, I found that about half the "scientists" I talked with were engineers (i.e., identified themselves as engineers even when their job title identified them as something else and sometimes even when they had an advanced degree in a science). What they reported doing varied considerably, too much for me to see a pattern. So, I was left asking what engineering's distinctive contribution to science is, if there is a distinctive contribution. That is a question that can only be intelligibly asked if a researcher can distinguish engineers in a science lab (whatever their official title) from others in the same lab. Much the same question can be asked about other engineers (and other professionals) in other technological organizations—from Boeing to the Illinois Commerce Commission.

Indeed, I would suggest that much may be learned about the ethics of technology from studying the distinctive role of each of the professions involved in an artifact's conception, development, manufacture, distribution, maintenance, use, and disposal, especially when the professions seem to have inconsistent commitments and must nonetheless work together. The positive role of conflict among technological professions seems to be a subject of study that, though inviting, remains largely unexplored (compare chapter 6).

4. ADVICE ON USING THE LENS OF PROFESSION

By now, it should be clear what the lens of profession is and why those engaged in the study of technology, especially, the study of its ethics, should find the lens useful. If so, then it is time to offer some advice concerning how to study the ethics of technology through that lens. The advice offered does not constitute a comprehensive handbook but merely seven "helpful hints" developed by considering mistakes that I have seen researchers make. While I have the impression that researchers today are less likely to make such mistakes than were researchers even a decade ago, I think it is still too soon for even one of these hints to be omitted because it is "too obvious to be worth mention." It is surprising how often "the obvious" is overlooked or forgotten—when unmentioned.

Each hint is stated as a way to avoid a certain mistake in the study of *engineering*. While I doubt the analogy with the study of other technological professions will always be clear, I shall say nothing about such analogies here. There are so many other professions that I must leave it to those who know the other professions better than I do to work out the analogies (and disanalogies) themselves.

1. *One cannot learn the role of engineers in technology simply by* observing *them*, that is, by studying them in the way we study bees or bonobos. Like all human activity, engineering has an "inside" as well as an "outside." Much of what is distinctive in engineering goes on "inside" individual engineers. Interviewing engineers, asking them for their reasons as well as for what they do, should be a part of every research plan for a study of engineering. The reasons that engineers offer for what they do can change our interpretation of what they do. For example, though

one might initially interpret the silence of engineers at a certain meeting as evidence that they approved the decision made at that meeting, interviewing them might reveal that they were silent only because they had already made their objections and been overruled. They were silent only because they believed repeating their objections would have been a waste of everyone's time. The decision taken at that meeting would then not count as a straightforward expression of the engineers' values.
2. *One cannot learn the role of engineers in technology merely by studying artifacts.* Different disciplines may produce similar artifacts (such as canals or bridges). The values seemingly encoded in an artifact may, or may not, be the values that the engineers brought to the artifact's conception, design, development, or the like. Sometimes engineers fail to embody their values because of mistake. Sometimes they fail because the decision was taken from them, for example, by "management" or a legislature. Most often, perhaps, they fail because they must compromise with others, some in other professions, or because the decision had to be postponed owing to an inability among engineers to agree on a change (though all agreed that some change was needed). To learn the role of engineers in a technology, one must study in detail the engineers involved, not only their acts and the arguments they made but also the thinking behind those acts and arguments. (For an example of how this might be done, both well and badly, see the substantial literature on the Challenger disaster, e.g., Vaughan, 1996.)
3. *One should not rely on institutional title or apparent function to identify engineers.* In general, engineers can be identified by their education and experience (whatever their institutional title). If they were trained as engineers and have long worked as engineers typically work, they are (all else equal) engineers (strictly speaking). There are, however, at least three exceptions to this rule that should not be overlooked.

First, there are some "engineers" who other engineers regard as incompetent, even though they have the appropriate degree, experience, and job title and claim to be engineers. They should be studied separately—much as one might study a broken radio to learn about how radios should work. What counts as incompetence can tell much about what it means to master the discipline in question. Something similar applies to engineers that other engineers regard as unethical or unsuited.

Second, there may be "engineers" who did not receive an engineering degree but instead a degree in chemistry, mathematics, physics, technical management, or the like or no degree at all but who claim to be engineers because they have worked beside engineers long enough to have learned the discipline. If the engineers they work with agree that they are effectively engineers, they are "engineers by adoption" and should be so counted.

Third, there may be "engineers" who, though educated as engineers and working as engineers typically do, do not consider themselves to be members of the profession but instead mere employees with certain skills—or members of another profession. My impression (derived largely from conversations with engineers in the United States) is that such "engineers" are extremely rare. Indeed, I can recall meeting only one, a member of a department of electrical engineering in a large university who considered himself "more physicist than engineer." But some researchers, especially those studying engineers in France or Japan, have a different impression (see, e.g., Didier, 1999). For that reason, I think it always worth asking early in an interview of someone whom one supposes to be an engineer both "Are you an engineer?" and "Why did you answer as you did?" (Davis, 1997). The existence in a country of many "engineers" who deny that "engineer" is anything more than a job description is evidence that engineering is not a profession there. It is, however, not decisive evidence. The engineers in question would also have to be indifferent concerning, for example, whether other "engineers" maintain engineering standards. If they understand themselves as working in cooperation with other engineers, not only others in their employer's organization but engineers outside of it (whether or not any of them are called "engineer"), they may still understand themselves in a way consistent with being members of the engineering profession.

4. *One should not only interview engineers about what they do but also ask them to evaluate each other and members of other professions with whom they work.* Part of being a member of a profession is expecting certain others, those one counts as members of one's own profession, to have certain abilities and to act in certain ways. Also part of being a member of a profession is being aware that those one counts as *not* belonging to one's own profession will typically lack some of those abilities and act in ways different from one's own. So, for example, an engineer might say that "any engineer worth his salt" would use such-and-such a safety factor in a certain design while doubting that a computer scientist or ordinary manager would do the same.

5. *One should not limit research into engineering ethics to what is in a code of engineering ethics.* A profession need not have a formal code of ethics to be a profession. It merely needs technical standards that satisfy the definition of ethics (morally permissible standards of conduct that every member of the profession, at his most reasonable, wants every other member to follow even if that would mean having to follow them too). The formal code of ethics may best be thought of as a convenience engineering can do without and still be a profession. Even if there is a code of ethics, it may be that few members of the profession actually have seen

it. A crucial question for such an occupation to be a profession is whether its members *nonetheless* rely on each other to act in accordance with the standards in the code—or, at least, when shown the code, can agree that it contains standards they want others sharing the occupation to follow even if the others following them would mean having to do the same. The code may simply document engineering's implicit standards.

6. *One should not assume that just any difference between engineering practice in one country and another (say, that engineers in one country typically present discoveries in tables while engineers in another country typically present them in equations) shows that the engineers lack a common profession.* Nor is it enough that philosophers or social scientists find a difference significant. The difference should be one that the engineers themselves regard as significant enough to divide them (for example, because it makes the work of the engineers of the other country unreliable or unintelligible to the other). To know that, a researcher must ask engineers about the *importance* of specific differences between the way that engineers in one country practice engineering and the way that engineers in another country do.

7. *Any social scientist organizing a research group to study a question concerning engineering ethics or the ethics of technology should consider including both at least one philosopher and at least one engineer.* It is good to have a philosopher in a research group from the beginning to help researchers define their terms, make sure they end up studying what they set out to study (or something at least equally worthwhile), and otherwise help researchers do something useful. (For examples of what can go wrong in research not philosophically informed, see Davis, 1996.) It is also good to have an engineer participate not only to help the social scientists understand the engineering in question, when engineering is in question, but also to point out that engineering is not in question, when it is not. An engineer may also have important insights relevant to interpreting interviews both of engineers and of those who work with engineers.

NOTES

Originally published as: "Profession as a Lens for Studying Technology," in *Methods for the Ethics of Technology*, edited by Sven-Ove Hansson (Rowman and Littlefield International, 2016): 83–96; a somewhat different version was published as "Profession as a Lens for Studying Technology" [以职业为棱镜研究技术], *Journal of Dialectics of Nature* 38 (July 2016): 1–8, DOI:10.15994/j.1000-0763.2016.04.001.

1. For a fuller exposition of this definition and a defense of the distinction between business and profession, see Davis, 1994.

REFERENCES

Bush, Lawrence, *Standards: Recipes for Reality* (MIT Press: Cambridge, Massachusetts, 2011).

Davis, Michael, "Thinking like an Engineer: The Place of a Code of Ethics in the Practice of a Profession," *Philosophy and Public Affairs* 20 (Spring 1991): 150–167.

Davis, Michael, "Is Engineering Ethics Just Business Ethics?" *International Journal of Applied Philosophy* 8 (Winter/Spring 1994): 1–7.

Davis, Michael, "Professional Autonomy: A Framework for Empirical Research," *Business Ethics Quarterly* 6 (October 1996): 441–460.

Davis, Michael, "Better Communications Between Engineers and Managers: Some Ways to Prevent Ethically Hard Choices," *Science and Engineering Ethics* 3 (April 1997): 171–213.

Davis, Michael, *Code Writing: How Software Engineering Became a Profession*, Center for the Study of Ethics in the Professions: Chicago, 2009b (http://ethics.iit.edu/sea/sea.php/9).

Davis, Michael, and Hengli Zhang, "Proving that China has a Profession of Engineering: A Case Study in Operationalizing a Concept across a Cultural Divide," *Science and Engineering Ethics* (December 2017): 1581–1596.

Didier, Christelle, "Engineering ethics in France. A historical perspective," *Technology in society* 21(1999): 471–486.

Hansson, Sven-Ove, "A note on social engineering and the public perception of technology," *Technology in Society* 28 (2006): 389–392.

Harris, Charles E. et al. *Engineering Ethics: Concepts and Cases* (Wadsworth: Boston, MA, 2014).

Herkert, Joseph R. "Ways of Thinking about and Teaching Ethical Problem Solving: Microethics and Macroethics in Engineering," *Science and Engineering Ethics* 11 (2005): 373–385.

Iseda, Tetsuji, "How Should We Foster the Professional Integrity of Engineers in Japan? A Pride-Based Approach," *Science and Engineering Ethics* 14 (2008):165–176.

Son, Wha-Chul, "Philosophy of Technology and Macro-ethics in Engineering," *Science and Engineering Ethics* 14 (2008): 405–415.

Vaughan, Diane, *The Challenger Launch Decision* (University of Chicago Press: Chicago, 1996).

Vincenti, Walter G. *What Engineers Know and How They Know It: Analytical Studies from Aeronautical History* (Johns Hopkins University Press: Baltimore, Maryland, 1990).

Weingardt, Richard, *Engineering Legends: Great American Civil Engineers* (Reston, VA: ASCE, 2005).

Wells, Paula, Hardy Jones, and Michael Davis, *Conflict of Interest in Engineering* (Kendall/Hunt: Dubuque, 1986).

Part III

PROFESSIONAL RESPONSIBILITY OF ENGINEERS

Chapter 9

"Ain't Nobody Here But Us Social Forces"

Constructing the Professional Responsibility of Engineers

A hobo is in the henhouse stealing eggs when the farmer, having heard the chickens complaining, comes out with a shotgun and asks, "Who's there?" The hobo responds: "Ain't nobody here but us chickens!"—old Vaudeville joke

There are many ways to avoid responsibility. Among them are attributing one's act or its consequences to society, market, culture, God, the devil, drugs, or drink. For engineers, "technology" or "the organization" might seem to serve quite well. I mock this response as, "Ain't nobody here but us social forces"—recalling the old Vaudeville joke. Of course, the Vaudeville joke depends on someone being there in addition to the chickens, someone who can properly take responsibility but tries to avoid it—by a device too desperate to succeed. The response thus mocked is not a joke, of course. It presupposes that certain facts can prevent assignment of responsibility. For example, some have claimed that engineers cannot be responsible for some consequences of their work because the technology in question is so complex that they cannot reasonably be expected to foresee those consequences.[1] A fact, the impossibility of reasonable foresight, rules out the responsibility in question.
What I propose to do here is argue that any such appeal to mere fact fails. The responsibility in question is something engineers can, and generally do, take on by their own voluntary acts. Engineers construct their responsibility, much of it at least, rather than discover it or have it imposed on them. Taking responsibility is an important but, it seems, overlooked way of becoming responsible. For engineers, the proper question is (primarily): *what*

responsibility have we *taken on*? The responsibility of engineers is, in large part, what, as a profession, they have made it.

I shall have nothing to say here about whether engineers have taken on too much, too little, or just the right amount of responsibility. That does not seem to be a question for philosophers, both because the question concerns what burdens *engineers* should bear and because proposed answers seem likely to be in terms of degree. This chapter has a different focus: arguments supposedly defeating responsibility claims altogether.

1. PRELIMINARY CAVEATS

Like "profession," "engineering," and "ethics," "responsibility" has too many senses for the term's safe use without distinguishing the more important ones, understanding their relationship, and then trying to keep them straight. We may distinguish the following nine interrelated senses as relevant here (all of which I will eventually use):

1. responsibility-as-simple-causation ("Katrina was responsible for the flooding in New Orleans"),
2. responsibility-as-faulty-causation ("He's responsible because he acted carelessly"),
3. responsibility-as-good-causation ("She's responsible for our success, give her the credit"),[2]
4. responsibility-as-competency ("a responsible person"—rather than, say, a careless or incompetent one—responsibility as a disposition or virtue),
5. responsibility-as-power ("Your skill gives you a responsibility"),[3]
6. responsibility-as-office ("the person responsible," the one in charge),
7. responsibility-as-domain-of-tasks ("These are her responsibilities," that is, what she is supposed to do),
8. responsibility-as-liability ("You should *pay* because you are responsible for what happened"), and
9. responsibility-as-accountability ("You should *explain* because you are responsible for what happened").[4]

This list has more structure than may be apparent at first. Five of the nine senses are (more or less) factual: the simple-cause, competency, power, office, and task senses. Two, in addition to the simple-cause sense, are entirely historical (or "backward looking"): the faulty-cause and the good-cause senses. When we are talking about responsibility in any of the three historical senses, we should (and generally do) use the past tense: "she *was* responsible for that." The remaining two (the liability and accountability senses) seem to be about the future, that is, they point to what should be done. In this respect, they resemble the power, office, and task senses. The responsibility that comes with power, office, or task is a responsibility to do something (to benefit humanity, to oversee this project, or to perform these tasks). The responsibility that comes with liability (whatever explains the liability) is also something to do, for example, accept blame, apologize, make compensation, or perform as promised. The responsibility that comes with accountability is to give an account of something, for example, offer a justification, excuse, or explanation, or answer pertinent questions. The standard way to talk about responsibility in one of these five future-oriented senses (power, office, task, liability, or accountability) is in the present tense: "She *is* responsible for that."

This difference in tense is only a general index. It is easy to find causal uses in the present tense (when there is a continuing or continual causing) as in, "He is responsible for all my troubles [that is, the cause of them all]." It is also easy to find uses of the future-oriented senses in the past tense (when the power has been lost, the office vacated, the task completed, the liability satisfied, or the account given) as in, "He *was* responsible for paying the damages [but paid them long ago]." Nonetheless, it seems important to stress the future-oriented aspect of all five senses, though each typically presupposes something in the past (the act assigning office or tasks, the event imposing liability or imparting power, and so on).[5] What is important about the future-oriented senses of responsibility is precisely that they have something to say about the future, something not true of competency or any of the three historical senses.

Neither technology nor organization seems to threaten responsibility in any of the five factual senses (cause, competency, power, office, or task). Whatever technology or organization becomes, engineers who help to design, build, distribute, maintain, or dispose of technology will remain causal agents, responsible in the first (simple-cause) sense of responsibility for whatever they bring about (whether anyone can trace their individual contributions or not). They may also remain responsible persons (agents to be trusted), experts (with the power that comes from knowledge, skill, or judgment), persons responsible (agents in charge), and persons with responsibilities (tasks to do). When technology or organization seems to threaten "responsibility," it is responsibility-as-liability or responsibility-as-accountability that is threatened. In particular, it might be (and sometimes is) argued that technology may allow engineers to do great harm without fault. We cannot (it is said) hold engineers liable or even accountable for the harm they do without fault because fault is a logical precondition of responsibility in the liability or accountability sense. What engineers do without fault is not their responsibility.[6]

Though strong, the relation between fault and liability (or accountability) is, I think, not that of *logical* precondition. The law is quite capable of holding persons responsible for outcomes when they were not personally at fault—without any incoherence. Among terms covering this sort of responsibility are "strict liability" (liability when there is causal agency but no fault), "vicarious liability" (liability when the fault is in some person other than the one held liable, such as when the principal or employer is held responsible for the faulty conduct of an agent or employee), and "surety liability" (when liability depends on contract rather than fault, causal agency, or relation of agent to principal, for example, when a person undertakes to pay a fine if another does not appear in court).[7]

Our question, though, is not what the law *can* do with responsibility-as-liability (much less with responsibility-as-accountability). Our question is in part what the law *should* do with responsibility—and, in part, what we should do with it outside the law, for example, when attributing professional responsibility. This "should" depends on both moral justification and rational justification.[8]

There is a weak, intermediate, and strong sense both of "moral justification" and "rational justification." I can, for example, morally justify my conduct by showing (a) that morality does not forbid it (a weak justification), (b) by showing that morality recommends it (an intermediate justification), or (c) by showing that morality requires it (a strong justification). We need not worry about this detail here. Weak moral justification is all we need for the arguments in question. Only at the end of this chapter will I have to say anything about rational justification.

While our question is (primarily) about moral responsibility rather than legal, these examples of legal responsibility-as-liability are worth keeping in mind. Since they seem not to stretch the concept of responsibility to the breaking point, or indeed to stretch it much at all, they demonstrate the range of that concept. Any claim that *moral* responsibility is not as flexible as legal responsibility will have to rely on moral arguments, not on appeal to the "logic of responsibility" (much less to "what 'responsibility' means"). I therefore reject John Ladd's claim that moral responsibility is separate from the nine senses just distinguished. Ladd defines moral responsibility as (roughly) "the concern of people for people."[9] Insofar as this concern constitutes responsibility at all, it seems to define a domain of (morally required or recommended) tasks (say, taking people into account when acting) also found in law. While I do not think moral responsibility and legal responsibility *necessarily* mirror each other, I see no reason to suppose that the term "responsibility" in morality has a special meaning that cannot be incorporated into law—or that "responsibility" in law has a special meaning not to be found in morality. The difference has certainly not been established.

There are certain factors commonly put forward as conditions *necessary* for the assignment of *moral* responsibility-as-liability—and of moral responsibility-as-accountability when distinguished. The four most common are: (1) cause (or causal factor), (2) rationality (or competency), (3) choice (or freedom), and (4) knowledge (or foresight). (See, e.g., Swierstra and Jelsma, 2006, esp. 312–313.) A person who has not caused an event cannot (it is said) be morally responsible for it, nor can a person who is not a "responsible agent," nor can a person who had no alternative or who did not know what would happen (or at least what she risked). Part of what makes this collection of factors attractive is that each connects responsibility-as-liability with another sense of responsibility. Thus, the first explains why we might use "responsibility" for both liability and cause (any liability presupposes cause); the second, our use of "responsible person" for someone who can be held liable (a competent agent); the third and fourth, with fault, since fault requires both choice (an exercise of deliberative power) and something the agent knew or should have known (an exercise of cognitive power).[10] Indeed, this way of thinking about responsibility puts liability at the concept's center, with the other senses of responsibility derivable by subtracting one or more conditions of liability.[11]

However attractive this structuring of the concept may be, it should not be embraced. *All* the claims on which it relies seem to be false or, at least, no more than roughly true. Consider the requirement of cause: it seems not to fit some *failures* to rescue. We sometimes hold a person morally responsible for the death of another (in the liability or accountability sense) when she could have, but knowingly did not, come to the aid of the other, for example, a drowning child she could have saved with little risk, cost, or trouble to herself and no breach of duty to another. We blame her for the death though it would have occurred even if she had not been there to help, indeed, even if she had died a decade before. Her inaction, though certainly an event, indeed, a conscious act, is no more a cause of the child's death (whether "cause" means the most important factor or just a necessary one) than the non-presence of the billions of other people who might have saved the child had they been there.[12] Her non-rescue of the child has no causal significance (no special role in the actual process leading to death). Only her *rescue* of the child would have had causal significance—and that did not occur. We nonetheless hold her morally (and, in some jurisdictions, legally) responsible for the child's death (in the liability or accountability sense). If responsibility-as-simple-cause matters here (as I think it does), it is not as cause in this world but as part of an explanation invoking possible worlds. The cause in question exists only in at least one counterfactual world (one reasonably close to the actual one). Such a counterfactual cause (the rescue that happened in that world but did not in this) cannot be the actual cause of the child's death. What is important to the responsibility of the bystander who did not help is that she *could* have saved the child but did not, that is, not faulty cause in this world (since she caused nothing) but good cause in another.[13]

Rationality (or reasonableness) is also said to be a necessary condition for moral responsibility. That is why (it is said) we do not hold the insane, the very young, or nonhuman animals morally responsible or, at least, why we do not *properly* hold them morally responsible. Of course, sometimes we do hold them morally responsible. For example, we may criticize even a young child for breaking a promise (saying something like, "You should have kept your promise"). Why we do that is open to discussion. My point now is merely that we *do* criticize the very young in this way; criticizing them in this way is blaming them (in a way criticizing a dog is not); blaming someone for breaking a promise is (normally) moral blame; and imposing moral blame on the very young at least seems to be holding them morally responsible (in the responsibility-as-liability sense). Certainly, when my son was four and broke a promise, I spoke to him as I would to someone who was morally responsible—and he responded accordingly, for example, by offering an excuse or justification, or otherwise accounting for his conduct, or by apologizing or suggesting some other way to satisfy the liability incurred (depending on the circumstances). Though I held him responsible in this way, I did not regard him as a competent adult (a responsible person). At four, he was far from that. On matters of safety, especially, I had more faith in our dog (who was then ten). If we in fact sometimes treat even the very young as morally responsible in this way without supposing them to meet the rationality condition (as I believe we do), the hard question is whether doing so is morally proper (and, if it is, why). I need not answer that question here (though it seems to me a proper enough question). My purpose now is merely to raise doubts about these four supposedly necessary conditions for responsibility-as-liability (and accountability). I have, I think, now said enough to raise substantial doubts about the second one ("agent rationality"). We often seem to act contrary to that condition without any feeling of doing something odd. Whatever this second condition is, it is not common sense. We can go to the third choice.

We cannot (it is said) have any moral responsibility for an outcome when we had no choice but to bring it about. Consider, then, a minor traffic accident. A cat runs out from between parked cars just as I am passing. I slam on the brakes as quickly as I can—but too late. I have not been drinking, missed a night's sleep, or otherwise done anything to impair my reactions. I was not speeding, using a cell phone, daydreaming, or otherwise failing to exercise reasonable care. The accident was, we might say, just one of those things—quite beyond my control—and therefore "definitely not my fault." Should I then drive on as if I were in no way responsible for the cat's death? I should not. I should stop, check to see if there is anything I can do for the cat, and perhaps call the police. I should do all that not to be a nice guy, a Good Samaritan, but because I am responsible (in some sense) for the cat's death. If I do not feel that responsibility, an obligation to respond in certain ways, I am a morally worse person than one who does.[14] I seem to owe the cat, its owner, or "the world" something for what I have done.[15] This owing seems like responsibility-as-liability. Of course, I may not owe what I would have owed if the cat's death had been my fault (say, the result of negligence). But I still owe something. I shall offer no defense of this leap from responsibility-as-simple-cause to responsibility-as-liability here. I merely point it out as a fact of moral life, one that at least seems to undercut the claim that choice is *required* for moral responsibility.[16]

The cat also provides a good example of the fourth condition for moral responsibility (knowledge) not being met. I could not know that the cat was there, between the two cars, much less that it would run out just as I was driving by. Of course, I did know, or at least should have known, that such accidents are possible. Cars are big, fast, and hard to stop; cats, not always as careful of traffic as they should be. Yet, the risk I ran, however knowingly, merely by driving, does not seem to be enough for moral fault. Such general knowledge of unlikely events may help to explain strict liability in law, but it is not the sort of knowledge commonly presented to justify moral responsibility-as-liability (or accountability). More specific knowledge is supposed to be required, at least knowing that there is a reasonable chance that the event will happen if I drive (and that there are reasonable alternatives such as a bus or bicycle instead). My knowledge fell well short of that.

I do not claim that these four conditions are irrelevant to responsibility-as-liability (or accountability), only that they are not *necessary* conditions. They seem to function more like considerations relevant to rejecting, imposing, or accepting responsibility (mere reasons rather than requirements). The more of those conditions that are met, the stronger the case for moral responsibility, all else equal. But context is important too. In the right context, even one factor, such as simple cause, can be sufficiently strong (more or less by itself) to decide the question (as perhaps it is in the cat case).[17]

We should keep all this in mind as we evaluate arguments for the claim that technology or organization is undermining the responsibility of engineers for much they do. In particular, we should be charitable in our interpretation of these arguments. We should interpret them, if at all possible, as relying on a less questionable claim than that any of these four conditions is necessary for responsibility (though many do in fact seem to rely on that questionable claim). Such claims fail even when they receive more charity than they deserve.

2. ENGINEERS' RESPONSIBILITY FOR WHAT THEY DO

There are at least seven arguments used to defend the claim that technology or organization is making, has made, or might someday make it impossible for engineers to be morally responsible for (at least some of) what they do. We may call them the argument from: (1) many hands, (2) many causes, (3) replaceability, (4) institutional constraint, (5) individual helplessness (lack of coordination), (6) individual ignorance, and (7) neutrality of technology. Each of these arguments suffers from one or more flaws. The responsibility in question is liability, accountability, or both.

2.1 Many Hands

We seem to owe the name of the first argument, if not the argument itself, to Dennis Thompson. His summary of it is worth quoting as much for what it does not say as for what it does say: "Because many different officials contribute in many different ways to decisions and policies of government, it is difficult even in principle to identify who is morally responsible for political outcomes" (Thompson, 1987, 41). For Thompson, the argument from many hands is primarily about large organizations ("governments"), not about technology. It is an epistemic argument, one about a problem more information might clear up, not a metaphysical argument (one relying on relations among things). In addition, the problem that the argument relies on makes identifying those morally responsible for an outcome "difficult," not impossible, and no newer than determining whether Henry II was responsible for the murder of Thomas Becket. Indeed, Thompson devotes the chapter with which this quotation opens to assigning responsibility to politicians and government officials who appeal to one version or another of the argument to avoid it. Thompson is not afraid to look into the henhouse.

For Thompson, as for most who discuss the argument from many hands, a crucial assumption is that responsibility is primarily about holding *others* responsible.[18] So, for example, Helen Nissenbaum (whose subject *is* primarily technology) notes, "Boards of directors, task forces, or committees issue joint decisions, and on the occasions where these decisions are not universally approved by all their members but are the result of majority vote, we are left with the further puzzle of how to attribute responsibility." Nissenbaum (1996, 29) Of course, Nissenbaum's "we" are left with the puzzle because "we" are not party to the "joint decisions" in question. "We" are trying to impose responsibility on others, not to decide our own responsibility. Were "we" among the decision-makers, we would at least know our part in the decision, for example, whether we voted for or against the measure in question. The problem of many hands (so stated) does not seem to be much of a problem for holding *oneself* responsible.[19] In any case, it is no problem for someone who *takes* the responsibility in question. I can take responsibility for what happened even if I have no idea how I contributed to the ultimate decision (as, e.g., President Truman did when he said, "The buck stops here").

2.2 Many Causes

For decision-makers themselves, the problem is (in part at least) many *causes* rather than many *hands*. There are two versions of the resulting argument against responsibility. One relies on the complexity of the causal story; the other, on the problem of identifying "*the* cause" (or "the chief causes") even when the causal story is relatively simple. Like the problem of many hands, the problem of complexity is epistemic, but the problem of identifying "the cause" is metaphysical (a problem about what is to count as "cause").

Nissenbaum's discussion of the THERAC-25 disaster (three patients dead and three more seriously burned by what should have been a relatively safe means of irradiating cancer cells) provides a good example of the complexity version of the argument from many causes:

> After many months of study and trial-and-error testing, the origin of the malfunction was traced not to a single source, but to numerous faults, which included at least two significant software coding errors ("bugs") and a faulty microswitch. The impact of these faults was exacerbated by obscure error messages, the absence of hardware interlocks, inadequate testing and quality assurance, exaggerated claims about the reliability of the system in AECL's safety analysis and, in at least two cases, negligence on the parts of the hospitals where treatment was administered.[20]

In its complexity, the THERAC-25 disaster differs little from most engineering disasters. Because engineers typically try to design considerable redundancy into their systems, an engineering disaster typically requires several independent failures. Nothing is *the* cause, but several factors together constitute "the chief causes." Responsibility-as-faulty-cause must then be divided among these—and, insofar as each factor is the work of a different agent, among different agents. But the resulting responsibility-as-liability (or as-accountability) is not like weight, a burden *necessarily* lessened when so divided. Some parts of responsibility may be lessened by being divided, especially financial liability, but the rest, such as the blame deserved or the obligation to explain one's own part, may not be.

Nonetheless, THERAC-25 seems to illustrate the argument that many causes can undercut responsibility. After all, how could an engineer involved in the disaster honestly take any responsibility for the disaster (such as blame or the obligation to explain) until a post-disaster analysis identified the chief causes—and, if the causal story is too complex to identify any engineer's part, how could an engineer take any responsibility whatever?

This question is not rhetorical. If it seems rhetorical, that is (in part at least) because we are not being careful enough about the sense of "responsibility" in question. If "responsibility" means responsibility-as-simple-cause or faulty-cause, then the engineer cannot honestly take responsibility (i.e., accept a description in which she is the cause) until she knows the causal story and how her faulty conduct, if any, fits into it. The same seems true if the engineer is to be called to account for the accident in question or held liable for it. How can an agent give an account or be held liable (even by herself) without knowing who was at fault?

Neither of these questions is rhetorical because both responsibility-as-liability and responsibility-as-accountability include more than responsibility for faulty conduct. Remember the moral responsibility arising simply from having caused the cat's death. We can see something like that in the THERAC-25 story. Confronted with the three deaths and three serious injuries, AECL (the manufacturer) investigated the THERAC-25's malfunction, eventually issuing a detailed post-disaster analysis. Undertaking that investigation meant accepting no more than that the THERAC-25 at least *seemed* to be a causal factor in the disaster. Undertaking such an investigation is something AECL's engineers should have urged. For engineers, even such a weak connection—being involved in the disaster as a *possible* causal factor—entails an obligation to prepare an accurate account of what happened. Preparing such an account is already taking (some) responsibility—and taking that responsibility (responsibility-as-accountability) is possible before anyone knows who was at fault—or even who caused the deaths and injuries.

This attribution of responsibility may seem like a small point, especially to engineers: What else would "we" (engineers) do when there has been a disaster and we, or our organization, may be involved in it? For many in management or politics, the answer seems to be: *Call in the "spin doctors."* Even among professionals, there seem to be answers other than the one engineers take for granted. Lawyers, for example, generally do not undertake anything like AECL's post-disaster analysis when they lose a major case. They "move on," each lawyer taking whatever lesson she may from the experience.[21]

Whatever is true of other professionals, engineers consider it their responsibility to study any disaster that seems to arise from what they did—and to report what they find. To commit a certain mistake once, even a serious one, is something engineers tolerate as part of advancing technology (provided the engineer in question exercised reasonable care). What engineers do not tolerate is that an engineer—this engineer or any other—should make the same mistake. Once a mistake has been identified, the state of the art advances and, all else equal, what was once tolerable becomes intolerable (a kind of incompetence). Because keeping good records is part of accountability, engineers have routines for recording what they do.[22] Indeed, anyone who watches engineers work will be surprised at how much of what they do is "documentation." Engineering is unusual among professions in recognizing an obligation to "acknowledge their errors"[23] (NSPE, 2010, III.1.a.). Keeping good records makes that easier.

So, one response engineers have to the problem of many causes is to *take* responsibility for enough investigation to assign further responsibility (both for past errors and for future tasks). They seek to disperse the fog of technology rather than viewing it as a cover under which they can escape responsibility. The argument from many causes overlooks the possibility of that response—indeed, of that sort of responsibility-as-liability (liability as a duty to investigate and report).

But once engineers have investigated, they will often—perhaps always—have at least three candidates for *the* cause (or the chief causes) of the same event: operator error (running the THERAC-25 contrary to instructions), organizational failure (inadequate training or supervision, inadequate quality assurance, or the like), and design flaw (faulty micro-switch or software bugs). There is a nice joke to illustrate how fundamental this trinity of candidate causes is (relying on the first instance of the problem of many causes):

> God says, "I told you not to eat the fruit of that tree. You have no one to blame but yourselves for your fall from grace."
> "But," Eve responds, "you should have explained better. You should have been more explicit about the consequences. You should have posted fiery angels beside the tree to defend its fruit."
> "And," Adam adds, "just what was that tree doing in the Garden, anyway? Would it not have been safer to plant it where we could not go?"

God is trying to treat operator error as *the* cause of the disaster. He accepts no responsibility for what happened. He has a point, of course: Adam and Eve did what they knew (or should have known) they should not do. Had they done otherwise, there would have been no problem. They are, in a sense, plainly the cause of their own fall from grace. But Eve also has a point. Had God been more explicit about the consequences, she might have been better able to resist the serpent's enticements. God certainly could have done a better job of organizing and supervising the Garden's workforce. The organization (God and his angels) failed to do all they reasonably could to prevent the disaster. But for that failure, Adam and Eve might still be in the Garden. And Adam also has a point. Had God designed the Garden differently, Eve could not have reached the tree. Why did God put the tree *in* the Garden? (And why, we might add, did God not fence the serpent out?) The Garden was a disaster waiting to happen—and God was responsible for that.

Generally, when something goes wrong with any technology, there will be this trinity of candidate causes. That fact yields a version of the argument from many causes independent of ordinary complexity. This version relies instead on a fundamental problem in all talk of causes, that is, distinguishing mere causal factors, always many, from the cause or chief causes (generally, a much smaller set).

What is *a* causal factor is (more or less) a simple matter of fact (or metaphysics).[24] What is *the* cause, or one of the *chief* causes, is not. As Joel Feinberg noted almost a half century ago, which causal factor we cite as cause depends on practical considerations, especially on the reason we are looking for a cause, what Feinberg called a "pragmatic criterion." Feinberg (1970) distinguished three such criteria (calling them): *explanation, engineering,* and *blame*. If our purpose is explanation, we look for the causal factor that helps us to understand why the event in question occurred. The serpent's presence in the Garden, and its cunning, may well be the best explanation of why Eve ate the apple. Without the serpent, the story might have been different. Adam and Eve might have lived in the Garden many happy years before they did anything wrong. They might be there still. The serpent was (in this respect) the cause of the fall from grace at the time it occurred.

Yet, if our purpose is "engineering" (in Feinberg's overly broad sense), then both Eve's organizational criticism and Adam's design criticism seem to pick out true causes. Both point to ways to prevent similar disasters. Eliminating the serpent would only eliminate "the immediate cause" (a particular temptation), not the "underlying" or "real" cause (as we might say), which is that the Garden has tempting fruit in it, something better design might eliminate by, for example, planting the fruit tree outside the Garden beyond the line of sight.

If, however, our purpose is simply to assign blame, then, and only then, does God seem to have a point. The serpent merely acted according to its serpentine nature; it lacked the free will Adam and Eve have. The Garden did not eat the fruit, Adam and Eve did, even though (thanks to God's warning), they knew better (or, at least, should have). They have no one but themselves to blame for their fall from grace.

The practical criterion most appropriate to engineering as such is neither explanation (which seems to belong to history or science) nor blame (which seems more appropriate to law or ordinary moral life), but to what Feinberg actually called "engineering."[25] The primary concern of engineers who have identified an engineering problem, whether or not of their own making, is—all else equal—to fix it if they can.[26] This is a corollary of the purpose of engineering (roughly, to improve the material condition of humankind). It is certainly the engineer's responsibility-as-domain-of-tasks. So, for example, though fools may be to blame for much they suffer, a routine part of engineering is to make designs "fool proof," that is, to protect even fools against foreseeable misuse (at least insofar as that is possible at reasonable cost).[27]

For engineers, then, the second version of the problem of many causes has a simple solution. Because the engineering criterion distinguishes cause from mere causal factor in a specific way, it forecloses the argument from many causes (in its second version). Indeed, engineers have constructed their profession to foreclose that argument. They make it their responsibility to solve certain problems, engineering problems. Among those problems is avoiding certain disasters (whether by design of the thing itself or by design of its organizational context). Solving such problems is something engineers are "liable to do" because they are engineers, not because they caused the problem. Here is a good example of (moral) responsibility-as-liability without responsibility-as-simple-cause.

This response to the argument from many causes may seem to dismiss the question of blame. It does not. Instead, it points out that engineers can, and often do, take responsibility where blame cannot be assigned. This response does, however, pose a question for those who find it unsatisfactory because it says nothing about blame: why is blame so important to them? Is it not more important to fix the problem, say, the malfunction in the THERAC-25, and to avoid similar malfunctions in the future, than to assign blame for the original malfunction? (Perhaps engineers and the public will answer this last question one way; lawyers and their clients, another.)

2.3 Replaceability

We turn now to the third argument against responsibility, the argument from replaceability. Like the first two arguments, this one depends neither on technology nor on any special feature of modern organizations. The argument is often used in some such form as this: "If I don't do it, someone else will, and since it makes no difference whether I do it or not, there's nothing wrong with me doing it" (Glover, 1975; Scott-Taggart, 1975; and Bayles, 1979). Though that is the common form of the argument, both its strength and weaknesses will be clearer if we state the argument more formally:

1. To be the cause, or a chief cause, a causal factor must be necessary to the event caused.
2. If I do not bring X about, someone else will.
3. If X will occur whether I bring it about or not, I cannot be necessary for X to occur.

4. If I am not necessary for X to occur, I cannot be the cause of X or even *a* chief cause.
5. No one is responsible for what he did not cause.
   Therefore: I am not responsible for X (whether I bring X about or not).

There are at least three flaws in this argument—at least as applied to engineers.[28] The first flaw concerns premise 5. It claims that no one is responsible for what he did not cause. We have just seen that that is not true, engineers can—and do—take responsibility for problems they did not cause (and may therefore be responsible for them—in both accountability and liability senses).

The second flaw is factual. The claim that someone else will do X if I do not (premise 2) is true only if the agent in question can in fact be replaced by someone who will do X. Having long ago identified replaceability as a problem, engineers have adopted standards of practice, including a code of ethics, to prevent employers from being able to make the replaceability argument to engineers in a wide range of circumstances. Within that range, the engineer can simply reply: "If you want to replace me with an unethical engineer, go ahead. There may be a few out there. But no ethical engineer will do what you are asking. Do you really want an unethical engineer working for you?"

Engineers have in this way largely foreclosed the argument from replaceability. Normally, the argument will be unsound. Except in extraordinary circumstances, there will be no engineer out there both able and willing to violate the engineering standard in question. And, in those extraordinary circumstances, the employer has a reason not to hire the person (for that sort of job): that person is either an "unethical engineer" or no engineer at all.

The third flaw in the argument from replaceability is logical. The argument assumes that there is only one criterion for cause when, as we have seen, there are at least three. Even if the engineer can be replaced, either with another engineer or with some other sort of technologist, the argument from replaceability carries weight only when we are considering what to fix. (If any operator would do X, the problem of X-ing is not solved by replacing this operator with another.) But to consider only what to fix is to rely only on (a narrow version of) the *engineering* criterion of cause. We might, instead (or in addition), be concerned with finding the cause because we are seeking to avoid deserved blame, damages, or punishment, or having to give an account. We would then appeal to a different criterion of cause. We might, for example, say that X, an act such as approving a flawed design, is wrong even if the approval would be given by a few other engineers if the engineer in question did not give it. Whoever does X is at fault and that individual's fault is the cause of any harm that actually follows. The possibility of alternate causal chains does not change the actual causal chain or make the cause in question less faulty. (The only question left open is whose fault it will be, X's or someone else's, assuming it is someone's.) Premise 3 seems to equivocate between a first-person version (my doing y caused X) and an impersonal version (someone's doing y caused X). The argument can block responsibility for what "I" in fact do only by adopting an interpretation of premise 3 in which "I" disappears from the causal chain. The argument from replaceability thus begs the question—whether *I* should contribute to the causal chain in question.[29]

2.4 Institutional Constraint

The argument from institutional constraint, though another argument (more or less) independent of technology, is not independent of organization. The argument seeks to undercut the claim that engineers have a choice. Lynch and Kline state it nicely:

> Most engineers operate in an environment where their capacity to make decisions is constrained by the corporate or organizational culture in which they work. Engineers are rarely free to design technologies apart from cost and schedule pressures imposed by a corporate hierarchy, a government agency concerned with its image, or market pressures. (Lynch and Kline 2000, 210)

Lynch and Kline are certainly right that most engineers do (and always have) worked in large organizations (that their profession is in this respect "captive"). With the possible exception of those engineers high in an organization, working in a large organization means working within a framework that others have constructed, one that necessarily drastically limits what an individual engineer can do. Of course, the other side of working within a large organization is that one's choices, while limited by cost, market, organizational politics, and the like, can have much larger effects than the same sort of choice outside the organization. To design a small plane in one's own little company is unlikely to affect many people. To design just the riveting for the wing of the Boeing 737 is likely to have a much larger effect. For engineers, working for a large organization is (as noted before) a Faustian bargain—except that Mephistopheles, if present at all, is largely in the details.

Can an engineer have responsibility for what her employer does—or only for what she does within the narrow bounds where she is free? I see no reason why the engineer cannot *take* responsibility for what her employer does, however large the employer. She has voluntarily accepted employment with that employer and can break off the association at any time just by giving notice. Certainly, engineers have long recognized that they should take responsibility for what their employers do—even when what the employer does is beyond the individual engineer's control. Among the provisions of one of the earliest codes of engineering ethics is this: "If after becoming associated with an enterprise he finds it to be of questionable character, he should sever his connection with it as soon as practicable" (American Institute of Electrical Engineers, 1912, A.2.). Many recent codes contain a similar provision. At a minimum, an engineer should be willing to give an account of an employer's conduct and accept some blame for what it does when what it does deserves blame (just as she might accept credit if the employer does something creditable). An engineer who declines to accept that minimum responsibility but continues to benefit from the employment will seem much like that great but greatly flawed engineer in Tom Lehrer's song:

"Once the rockets are up, who cares where they come down?
That's not my department," says Wernher von Braun.

While engineers work *in* departments, they work *for* an organization. If the organization's name is on the rocket (literally or figuratively), the engineers may have something to answer for when it comes down—even if their department has nothing to do with sending it up.

That, however, is not the end of the argument from institutional constraint. The main point of the argument is to free engineers from responsibility for what *is* "their department," that is, for what they in fact help to bring about. The engineers are causal factors but (the argument runs) the organization so hems them in that they are not free enough to be *faulty* causes. The engineer who prepares a plan for a project is not responsible for it because she could have prepared no other. She had "no choice."

This version of the argument from institutional constraint has at least three flaws (beside the one already discussed). Each flaw is serious in itself. One is logical. The engineer always has at least one other choice, that is, not to prepare the plan. That other option may be literally suicidal, but—as a matter of logic—it always exists. Morally, there is an important difference between "no choice" and "no attractive choice" (and even "no reasonable choice").

The second flaw is conceptual. Engineers are hired to exercise engineering judgment on behalf of their employer. Where judgment is necessary, there must be at least two (significantly different) options, aside from preparing no plan. If there is only one option (with no judgment required), a *technician* can prepare the plan. Even if engineers do not make the final decision, they must make the initial decisions on almost any project they undertake. That decision-making power, that freedom to choose, is what, as a conceptual matter, distinguishes an engineer from a mere technician. The argument from institutional constraint is not a reason why engineers cannot be responsible for how they exercise that decision-making power—in the faulty cause, accountability, or liability sense of responsibility. They have the necessary freedom to choose.

The argument's third flaw is empirical. Those of us who have watched engineers at work have noticed that many of the decisions they make are *in fact* final. Superiors review them but, unless something is clearly wrong, the superior will not enquire deeply into the decision. Going along with engineering decisions is generally more efficient than getting a second opinion or overruling the engineer "on principle." As a matter of fact, *most* engineers in large organizations have considerable decision-making power. Whenever that is true, the argument from institutional constraint simply does not apply.[30]

2.5 Individual Helplessness

But, it may be said, the problem is not so much freedom but the need to cooperate with others to accomplish the task in question. One engineer cannot, for example, do anything to help her employer adopt sustainable practices if adopting such a practice would raise the price of the product in question. The market blocks the way. The individual engineer is helpless. Hence, even though the engineer's code may, as the code of ethics of the American Society of Mechanical Engineering (ASME) or of the American Society of Civil Engineering (ASCE) does, require engineers to "strive to comply with the principles of sustainable development in the performance of their professional duties," an individual engineer can in fact do nothing about sustainable development. Lacking any power to follow the code in this respect, the engineer can have no responsibility to do so (See, e.g., Miller, 2005).

So long as we think of engineers as individuals without the power to coordinate their conduct, this argument from helplessness is sound—at least for many engineering decisions. The flaw in the argument is that engineers, as members of a profession, are never mere individuals. They are always, in addition, members of the engineering profession. That profession may adopt standards of practice which, applying across all employers in a market, can eliminate market pressures that might otherwise block adoption of the standards. If the market still resists adoption of the standards, the engineers may, through professional organizations, appeal to insurance companies, other private entities, or the government to enforce the standards. (Which entity they should appeal to will, of course, depend on who would benefit from the standard's adoption.) American engineering societies have been developing technical standards of this sort for more than a century. There is a more recent history of similar standard-setting by international bodies, such as the IEEE or International Organization for Standardization (ISO).[31]

2.6 Individual Ignorance

But, it may be said, the problem is not so much freedom or power as knowledge. The complexities of any large project make it hard for most engineers to know the whole. On many large projects, engineers literally do not know what they are doing. And, because they do not, they cannot be responsible in either the liability or accountability sense for what happens. That is the sixth argument against responsibility, the argument from individual ignorance.[32] It has at least two flaws.

The first flaw should be obvious from what we have already said. Engineers generally try to understand how what they are working on fits into the world. That is part of good engineering. An engineer who literally does not know what she is doing has a lot to account for. There may, it is true, be some contexts where engineers are kept ignorant of the big picture. Some military work is said to be organized to keep most participants in the dark about the ultimate purpose. And some projects, especially those with large software components, are unusually hard to understand. Like the THERAC-25, they end up behaving in ways that no one foresaw—some, perhaps, in ways no one could have foreseen given the state of the art at the time. But even such unusual contexts leave engineers with a responsibility to try to find out what they do not know but need to know to do a good job; it does not excuse them from trying to do a good job, for example, by consulting the appropriate experts. The argument treats the elimination of some responsibility-as-liability as the elimination of all liability.

Ignorance of outcome is, however, rare—and, in any case, only frees engineers from blame. That is the second flaw in the argument. Whatever force the argument from ignorance has, the engineers will still owe an accounting for what happened. And, if anything went wrong, they will also have a responsibility (a liability) to figure out how to prevent it from happening again. The argument from ignorance only defends against one sort of responsibility-as-liability, the sort engineers as such are least interested in (blame).[33]

2.7 Neutrality

That brings me to the last of my seven arguments, the argument from technology's neutrality. While it is true that engineers produce technology, they cannot, it is said, control what is done with it. Any piece of technology, even something as innocuous as a scissors, can as easily be put to a bad use as to a good one (putting out the eye of an innocent person). Engineers should not be held responsible for what is done with their work.[34]

This argument does not, it should be noted, deprive engineers of all responsibility, for example, responsibility for flaws in design or manufacture over which they actually have control or, indeed, responsibility for uses for which the technology was designed. All it can deprive engineers of is responsibility for *misuses* of their work *beyond their control*. And, depending on the kind of liability or accountability in question, it may not be able to do even that. Consider the THERAC-25 disaster again. While engineers are certainly not to blame for what they could not control, their response to the disaster was not to crow about how they were not to blame. Instead, they viewed themselves as having a professional responsibility to prevent similar disasters once they knew what in particular had gone wrong with the machine. For engineers, what they create should not be neutral. It should be designed, built, distributed, maintained, and even disposed of in a way at least consistent with the public health, safety, and welfare. Technology as such may, or may not, embody certain values. That (as noted earlier) is a deep question for the philosophy of technology.[35] *Engineered* technology, in contrast, necessarily embodies (or at least should embody) certain values. An engineer who produces a product that, however clever, fails to benefit humankind has failed—as an engineer. Engineers have been making that point about engineering at least since civil engineering separated from military engineering early in the nineteenth century (Davis, 1998, 12–15). The argument from neutral technology simply overlooks the distinction between technology as such and engineered technology. For that reason, as much as because of government regulation, today's scissors generally lack the sharp tips they once all had. Part of good engineering is designing to prevent foreseeable misuse. Another part of good engineering is seeking to add to the misuses that can be foreseen, for example, by tracking products in use to discover how they have been misused.

3. THE RATIONALITY OF TAKING RESPONSIBILITY

I have canvassed seven arguments that seek to relieve engineers of moral responsibility for what they do (or fail to do). These seven are, I believe, all the arguments now in the literature. I have dealt with them in different ways. Some I disposed of by detailed analysis of the key concepts (such as "cause") or by pointing to logical mistakes. Some I could dismiss because they relied on a claim for which there was substantial adverse evidence. But, for most, one step was to point out how engineers have *taken* the responsibility in question (liability or accountability). Some responsibilities of engineers arise from their own voluntary acts (whether as individuals or as the engineering profession). Though people are generally thought to shy away from responsibility-as-liability and responsibility-as-accountability, engineers do not seem to. Instead, they seem to claim certain responsibilities most of us, even technically trained managers and other technologists, try to avoid.[36] Some might reasonably wonder whether I have described a mass pathology rather than explained why engineers can have such moral responsibilities. Why take on responsibilities others do not want?

That question, one concerned with rational justification, is one I must now answer. It is, I think, a question easily answered in a way preserving the moral responsibilities in question. That answer is that engineers, as a group, gain more by taking on the responsibilities in question than they lose. Consider a joke engineers like to tell:

> A priest, a lawyer, and an engineer are about to be guillotined in one of the great squares of Paris during the Reign of Terror. The condemned must lie on a bench with his head set in braces under a huge angled blade. The priest is first. Asked whether he would like to lie face up or face down, he chooses face up—adding, "I want to look where I hope to go." He says a prayer and lies down. The blade drops—but just as it reaches his throat, it stops. The executioner declares "divine intervention," unties the priest, and lets him go. The lawyer is next and, given the same choice, chooses to "follow precedent—and hope for the best." Again the blade stops just short. The lawyer is also let go, but this time the executioner checks the machine before proceeding. He finds nothing amiss. The engineer is brought forward. He too chooses to lie face up. Just as the blade is about to drop, he shouts "*Stop*! I think I see the problem!"

This is a good joke and, like most good jokes, has several uses. I first heard it from an engineer making the point that not every problem an engineer can solve is a problem an engineer should solve. I have also heard it used to make the point that engineers see problems others do not. Here, however, I tell it to illustrate how and why engineers might take on responsibilities not otherwise theirs. The engineer under the blade was not (as far as we know) even a causal factor in the design, production, installation, or maintenance of the guillotine, or under contract or otherwise obliged to take on its repair or even to investigate its malfunction. He might have kept silent without blame—and thereby protected his neck. Fixing the guillotine definitely was not "his department."

Yet, whatever we think of this engineer's prudence, we recognize in him someone we can trust with the world's machinery, someone quite unlike the hobo in the henhouse. Avoiding responsibility has many advantages, but winning the trust of others is not one of them. What engineers gain by taking on responsibilities that others avoid are several tasks or offices that, as a matter of fact, have combined to become a relatively lucrative occupation. Creating a lucrative occupation is reasonable enough to justify accepting the moral responsibility (both accountability and liability) that makes the occupation possible (provided the tasks accepted are also morally permissible—as engineering's tasks and offices seem to be). The very barriers to moral

"Ain't Nobody Here But Us Social Forces" 143

responsibility that technology and organization throw up are opportunities for a profession willing to make overcoming them their responsibility.[37]

NOTES

I should like to thank Ibo van de Poel and Vivian Weil for many helpful comments on one or another earlier drafts of this chapter. I should also like to thank those who responded to the paper during a plenary session of the *International Conference on Moral Responsibility, Neuroscience, Organization, and Engineering*, Delft Technological University, The Netherlands, August 27, 2009. Originally published as: "'Ain't no one here but us social forces': Constructing the professional responsibility of engineers," *Science and Engineering Ethics* 18 (Winter 2012): 13–34

1. See, for example, Nissenbaum (1996). Though her subject is, as her title suggests, computer scientists, her argument fits engineers as well—and some of her examples include engineers.

2. Though responsibility-as-good-cause seems the fraternal twin of responsibility-as-faulty-cause, it is overlooked in most discussions of responsibility—perhaps because there is no obvious credit version of responsibility-as-liability or responsibility-as-accountability. The term for this missing sort of responsibility might be "responsibility-as-reward" (or perhaps "creditability").

3. Responsibility-as-power often appears in discussions of professional ethics. See, for example, Alpern (1983), for an early example applied to engineering.

4. Hart (1968, 210–223) seems to have been the first to distinguish four of these senses: responsibility-as-role (which I have divided into "office" and "task"), responsibility-as-simple-cause, responsibility-as-liability, and responsibility-as-competency ("capacity"). Ladd (1982, esp. 64–65) seems to have been the first to distinguish accountability (which, unfortunately, he treats as a merely organizational notion—as if a perfect stranger cannot "call me to account" for something untoward I did as a mere individual). Kuflik (1999, esp. 174–175) offers six senses—some significantly different from mine. For example, he distinguishes responsibility-as-simple-cause (using a hurricane example) from responsibility-as-function ("the heart is responsible for pumping blood"). Though a useful distinction for some purposes, it would only complicate discussion here. Nevertheless, that distinction is a reminder that my list is neither exhaustive nor canonical.

5. Those who distinguish between "backward-looking liability" and "forward-looking liability" seem to have confused responsibility-as-liability with responsibility-as-simple-cause or responsibility-as-faulty-cause. The reason for that confusion seems to be that blame is (generally) backward-looking. We generally blame people for what they have done, not for what they will do, and liability (it is thought) must work in the same way. The mistake is to assume that other blame-related concepts must be just as backward-looking as blame itself. While it may be true that blaming someone for doing such-and-such is to say of her that she *did* something she should not have (or failed to do something she should have), that is, to refer to the past, to

say of someone that they *deserve* blame (or that they are blamable) is not only another way to blame them; saying that also sets out a set of tasks (blame them in various ways until the reason for blaming is gone), that is, says something about the future (as blaming does not). Deserving blame differs from responsibility-as-liability only in not assigning the tasks to anyone, though it gives each of us a reason and a right to do the tasks.

6. There should be a corresponding problem about engineers and good causation, but no one seems to want to deny engineers credit for their good works even though the same arguments that undermine liability and accountability would seem to undermine credit. Where there can be no blame, there can be no praise (or so we generally think).

7. Indeed, holding *persons* responsible without fault is not the limit of what the law can do. The law can also hold non-humans responsible without fault. For example, in 1595, in Leiden, Provetie, an ordinary dog, was tried for the murder of a child whose death it caused. The dog nipped the child's hand while taking meat from it, the wound became infected, and the child died as a result. All this having been proved, Provetie was found guilty and publicly hanged. There was no claim that Provetie intended to harm the child, that it foresaw the harm, or even that it should have foreseen the harm. It was enough to hold the dog responsible for the death that it did what it should not have, nipped the child's hand, however innocently, and the child died as a result. (Arthur and Shaw, 2006, 245–246).

8. So, for example, morality is silent concerning the legal punishment of dogs. Since reason no longer seems to recommend such punishment, we may wonder *why* it ever did (if it did). One answer at least is obvious: Going through the forms of criminal trial may have been the simplest way to deal with an unusual case. There is some evidence that this obvious answer may also be the right one. The official summary of the case includes language probably standard at the time but wholly inappropriate to this particular case, for example, "all of which appears from the prisoner's own confession, made by him without torture or being put in irons." Arthur and Shaw, 2006, 246. How could the dog "confess?"

9. Ladd (1982), 67. Note that Ladd explicitly connects moral responsibility with power (my fifth sense of responsibility), for example, "one of the principal factors that creates moral responsibilities for one person rather than another is a difference of *power*, which usually consists of superior knowledge and ability to affect outcomes."

10. Those who are not competent (in the appropriate respect) cannot (it is assumed) be at fault (with respect to that). Neither can those who could not have foreseen the harm they caused be at fault in that respect. Competence and knowledge (of the appropriate sort) are (it is assumed) at least preconditions of fault.

11. For those who think that I should have included "control" among the list of necessary conditions, I recommend Sher (2006). As will be plain, nothing important turns on this omission.

12. Her inaction is, in this respect, quite different from the inaction of a lifeguard assigned to watch a stretch of beach. His inaction is a causal factor in any drowning death he might have prevented precisely because another lifeguard (his "relief" would have been there had he not been. The beach would not (we may suppose) be open for swimming if a lifeguard were not present, guaranteeing the safety of swimmers.

13. Lawyers like to interpret "cause" as a necessary condition ("but for"). The problems with so interpreting "cause" are well-known (e.g., an inability to make sense of redundant causes or to limit the number of causes to a manageable few). We need not concern ourselves with those problems here. The objection made here stands even if those problems can be solved. For recent work on omissions as causes (more or less consistent with what I say here), see: McGrath (2005), Dowe 2004), Boniolo and De Anna (2006), Pundik (2007), and Baumgartner (2008).

14. Curtis Forbes tells me that this comment may reveal that I am an urbanite. He has noticed that those who grew up in the countryside, as he did, are much less likely to feel this responsibility. He may be right, but even if this sense of responsibility (and the moral judgment of "worse" it supports) is "cultural-centric," it relies on a conception of responsibility independent of choice (and does not appear to be incoherent because it does), which is all I need to make my point.

15. Now, it might be objected that this example is really about a duty to aid because one can do so, rather than a duty to aid because one caused the problem. I do not think there is a *decisive* way to disprove this objection. But I do think there are good reasons to dismiss it. The most important is that the duty to aid usually involves serious danger to another—not, as in this case, merely cleaning up a (very small) mess. The only people who are likely to see much force in this duty to aid, apart from the causal connection, are act utilitarians, supposing I have no better way to use my time than looking after this dead cat.

16. I am not the only one to notice this sort of responsibility. See, for example, Kutz (2002), 558: "Sometimes an agent's mere causal linkage with harm may warrant a response ... agents can reproach themselves for faultless conduct that causes a harm, even when their victims, and onlookers, do not reproach them." The first half of Kutz (2002) is worth reading for the way his "Strawsonian" interpretation of responsibility tends to reach the same conclusions that I do in this section.

17. To see how important context is, suppose the cat had been a rat or other small non-pet. Would we feel the same obligation to stop? Should we feel any? Then suppose the cat had been a child or youth. Would we not feel an even stronger obligation to stop and help?

18. There is, I think, a close analogy between this view of responsibility as primarily imposed and what is called "the external" perspective on law. The external perspective on law misses essential features of law that the internal perspective includes, especially its authority. See Hart (1961, 101–102). From the outside, responsibility may seem to be primarily about having to answer to others for what one does (what Kant would call "heteronomy"). From the inside, responsibility may seem to be primarily about being trustworthy, someone capable of answering in a reasonable way for what he does, a certain (virtuous) disposition (what is recognizably Kant's "autonomy").

19. I am inclined to think that discussions of responsibility suffer from supposing, however implicitly, too close an analogy between moral and legal responsibility. While legal responsibility is primarily about holding others responsible, moral responsibility is at least as much about holding ourselves responsible as holding others responsible. The structure of moral responsibility may (despite many overlaps

with legal responsibility) have far less to do with punishment, compensation, or even blame than legal responsibility has. For more on this, see Cane (2002).

20. Nissenbaum (1996, 30) has therefore confused the problem of many hands with the problem of many causes. She is not the only one. See, for example, Bovens (1998, esp. 45–50), a section titled "Accountability: the problem of many hands" and beginning with the Thompson quotation used above, but consisting of several interesting examples all of which illustrate the problem of many causes rather than many hands.

21. The failure to take account of such differences between professions is one reason why Bovens' treatment of professional responsibility (Bovens 1998, 161–163) is—like that of most social scientists and the philosophers who follow their lead—decidedly misleading.

22. The obligation to keep good records, though helpful to accountability (should giving an account become necessary), is not itself accountability but a liability (an obligation to do something other than *give* an account).

23. Some other engineering codes have similar provisions. See, for example, IEEE (2006): "7. to seek, accept, and offer honest criticism of technical work, to acknowledge and correct errors, and to credit properly the contributions of others." (I cite codes of engineering ethics in this book to demonstrate that I am not making up what I claim to be facts about engineering. The codes constitute evidence, though not decisive evidence, of what responsibilities engineers have taken on.)

24. Well, relatively simple. There are conceptual problems about how to treat redundant causes, alternate causes, and the like (as noted already)—and these affect (and are affected by) how we conceptualize cause itself. See especially, Hart and Honoré (1973).

25. Of course, engineers do blame one another sometimes, but there is nothing especially interesting about that. Their blaming will have much the same structure as ordinary blaming—though it may consider the special responsibilities engineers have taken on.

26. Interestingly, Feinberg (1970), 20 eventually calls the engineering criterion "the handle criterion," having in view the expression "to get a handle on it."

27. Whether this is a good idea is a separate question. Compare Bucciarelli (2002).

28. There is a fourth flaw that raises deeper questions about causation. Premise 1 also seems to be false because redundant causes can be causes. Suppose that two hunters shoot the same bystander at the same moment (because the bystander has come between them and a duck that they are both aiming at), that both their bullets enter his body at the same time, and that either bullet would be sufficient to cause death. Neither hunter is necessary for the bystander to die. The death is "over-determined." Yet, it would seem odd to let both hunters off because neither was the cause, or even a cause, of the bystander's death. I ignore this flaw (and several related ones) here because I promised charity on the subject of necessary conditions for responsibility.

29. For a more detailed discussion of this argument, see the exchange between: Davis (1986a, b) and Bayles (1986).

30. Swierstra and Jelsma (2006) draw this conclusion from their study of European engineers. I came to the same conclusion from a study in the United States. See Davis (1997).

31. So, for example, a claim like the following seems to me clearly mistaken: "As an engineer, I think we take ourselves too seriously if we think we are ever going to influence what society decides to do by looking at the problem of ethics within the context of technology itself." Wiseman (1980), 166. This argument has an extension to society as a whole not relevant here. See, for example, Winner (1995).

32. See, for example, Florman (1978, 323): "An engineer designing a rapid transit cannot become expert in acoustics, urban planning, and the habits of woodland birds, and at the same time be an expert in the design of monorails. Nor can he do his best work if he is excessively apprehensive about the consequences of his every move." Note that Florman seems to be thinking of the engineer as working alone, not in teams that would include experts of the appropriate sort.

33. Of course, engineers are interested in avoiding blame, just as they are interested in avoiding legal liability. All I claim is that, as engineers, that is (and should be) what they are *least* concerned about—just as lawyers, as such, are (or at least should be) most concerned with legal liability. As individuals, avoiding blame may be their chief concern.

34. See for example, Schnädelbach (1980, 28): "'We dealt only with the technical problems and had no influence on the determination of goals'—this is a type of excuse frequently advanced in order to separate technical from ethical responsibility after political and moral catastrophes." While Schnädelbach does a good job of stating the argument from neutrality, he in fact rejects it—even for applied science. I am still looking for a contemporary thinker who endorses it.

35. For a good survey of this debate, see Sundström (1998). By "technology," I simply mean all those systems of people and things that constitute the world humans have made, everything from bridges, computers, butter knives, programs, and dictionaries to hybrid corn, hot dogs, and Labrador retrievers.

36. Think, for example, of programmers who claim to be "artists rather than engineers" and therefore need not "look back"—or scientists who rely on the neutrality argument discussed above in Section 2.7.

37. Compare Ladd (1982, 66): "responsibilities are not incurred or acquired like obligations. Rather one finds oneself responsible for something or other as a result of being in a certain position, e.g. of power." The responsibility that the engineer took here is (Ladd notwithstanding) incurred like an obligation, not by having the power to help but by a specific voluntary act, claiming to "see the problem." Had the engineer remained silent, he would have had no responsibility to fix the guillotine.

REFERENCES

Alpern, Ken, Moral responsibility for engineers. *Business and Professional Ethics Journal* 2 (1983): 39–48.

American Institute of Electrical Engineers, *Code of ethics* (March 8, 1912), http://ethics.iit.edu/publication/CODE–Exxon%20Module.pdf. Accessed July 22, 2010.

Arthur, John, and William H. Shaw (eds), *Readings in the philosophy of law*, 4th ed. (Upper Saddle River, NJ: Pearson Prentice Hall, 2006).
Baumgartner, Michael, "Regularity theories reassessed," *Philosophia* 36 (2008): 327–354.
Bayles, Michael D., "A problem of clean hands," *Social Theory and Practice* 5 (1979): 165–181.
Bayles, Michael D., "Reply to Davis," in Davis & Elliston (1986, 458–460).
Boniolo, Giovanni, & De Anna, Gabriele, "The four faces of omission: Ontology, terminology, epistemology, and ethics," *Philosophical Explorations* 19 (2006): 277–293.
Bovens, Mark, *The quest for responsibility* (Cambridge: Cambridge University Press, 1998).
Bucciarelli, Louis L. "Is idiot proof safe enough?" *Applied Philosophy* 2 (2005): 49–57.
Cane, Peter, *Responsibility in law and morality* (Oxford: Hart Publishing, 2002).
Davis, Michael, "The right to refuse a case," in Davis & Elliston (1986a), 441–457.
Davis, Michael, "Rejoinder to Bayles," in Davis & Elliston (1986b), 451–464.
Davis, Michael, "Better communications between engineers and managers: Some ways to prevent ethically hard choices," *Science and Engineering Ethics* 3 (1997): 171–213.
Davis, Michael, *Thinking like an engineer: Essays in the ethics of a profession* (New York: Oxford University Press, 1998).
Davis, Michael, and Frederick Elliston, eds, *Ethics and the legal profession* (Buffalo, NY: Prometheus Books, 1986).
Dowe, Phil, "Causes are physically connected to their effects; Why preventers and omissions are not causes," in C. Hitchcock (Ed.), *Contemporary debates in philosophy of science* (pp.189–196). Oxford: Wiley-Blackwell, 2004.
Feinberg, Joel, *Doing and deserving* (Princeton, NJ: Princeton University Press, 1970).
Florman, Samuel C., "Moral blueprints," *Harper's* 257 (1978): 311–323.
Glover, Johnathan, "It makes no difference whether or not I do it: Part 1," *Aristotelian Society: Supplementary Volume* 49 (1975): 171–190.
Hart, H. L. A. *The concept of law*. Oxford: Oxford University Press, 1961).
Hart, H. L. A. *Punishment and responsibility* (New York: Oxford University Press, 1968).
Hart, H. L. A., & Honore´, A. M. *Causation in the Law* (Oxford: Oxford University Press, 1973).
IEEE, *Code of ethics* (2006). http://www.ieee.org/ membership_ services /membership/ethics_code.html. Accessed July 22, 2010.
Kranzberg, Melvin (Ed.), *Ethics in an age of pervasive technology* (Boulder: Westview Press, 1980).
Kuflik, Arthur, "Computers in control: Rational transfer of authority or irresponsible abdication of autonomy?" *Ethics and Information Technology* 1 (1999): 173–184.
Kutz, Christopher, "Responsibility," in Jules Coleman & Scott Shapiro (eds), *The Oxford handbook of jurisprudence and philosophy of law* (Oxford: Oxford University Press, 2002).

Ladd, John "Philosophical remarks on professional responsibility in organizations," *The International Journal of Applied Philosophy* 1 (1982): 58–70.
Lynch, William T., & Kline, Ronald, "Engineering practice and engineering ethics," *Science, Technology, & Human Values* 25 (2000): 195–225.
McGrath, Sarah, "Causation by omission: A dilemma," *Philosophical Studies* 123 (2005): 125–148.
Miller, David, "Distributing responsibilities," *The Journal of Political Philosophy* 9 (2005): 453–471.
Nissenbaum, Helen, "Accountability in a computerized society," *Science and Engineering Ethics* 2 (1996): 25–42.
NSPE, *Code of ethics for engineers* (National Society of Professional Engineers, 2010), http://www.nspe.org/Ethics/CodeofEthics/index.html. Accessed July 22.
Pundik, Amit, "Can one deny both causation by omission and causal pluralism? The case of legal causation," in Frederica Russo & Jon Williamson, eds, *Causality and probability in the sciences* (London, UK: College Publications, 2007), 379–412.
Schnädelbach, Herbert, "Is technology ethically neutral?" in Kranzberg (1980).
Scott-Taggart, M. J. "It makes no difference whether or not I do it: Part 2," *Aristotelian Society: Supplementary Volume* 49 (1975): 191–209.
Sher, George, "Out of control," *Ethics* 116 (2006): 285–301.
Sundstro"m, P. "Interpreting the notion that technology is value-neutral," *Medicine, Health Care and Philosophy* 1 (1998): 41–45.
Swierstra, Tsjalling, and Jaap Jelsma, "Practice responsibility without moralism in technoscientific design," *Science, Technology, and Human Values* 31 (2006): 309–332.
Thompson, Dennis E., *Political ethics and public office* (Cambridge, MA: Harvard University Press, 1987).
Winner, Langdon, "The enduring dilemmas of autonomous technique," *Bulletin of Science, Technology & Society* 15 (1995): 67–72.
Wiseman, Gdalyah, "The powerlessness of engineers," in Kranzberg (1980).

Chapter 10
Engineering Ethics, Individuals, and Organizations

This chapter evaluates a family of criticism of how engineering ethics is now generally taught. The short version of the criticism might be put this way: *Teachers of engineering ethics devote too much time to individual decisions and not enough time to social context.* There are at least six versions of this criticism, each corresponding to a specific subject omitted. Teachers of engineering ethics do not (it is said) teach enough about: (1) the culture of organizations; (2) the organization of organizations; (3) the legal environment of organizations; (4) the role of professions in organizations; (5) the role of organizations in professions; or 6) the political environment of organizations. (By "organization," I mean any employer of many people, whether government, business, or non-profit.)[1]

My conclusion is that, while all six criticisms recommend worthy subjects, there is neither much reason to believe that any of those subjects are now absent from courses in engineering ethics nor an obvious way to decide whether they (individually or in combination) should (or should not) have their share of the course augmented. What we have here is not some well-defined either-or but a dispute about how much is enough. How-much disputes are not to be settled without agreement concerning the method by which we are to tell whether we have enough of this or that.[2] Right now we seem to lack such a method—and not to have much reason to expect one any time soon.[3]

1. CULTURE, ORGANIZATION, AND LAW

By "culture," I again mean a distinctive way of doing certain things (including standards for evaluating success and failure).[4] Lynch and Kline (2000), two historians of technology we have met before, have offered the following argument for including more about *organizational* culture in courses in engineering ethics: "Most engineers operate in an environment where their capacity to make decisions is constrained by the corporate or organizational culture in which they work." Engineers, "even public-spirited, highly ethical engineers," do not "spontaneously and infallibly know what the public interest demands [or how to achieve it]." To know what the public interest is and how to achieve it requires detailed knowledge of circumstances, including the culture of the organization (and the larger society). A course in engineering ethics should alert students to the need for such knowledge.

Lynch and Kline use what has come to be known as "the Challenger case" to illustrate this point, contrasting the "typical" approach that emphasizes the heroic (but ineffective) dissent of Roger Boisjoly (and a few others) with Vaughan's book-length reconstruction that makes the decision to launch on January 28, 1986, seem inevitable well before Boisjoly's heroic dissent, a product of the culture that NASA developed over many years. To stop the launch, Boisjoly would have had to have begun to change that working environment months or even years before. The night of January 27–28 was too late. Certain ways of working had left too few options. There was too little information available to make the case that the engineers would have had to make that night to get the decision they believed right. They were in no position to overwhelm their managers with facts.

The trouble with Vaughan's approach, apart from the fact that it takes a book to make the argument, is that it effectively frees the decision-makers at Morton-Thiokol of responsibility for the decision they made the night before the launch. If anything was to be done to prevent the disaster, it would (she argues) have had to be done earlier—much earlier—and (in large part) by others. While enriching our understanding of the context in which engineers work, Vaughan's book also tends to elide the ability to decide that is a precondition for any discussion of engineering ethics. Still, the elision might be allowed if the engineers and managers assembled on the night of January 27–28 to decide whether to launch felt they had no decision to make, but all seem to have felt otherwise. Both engineers and managers seemed to agree that the responsible decision-makers at Morton-Thiokol might have stood by the decision not to launch that they had made that afternoon. The launch would then (probably) have been postponed. Morton-Thiokol might have lost its NASA contract as a result and the managers who made the decision might then have lost their jobs. Morton-Thiokol was certainly under pressure from "Houston" (NASA)—pressure generated by earlier decisions, for example, the decision to assign budget to certain problems rather than to others. But resisting pressure, however great, is not like trying to push back the sea with a broom. Pressure does not rob decision-makers of the power to choose otherwise than they in fact did—or of the power to affect events by their choice. It just makes the decision harder. Decision-makers who "can't stand the heat" are supposed to "get out of the kitchen." They should not stay in the "kitchen" while pleading, "I had no choice."

I do not deny that the way NASA and its contractors were then working made a disaster like that of January 28, 1986, likely—perhaps even certain "in the long run" (which might be measured in years, not days). What I do deny is (in part) that the disaster had to occur when it did. Had certain engineers or managers thought about what they were doing somewhat differently, something well within their power (given that they did think about it that way a few hours earlier), they could have postponed the launch to a warmer day—and avoided *this* disaster. I also deny that the disaster, whenever it occurred, had to be a scandal (an ethical disaster as well as a technical one). The scandal was the result of overruling the engineers, not of the shuttle's disintegration soon after launch.

The history of engineering is, in part, a history of disasters. Engineers are supposed to produce new technology. Insofar as technology is new, it goes beyond what we know—including what we know to be safe. Whatever care engineers take, they will now and then go too far and their products will fail, sometimes disastrously. Engineers have developed (and are still developing) routines for keeping failure to a minimum. That is why engineers are so central to the development, testing, operation, distribution, maintenance, and disposal of so much technology (Petroski, 1985). When engineers follow those routines, the failures, however disastrous, are not scandals. When they do not follow those routines, the failure, if it occurs, becomes a scandal. One of those routines is working by consensus. It is that routine that the managers violated on the night of January 27–28, 1986.

While the sociological approach need not make decisions seem inevitable, that is its tendency. Sociologists generally try to understand events as linked by "social forces" rather than by individual decisions. Social forces are said to "determine" individual decisions, explaining the decisions by explaining them away. Decision-makers are understood to have "no choice" (hence, no [real] decision). Teaching engineering ethics that way would, it seems to me, not teach it at all. Telling students that they will have no choice but to do what "the culture" asks cannot help them make a good decision. Indeed, it seems likely to have the opposite effect. Yet, I agree that, like all other decisions, engineering decisions are, though free, constrained—and culture is both an important constraint and one easily overlooked (especially, the small-scale structuring of work embedded in budgets, specifications, work rules, and even the location of offices). We must understand the cultural context of engineering decisions to understand what options are actually available to engineers and how engineers should choose among them. We should therefore find a way to have the insights of sociology without its excesses (one or another form of determinism). Sociology gives us tendencies or probabilities, not certainties.

Having agreed that we should try to include the insights of sociology in any course in engineering ethics, I need not conclude that those teaching engineering ethics should take *more* time to teach those insights. There are at least two reasons I need not. The first is "economic." Decisions about what to cover in a course are inherently comparative, a matter of budgeting time. Before we can decide that we should include this or that in a course, we must know what we are to exclude to do it; yet Lynch and Kline do not tell us what to leave out—or even how to go about comparing sociology's contribution to that of other subjects already in the course. Their argument does not even rule out the possibility that everything now in a typical course in engineering ethics contributes to "addressing moral responsibility" more than what they propose to replace it with. Lynch and Kline need a comparative argument to make their case, an argument they do not even hint at.

That is one reason that I need not conclude from the argument that Lynch and Kline make that we should devote *more* time to organizational culture. There is another: No one teaching engineering ethics seems to think that teaching organizational culture is irrelevant. The importance of organizational culture seems to have been admitted since the beginning of philosophical courses in engineering ethics in the late 1970s. Consider, for example, this (partial) description of one of the first philosophical courses in engineering ethics:

> Undeniably, however, moral conduct demands effort. One must be alert, reflect, plan ahead, anticipate likely consequences, and exert oneself at opportune moments. Some like to dismiss the study of moral issues saying, "A person is either decent or a scoundrel, and a formal course in ethics will not change that." We answered by showing the kind of thought and effort needed to engage in morally acceptable behavior in complex situations. Case studies impressed our students with the potential for catastrophe in some of these intricate circumstances. (Weil, 1977, 36)

The emphasis on "intricate circumstances" rather than "culture" suggests that one may be able to work one's way through the intricacies, a suggestion useful if one is to teach practical ethics rather than social science. But the emphasis on circumstances ("potential for catastrophe") must include an understanding of how organizations work. In practice, then, what may—on paper—be a "short case" (with no cultural, legal, or other background information) may, by the end of class, work exactly like a much longer case—with the background printed. The case discussion may include "mini-lectures" or student responses providing the relevant background information.

Everyone I know who teaches engineering ethics today already devotes a significant part of the semester to teaching students how organizational culture constrains engineering decisions. Lynch and Kline offer no procedure for determining that what is already being taught is insufficient. They do not even offer a syllabus to suggest what "doing enough" might look like. From the evidence they present, we cannot conclude that the engineering ethics course they recommend would be much different from those they criticize. Perhaps they are "just bringing coals to Newcastle." How are we to tell?[5]

For similar reasons, I agree that we should devote some time in engineering ethics to explaining the structure of large organizations and the legal constraints under which they operate. Engineering students should know, for example, both that many organizations have procedures for taking ethical issues out of channels and that going over a boss's head is generally imprudent. Students should also know something about product liability, the Federal Occupational Safety and Health Administration, patent law, and other ways in which government constrains what engineers or their employers can legally do. Such constraints are part of the context of engineering decision (in the United States), something an engineer must take into account to make a good decision (whether the decision concerns ethics or not).[6] But, again, the question is not whether we who teach engineering ethics should do something about such things. The question is *how much* we should do—or, rather, whether we are not *already* doing enough (or too much). I see no reason to think that we are not already doing enough—given the length of the semester and the other topics we should cover. So (again I ask), how are we to tell that teachers of engineering ethics are not (generally) doing enough about teaching organizational culture already?

2. PROFESSIONS, ORGANIZATIONS, AND LEGISLATION

Organizations have important roles in engineering. For example, an engineer may not be able to take time to attend a professional meeting without permission from a superior—and may not be able to afford to attend unless the employer pays. Engineering also has an important role in setting organizational culture. Working through various standard-setting bodies, engineers have much to say about what they will and will not do for their employers— and how they will do it. So, for example, many of engineering's organizations (such as IEEE or ASME) have standard-setting bodies. The standards those bodies adopt become *de facto* standards of competent work. And often, they become legal standards as well because local, state, and even national governments incorporate them by reference into safety codes, building codes, procurement contracts, and the like. Sometimes the best way for individual engineers to solve an ethical problem is to get their technical society to change the standards under which they are supposed work—or to do something similar through a governmental body such as the Environmental Protection Agency or a local zoning commission.

That, I think, is the point that the three Dutch critics of today's engineering ethics courses had in mind when they argued:

> To our way of thinking, engineers will only be able adequately to solve or diminish the ethical problems they have already identified by understanding the broader context from which they originate and by accepting that they must play an active role in helping to reshape that context whenever that may be necessary. In order to prepare appropriately future engineers for this, a multi-disciplinary approach to ethics and engineering is required, which will involve paying systematic attention to the actual and possible role of law, organizations, and procedures for collective decision-making. (Zandvoort, van de Poel, I., and Brumsen, 2000, 297)

Beside philosophy and sociology (or, at least, the social study of technology), these Dutch critics seem to think at least two other disciplines ought to have a place in engineering ethics: history and law. Who could disagree?

The problem is determining how this description of what engineering ethics courses *should* contain differs from what most engineering ethics courses *now* contain. In 2001, I made my first visit to Delft University of Technology where these three Dutch critics were teaching, discussed their engineering ethics course with two of them for more than an hour, and concluded that it differed in no important way from the course long taught at IIT (where I was teaching). There were differences in emphasis, a few hours more on this, a few hours less on that, but nothing that struck me as beyond what I ordinarily change when I teach a course again. The two Dutch critics were sure I was wrong—or, at least, that IIT was an exception. But they could provide no way to determine that I was wrong—except by quoting other critics of "American-style engineering ethics" who also provided no evidence that showed what they claimed. How is it possible to be so sure—without evidence?

Those Dutch critics might reply that the argument I am making here unfairly assigns the burden of proof to them. Since we are unlikely to get much more information any time soon about what actually goes on in most of the hundreds of engineering ethics courses offered in any particular semester across North America, assigning burden of proof is virtually decisive. Am I being unfair when I place the burden of proof on the critics of today's American-style engineering ethics? I think not. My argument would have much the same force even if I simply agreed that assigning the burden is itself controversial (as it seems to be). Criticism of a widespread and longstanding practice should not itself rely on little more than a controversial assignment of burden of proof. It is, I think, fairer to assign the burden of proof to those who argue for change.[7]

There were, I admit, two problems of language that complicated discussion with my Dutch friends. The first was that all materials for their course (syllabus, text, and handouts) were in Dutch. (So, as the saying goes, "It was all Dutch to me.") I had to depend on my friends' description of the material (given in perfect English). I could not make my own direct assessment.[8] Second, and more important, my two Dutch friends did not use the term "engineer" in quite the way that I (and most other North Americans) do. All graduates of Delft, including architects, industrial designers, and policy analysts, are said to be (and have a degree to prove that they are) "engineers." They are entitled to put "Ing." after their name—much as I am entitle to the title of "Dr." even though I know nothing of medicine. For my friends at Delft, there is less (but not much less) difference between the ethics of technology (or technology assessment) and engineering ethics than there is for me. Engineering was not for them a distinctive way of carrying on technology (though, except for Informatics, all the departments actually directing students to their course in engineering ethics were what North Americans would unhesitatingly identify as engineering).[9]

3. A POSSIBLE EXPLANATION OF THE QUESTION

My experience at Delft, along with my reading of the relevant literature (and discussions with other critics of how engineering ethics is now taught), has led me to wonder whether the family of criticism addressed here might not itself have a cultural explanation. The critics of "American-style engineering ethics" seem to have started their careers either in philosophy of technology or in science and technology studies (STS). These fields have never had much interest in individual decision-making. Their chief interest has always been *social* policy—when it has not simply been understanding (or criticizing) technology. The sort of course that grew out of their concerns was *X and Society* (Computers and Society, Science and Society, or the like). Anyone coming from such a field to engineering ethics must initially find the field's emphasis on individual decision-making surprising. With more experience, they may come to see the practical advantages of that emphasis. After all, their students, like mine, will often not have time to wait for a social or even organization-wide solution to the problem that arose at 10:17 this morning. They will have to act as an individual or part of a small group in the next few hours—or minutes.

Teachers coming to engineering ethics from philosophy of technology or STS may also have been misled by a certain custom within engineering ethics. As in other fields of practical or professional ethics, many (but not all) textbooks in engineering ethics still have a section on moral theory.[10] And many new teachers of the subject doubtless begin their career assuming that they should follow the text—giving three to six hours of class time to moral theory. Generally, as they gain experience, they drastically cut back on moral theory—and a few of us, even those who (like me) also regularly taught a course in moral theory, actually ceased to teach moral theory in the engineering ethics course (or teach moral theory only "on demand")—to make room for history, sociology, law, and the like. We make room for history, sociology, law, and the like because we quickly see that these subjects are necessary to provide context for the decisions in question—and are more useful to our students than a small amount of moral theory. That, anyway, is the impression I have after more than three decades in the field.

It is hard to find anything in print about this cutting back on moral theory. Why? I don't know. I also don't know why so many texts in engineering ethics continue to include a large section on moral theory. Perhaps there is a fear that publicly admitting how little moral theory one needs for engineering ethics would make the field look less like legitimate philosophy than is prudent. The course in engineering ethics is secure in only a few philosophy departments. Most of us recall colleagues asking questions like, "Is this a real philosophy course? Isn't it too practical? Doesn't it belong in the engineering school—or, at least, in the Department of Social Sciences?" There is indeed a risk that a philosophy department here or there will disown a course in engineering ethics if it contains no moral theory. But I think that risk better dealt with by pointing out that immersion in a subject outside philosophy is almost a precondition for competently teaching what most philosophers now consider legitimate philosophy courses, everything from philosophy of law to philosophy of physics. Such courses should not be abandoned for "real philosophy" just because they do not include much *traditional* philosophy. Not only are philosophy-of courses real philosophy, they also seem to have saved philosophy from the fate of classics, a field now in danger of disappearing from most universities. Philosophy has never been just metaphysics, epistemology, and moral theory. Philosophy departments that retreat to such a "core" will, in all probability, eventually become too small to survive.

NOTES

Thanks to Michiel Brumsen, Bill Lynch, and an anonymous reviewer for *Science and Engineering Ethics* for comments on one or another draft of this chapter. An earlier, and longer, version (focused on teaching professional ethics generally) appeared as Davis, 2004. This article was originally published as: "Engineering Ethics, Individuals, and Organizations," *Science and Engineering Ethics* 12 (April 2006): 223–231.

1. A profession is not an organization in this sense, though a professional society (that is, a society claiming to represent a profession) is—if it itself employs many people, as some do.

2. Compare Hume, 1948, 85: "[There] is a species of controversy which, from the very nature of language and of human ideas, is involved in perpetual ambiguity, and can never, by any precaution or any definitions, be able to reach a reasonable certainty or precision. These are the controversies concerning the degrees of any quality or circumstance."

3. Because professors tend to appeal to academic freedom to defend conduct in the classroom for which they can think of no better defense, let me add that I do not think academic freedom an issue here. Those who criticize what is now done or not done in engineering ethics recognize the academic freedom of those they address. For them, the issue is what would be a wise use of that freedom.

4. This definition, though simple, will be sufficient for our purposes. For my reasons for preferring it to some that others might prefer instead, see Davis, 1996.

5. A web survey may seem the solution. I doubt it. My quick search using "engineering ethics course" (February 13, 2006) produced 486 items. Most of these provided nothing like a syllabus. When I searched using "engineering ethics syllabus," the number of items produced became 63. Most of these clearly contained at least one reading I could identify as concerned with organizational culture. When I repeated the two searches fifteen years later (February 24, 2021) using Google, the returns were too numerous to study quickly: about 228,000,000 for "engineering ethics course"; about 12,000,000 for "engineering ethics syllabus."

6. These topics do seem to appear even in textbooks that Lynch and Kline (2000) identify as not having the right emphasis. Consider, for example, Harris et al, 2000, one of the best-selling texts in engineering ethics at the time Lynch and Kline were complaining. Of its eleven chapters (377 pages), two, Chapter Seven (145–172) and Chapter Nine (206–232) deal in large part with the legal environment and the chapter between these two (173–205) concerns "Engineers as Employees." (These proportions have not changed much in later editions.) I leave it to the reader to confirm that much the same is true of the other textbooks that Lynch and Kline mention.

7. Below I consider and dismiss the only other evidence on which my Dutch friends rely, that is, the chapter on moral theory or moral decision-making in most textbooks that Americans use to teach engineering ethics.

8. They have since described their course in print. See van de Poel, Zandvoort, and Brumsen (2001), especially the list of eleven "subjects" (in effect, a short syllabus) on 274. Significantly, the only subject I would not include in my own course on engineering ethics is—*moral theory*! Delft's recent contribution to the teaching of engineering ethics, the very interesting computer program *Agora*, is almost entirely about moral theory. From this, I conclude that "the European approach to engineering ethics" is more individualistic (insofar as focused on moral theory rather than political

philosophy) than is its American counterpart. Not the conclusion they would draw but nonetheless one the evidence then supported at least as much as it supported theirs.

9. Since de Poel, Zandvoort, and Brumsen now distinguish their course from "traditional American" courses in engineering ethics (rather than just "the standard American course"), they implicitly pose yet another empirical question: What in fact is that American tradition (if there is one—and only one)? I do not claim to know, but I do claim to be part of at least one American tradition (in virtue of being an American sharing my way of teaching engineering ethics with a significant number of other Americans going back several decades). I therefore constitute evidence of what the American tradition must be (or, at least, include)—something they cannot claim.

10. Deborah Johnson's popular anthology (1991) seems to be the chief exception. Significantly, Johnson, though a philosopher by training, has had a career (almost) entirely in STS, not philosophy proper. She did not begin with the prejudice that moral theory was (more or less) all that philosophy has to offer engineering ethics. For a brief biography, see http://www.onlineethics.org/Connections/Community/DJohnson.aspx (access May 23, 2020).

REFERENCES

Davis, Michael, "Second Thoughts on Multi-Culturalism," *International Journal of Applied Philosophy* 11 (Summer/Fall 1996): 29–34.

Davis, Michael, "Teaching Moral Responsibility Within Organizations: Are We Doing What We Should?" *Business and Professional Ethics Journal* 12 (Fall 2004): 77–91.

Charles E. Harris, Jr., Michael S. Pritchard, and Michael J. Rabins, *Engineering Ethics: Concepts and Cases*, 2nd edition (Wadsworth: Belmont, CA, 2000).

Deborah Johnson, *Ethical Issues in Engineering* (Prentice Hall: Englewood Cliffs, NJ, 1991).

Hume, David, *Dialogues Concerning Natural Religion* (Hafner Press: New York, 1948).

Lynch, William T. and Ronald Kline, "Engineering Practice and Engineering Ethics," *Science, Technology, and Human Values* 25 (2000): 195–225.

Petroski, Henry, *To Engineer Is Human: The Role of Failure in Successful Design* (St. Martin's Press, New York, 1985).

van de Poel, H. Zandvoort, and M. Brumsen, *Science and Engineering Ethics* 7 (April 2001): 267–282.

Vaughan, Diane, *The Challenger Launch Decision: Risky Technology, Culture, and Deviance at NASA* (University of Chicago Press, Chicago, 1996).

Weil, Vivian M. "Moral Issues in Engineering: An Engineering School Instructional Approach," *Professional Engineer* (1977): 35–37.

Zandvoort, H., van de Poel, I., and Brumsen, M. "Ethics in the Engineering Curriculum: Topics, Trends, and Challenges for the Future," *European Journal of Engineering Education* 25 (2000): 291–302.

Chapter 11

"Social Responsibility" and "Social Justice" for Engineers?

Most engineers I know dislike talking about "the social responsibilities of engineers." They have a similar reaction to talk of "social justice" as a criterion for evaluating engineering work. They consider such uses of "social" to be overly political, faddish, fuzzy, or otherwise inappropriate in any discussion of engineering. Many non-engineers take these reactions to confirm their stereotype of engineers: politically conservative, socially backward, intellectually stunted. I think there is a better explanation, one that may help us—both engineers and non-engineers—understand engineering better.

1. ORIGIN OF TERMS

The term "social responsibility" comes to engineering from business. There it responds to the thesis—associated with the economist Milton Friedman since 1970—that the only responsibility of business is to make as much money as legally possible without fraud or coercion. Those who appeal to "the *social* responsibilities of business" do so to remind business that it has responsibilities beyond the *economic*, for example, to contribute to local charities or to take account of worker safety in an overseas plant even when local law does not require it. The appeal to social responsibility is supposed to *encourage* acts beyond the moral (and legal) minimum.

The term "social *justice*" has a different origin. It first appeared in Catholic social teaching about 150 years ago. Distributive justice, corrective justice, and criminal justice apply to everyone. "Social justice" was meant to emphasize the special needs of the poor (and others less able than most to protect themselves). It was to suggest not mere equality of rights or fair treatment of everyone but "solidarity with the downtrodden," programs for improving the condition of "the least among you" (Cullen et al., 2007).

Both terms, "social responsibility" and "social justice," are controversial within their normal domain. That is one reason *not* to bring them into engineering. Engineering has enough controversy of its own (concerning quality, risk, sustainability, and so on). But there is a better reason to bring neither into engineering: bringing them in would add nothing to the responsibilities that engineers already accept as part of their *professional* responsibilities. Indeed, bringing those terms into engineering would instead suggest that engineers should do less than they are already required to do; that is, that the responsibilities of engineers are no greater than the responsibilities of the organizations for which they work.

2. DETERMINING ENGINEERS' RESPONSIBILITIES

How are we to determine the professional responsibilities of engineers? For the sake of variety, let us ignore the well-known codes of engineering ethics we have been referring to so far. Let us start instead with the Code of Ethics of Professional Engineers Ontario's (PEO's; which is not only a code of ethics but also an Ontario statute governing all the province's practicing engineers). The obvious provision relevant to social responsibility and social justice there is the familiar language of subsection 2.i: "[A practitioner shall] regard the practitioner's duty to public welfare as paramount." This subsection is more demanding than any statement of business's social responsibilities that I know of. It *requires* engineers—in the practice of engineering—to give priority to the public welfare. An engineer who fails to put the public's welfare first in professional work fails to satisfy a minimum requirement of engineering. A social responsibility (whatever it is exactly) is something less than a "duty" or requirement. (See chapter 9)

Turning to the Code of Ethics of Engineers Canada, we find that its first principle, though similar to the PEO principle just quoted, is even more demanding. It requires professional engineers not only to "Hold paramount the safety, health and welfare of the public" but also to "[hold paramount] the protection of the environment and promote health and safety within the workplace." If we assume (as I think we should) that the public's safety and health are part of what the PEO's code means by "welfare," the first principle of the Engineers Canada code adds to the obligations the PEO identified at least two others: first, protection of the environment (another paramount requirement); and, second, promotion of workplace safety.

That is not all the Code of Ethics of Engineers Canada has to say that is relevant to social responsibility or social justice. Its Principle 5 requires professional engineers to "Conduct themselves with equity, fairness, courtesy and good faith towards clients, colleagues and others..." Whatever "equity" is, it is more than mere "fairness," and whoever "others" covers, it covers more than "clients" and "colleagues." Principle 8 adds that engineers shall also "[be] aware of and ensure that clients and employers are made aware of societal and environmental consequences of actions or projects."

Much the same requirements can be found in engineering codes outside Canada, for example, south of the border in the NSPE's Code of Ethics (NSPE, 2007), across the Atlantic in the Code of Conduct of the European Federation of National Engineering Associations (FEANI, 2009), and even across the Pacific in the Asian Declaration of Engineering Ethics (2004). There are, of course, differences between these codes—depending in part on when the code was last revised, on the state of ongoing discussions within engineering, on whether the code is to be enforced by law, and so on. So, for example, the NSPE code includes a provision (III.2.d) that encourages engineers "to adhere to the principles of sustainable development in order to protect the environment for future generations" (a provision that began entering engineering codes in the 1990s). The Asian Declaration has something similar. But, like the Canadian codes, FEANI's does not.

3. RELATION TO SOCIAL JUSTICE AND SOCIAL RESPONSIBILITIES

On any reasonable reading of the paramount provisions, all the codes just cited cover most, if not all, the subjects that "social responsibility" is supposed to cover. What is "paramount" for engineers is not profit, whether their client's, employer's, or their own, but the "public welfare." While "public welfare" is a very general term, allowing much room for interpretation, it certainly includes health, safety, and other material conditions in society at large.

The relation of all these engineering codes to social justice is less obvious. So, assume (as seems probable) that, all else equal, a dollar spent improving the welfare of the downtrodden is likely to add considerably more to overall welfare than an equal amount spent on anyone else. Looking after the least well-off would, then, all else equal, be the most *efficient* way to improve the *public* welfare. For example, a small improvement in the safety of the cheapest cars should, all else equal, save many more lives than an equal improvement in the safety of expensive cars (i.e., equal in total social cost)—in part at least because there are many more cheap cars and many more ways to improve them at little cost.

This point is not merely hypothetical. If we consider the material well-being of the least well-off over the past 150 years, there is no question that it has improved considerably (and more than the welfare of the best-off has): life span is longer, health is better, hunger is rarer, and so on. Much remains to do but much of what has been done is in large part the work of engineers, the result of fewer industrial accidents, better transportation, cleaner air, improved water supply, and so on.

4. CONCLUSION: ENGINEERS AS REVOLUTIONARIES?

Measured by achievement, not intention, might we not say that engineers are the true social revolutionaries of the past 150 years? Indeed, should we not say that social revolution (as measured by improved conditions of life at least) is part of the ordinary work of engineers?

Of course, the way engineers have made their social revolution does not look much like revolution. Engineering tends to change life slowly and in small ways, for example, by increasing the speed with which boilers shut down when the water level drops too low. Engineering changes also tend to be incorporated into technical designs, technical standards, and technical procedures rather than in the memorable phrasing of a public declaration or the bloody language of a show trial. The guillotine was the invention of a physician, not an engineer.

So, those who call on engineers to exhibit social responsibility or to contribute to social justice seem to make at least two mistakes. The first is overlooking how much engineers are already doing. The second is failing to understand that engineers, though already committed to socially responsible engineering and social justice, probably cannot do much more about those commitments without better tools. Any engineer worthy of the name would, all else equal, be happy to invent something to improve the welfare of the poor or protect the public from whatever threatens its safety, health, or welfare. The reason most engineers do not (when they do not) is that they lack the tools necessary for it. For example, an engineer who wants to design an environmentally neutral component for a cell phone needs to know not only the environmental effect of how the materials for that component are mined, shipped, and shaped, but also the environmental effect of how the component will be used and disposed of. That is, she needs a system she can rely on to track such information, evaluate it, and rate cell phone components accordingly, not only for the cell phone she is working on but also for those it competes with or replaces, a system that standardizes cell-phone information in the way much safety or environmental information in now standardized.

Such a system is never the work of *one* engineer. Some standards are the work of government agencies; some, the work of the standard-writing bodies of national or international engineering societies; some, the work of nongovernmental interdisciplinary bodies (such ISO); and some the work of a large corporation that has a near monopoly on some new technology. In short, the work of making it possible for engineers to show more social responsibility or to do more for social justice is itself a social undertaking, not something usefully assigned to an individual engineer. Those who want engineers to be more socially responsible or to contribute more to social justice should focus on providing engineers with better tools for that work. Engineers can certainly use help in carrying out their professional responsibilities. No engineer can do much alone.

NOTE

This chapter was originally published as: "'Social Responsibility and 'Social Justice' for Engineers?" *Engineering Dimensions* (March/April 2013): 39–41.

REFERENCES

Asian Declaration of Engineering Ethics (2004), http://ethics.iit.edu/ecodes/node/5076 (accessed August 10, 2020).

Cullen, Philomena, Bernard Hoose, and Gerard Mannion (eds), *Catholic Social Justice: Theological and Practical Explorations* (T. &. T Clark/Continuum, 2007).

Engineers Canada, Guideline on the Code of Ethics, http:/ https://engineerscanada.ca/publications/public-guideline-on-the-code-of-ethics/ethics.iit.edu/ecodes/node/5076 (accessed August 9, 2020).

European Federation of National Engineering Associations (FEANI), Code of Conduct (2009), https://ethics.iit.edu/ecodes/node/6410 (accessed August 10, 2020).

Friedman, Milton "The Social Responsibility of Business is to Increase Its Profits," *The New York Times Magazine* (September 13, 1970). (Accessed November 9, 2012).

National Society of Professional Engineers (NSPE), Code of Ethics for Engineers (2007), http://www.mtengineers.org/pd/NSPECodeofEthics.pdf (accessed August 10, 2020).

Professional Engineers Ontario, Code of Ethics (1990), http://ethics.wikia.com/wiki/The_Code_of_Ethics_of_Professional_Engineers_Ontario (accessed August 10, 2020).

Chapter 12

Macro-, Micro-, and Meso-Ethics

There has recently been renewed interest among many concerned with social policy, philosophy of technology, and the social study of technology, in the distinction between macro-ethics and micro-ethics (see, e.g., Son, 2008, and works cited there and below). Those so interested have generally been critical of engineering ethics, especially the classroom and textbook versions, as too much concerned with micro-ethics. Much more should (they claim) be said about macro-ethics. Sustainable development seems to be just the sort of issue for which engineering ethics might deserve that criticism. Courses and texts in engineering ethics still include little about sustainable development as such, though most include something about protecting the environment.

By "sustainable development," I mean (roughly) improvement in material conditions of humanity meeting "the needs of the present [justly] without compromising the ability of future generations to meet their own needs [as justly]" (Kates et al., 2005, 9–10.). What could be more "macro-ethical" than problems concerned with social justice, the environment, and the material welfare of all humanity, now and in the future as far as humans can plan? Would not engineering ethics (the course) have to change dramatically to deal even reasonably well with a "macro-subject" like sustainable development?

My answer to the first is that sustainable development is no more or less macro-ethical than many other questions now a routine part of courses in engineering ethics. My answer to the second question is that engineering ethics can easily deal with sustainable development—*insofar as it involves questions of engineering*—without substantial change (no more than a few new problems and a paragraph or two giving background information). Both these answers are controversial. I shall defend them here by arguing (1) that the micro-macro distinction misses an important intermediate domain in ethics (the "meso"), (2) that engineering ethics, at least when taught in the standard way, that is, as professional ethics, belongs to that intermediate domain, and (3) that what the "macro-ethics" advocates want to include in engineering ethics courses does not seem to be ethics at all (in any interesting sense) or, while ethics, does not seem to be engineering or, while engineering, does not seem to be "macro." Any problem of sustainable development an *engineer* might address *as engineer* belongs to the intermediate domain rather than to either micro-ethics or macro-ethics; the same for any engineering organization addressing the problem as an organization of engineers. Those advocating more "macro-ethics" in courses in engineering ethics are confused about what engineering ethics (the study) is. It is not about technological decisions as such but about decisions engineers make as engineers (an agent-centered study rather than an object-centered one). The term "macro-ethics" need not appear in discussions of how to incorporate sustainable development into courses and texts in engineering ethics. Indeed, given the conceptual confusion its application to engineering has involved, the term should probably be avoided altogether.

1. MICRO, MACRO, AND THE GREAT IN-BETWEEN

The distinction between micro-ethics and macro-ethics seems to have been constructed on the model of a fundamental distinction in economics. (Indeed, the source of the distinction typically cited, Ladd, 1980, 156, explicitly claims to be adopting the distinction from economics.) Micro-*economics* is the study of markets. Its subject is the making, selling, and buying of goods by individuals, households, partnerships, corporations, and other market agents. Macro-economics is, instead, the study of the economy of a state or geographical region; its concern is national or regional income, money supply, taxation, balance of payments, government expenditure, and the like. Micro-economics is treated in one set of economics courses; macro-economics in another. The distinction between macro and micro in economics seems to date from the 1930s.[1]

When Ladd brought the micro-macro distinction into applied ethics in 1980, there was a closer analogy between economics and ethics than there is today. In the 1980s, (philosophical) ethics was still largely concerned with decisions of mere individuals; political philosophy, with decisions of government; and other sorts of decision (most of what we now call "applied ethics") were only beginning to win much attention—within philosophy or outside. Ladd himself had argued strenuously against the possibility of organizations, especially corporations and bureaucracies, being either moral agents or owed moral obligation. For Ladd, ethics was about what individuals should do; ethical standards, the same for each individual whether acting alone or in concert with others (Ladd, 1970). Today most of us recognize families, businesses, trade associations, professions, religions, charities, private universities, and other voluntary groups as distinct moral entities. The collective term now in vogue for these non-governmental entities is "civil society."

Since 1980, civil society has become increasingly important in our thinking about "society"—that is, the largest and most inclusive collection of human beings living together for mutual benefit. That thinking has concerned both what society is and what it should be. Consider two recent articles by one of the strongest advocates of macro-ethics, Joseph Herkert. Herkert (2001) offered a table listing five versions of the micro-macro distinction—including Ladd's. By Herkert's count, three of the five recognized an intermediate category between individual ethics and "social ethics," though each did it differently (Herkert, 2001, 405). Herkert counts Ladd as one of the three who recognize an intermediate category. That, I think, is a mistake. Ladd (1980), 155, is quite clear that "there is no special ethics belonging to professionals."[2]

Herkert (2005), 374, proposed Herkert's own version of the distinction. It divided civil society down the middle:

> Engineering ethics can be viewed from three frames of reference—individual, professional and social—which can be divided into "microethics" concerned with ethical decision-making by individual engineers *and* the engineering profession's internal relationships, and "macroethics" referring to the profession's collective social responsibility *and* to societal decisions about technology. (*Italics mine*)

Though Herkert clearly is aware of the importance of civil society (or, at least, of the engineering profession), just as clearly he has a problem making civil society fit the micro-macro distinction. One sign of that difficulty is that, in his version of the distinction, part of professional activity (the "internal") ends up on one side of the divide while the rest (the external or "social") ends up on the other.

When (in an email) I pointed out to Herkert how arbitrary it seemed to divide civil society in this way, comparing his approach unfavorably to King Solomon's *threat* to cut the disputed baby in half, Herkert responded:

> I don't have any difficulty at all in making this distinction. The internal and external relations of the engineering profession are very different. In fact, it is this difference that drew me to the micro/macro distinction in the first place. (email, April 13, 2009)

Herkert went on to link his attraction to the distinction to his experience working on "macro issues" within the IEEE. That experience is substantial.

See, for example, Herkert (1998). So, his response is not to be taken lightly. The response relies, however, on observations concerning how professional societies (sometimes) conduct themselves, not on judgments concerning how they *should* conduct themselves, that is, on fact, not ethics. Herkert (1998) makes a good case for his conclusion that the IEEE and other engineering societies have in fact failed to support sustainable development but, more relevant here, makes that case without any use of the term "macro-ethics."

The chief problem with the micro-macro distinction in ethics is not that the analogy with the similarly-named distinction in economics is now too distant (though that is a problem). Nor is the chief problem that the distinction does little or no useful work (though that too is a problem). The chief problem is that the distinction tends to hide an important fact: the crucial role of civil society in defining what we mean by engineering *ethics*. "Ethics" has, as already noted here, many senses in English. Four seem relevant here: ethics-as-ordinary-morality, ethics-as-moral-theory, ethics-as-theory-of-the-good-society, and ethics-as-special-standards (that are morally binding). Among interesting senses that do not seem to be useful here are: ethics-as-domain-of-problems (those problems that someone might propose that a new moral standard should resolve); ethics-as-actual-moral-practice (positive morality); and ethics-as-moral-ideal (aspirational ethics). Which of the relevant senses is (or should be) primary when we speak of "engineering ethics?"

Ordinary morality consists (more or less) of those standards all reasonable persons (at their most reasonable) want all others to follow even if that would mean having to do the same: don't lie, don't cheat, keep your promises, help the needy, and so on (rules, principles, ideals, and the like). Ethics-as-ordinary-morality is about what individual moral agents should or should not do, the domain of *micro*-ethics.

Ethics-as-moral-theory (moral philosophy) is the attempt to understand morality as a reasonable undertaking. Its focus is therefore also micro-ethics. Ethics-as-theory-of-the-good-society is, in contrast, about how *society*—in its widest sense—should be organized to achieve the good. It may go beyond what ordinary morality requires, recommends, or forbids. It is, therefore, an undertaking distinct from ethics-as-ordinary-morality. Indeed, the attempt to define the overall organization of society, to make recommendations concerning international relations, constitution, government, and laws is usually called "political philosophy" (or "political theory"). Every definition of macro-ethics includes this political domain (whether or not it includes any part of civil society).

The division between micro and macro is well-established in philosophy (even if the terms are not). It corresponds (more or less) to the division between (what we now call) Aristotle's *Nicomachean Ethics* and his *Politics* (which, for Aristotle, was one work, not two). Aristotle has, I believe, almost nothing to say about civil society even though his own Lyceum is a good example of the sort of institution that might have made up the civil society of ancient Greece. Indeed, Aristotle would probably have rejected the micro-macro distinction—but for the opposite reason I have. For Aristotle, the micro-macro distinction divides what should be treated together. Individuals do not exist except in society, and society does not exist without individuals. For Aristotle, morality cannot be a matter of individual decision alone, since mere individuals do not exist alone (except as gods or beasts).

Ethics-as-special-standards, the last of my four relevant senses of "ethics," consists of those morally permissible standards of conduct all members of a *group* (at their most reasonable) want all others in the group to follow even if that would mean doing the same. Ethics-as-special-standards resembles ethics-as-theory-of-the-good-society insofar as it concerns more than the conduct of individuals. It is, however, different from ethics-as-theory-of-the-good-society insofar as the groups in question are not "political societies" but certain members of civil society, that is, those organizations (associations, institutions, corporations, or the like), including political parties and special interest groups, free to exist under a constitution, government, and legal system but not required to. Ethics-as-special-standards stands between micro-ethics (standards for mere individuals) and macro-ethics (standards for society at large). For purposes of brevity and to bring out its intermediate status, I shall hereafter refer to ethics in this fourth sense as "meso-ethics."

Meso-ethics is part of morality but not part of *ordinary* morality. How is that possible? Here is a proof that it is possible: If I join a chess club having certain (morally permissible) rules, I have (all else equal) implicitly (and perhaps even explicitly) promised to follow those rules. Since ordinary morality includes a prima facie obligation to keep promises, I have a prima facie moral obligation to follow the club's rules. Insofar as the club's rules are morally binding on me, they are now part of morality (as it applies to me). But insofar as the rules do not apply to everyone, only to members of the club, those rules are special standards. In this way (and perhaps others), meso-ethics can be both part of morality (because morally binding) and distinct from ordinary morality (because the standards are special).

2. ENGINEERING ETHICS AS MESO-ETHICS

Engineering ethics is a kind of meso-ethics even when concerned with a large subject such as sustainable development—as I shall now show. Consider this typical engineering problem:

You, a mechanical engineer, are helping to design an office printer (with copier, scanner, and fax included). Sales are expected to be 10,000 or so. The specifications require that the device be able to print on one side or two but not which side should be the default setting. Single-side is the customary default, but that default seems to you an invitation to waste paper. Should you recommend two-sided printing as the default? (Anke Van Gorp, 2005, 16)

The decision "you," the individual engineer, will make is whether to *recommend* one design or another, that is, it is a decision within an organization (a part of civil society), as most engineering decisions are. You may have to defend the recommendation at higher levels. You will certainly have to win the organization's cooperation to build the printer as you wish. That is one respect in which the decision in question is meso rather than micro: the engineer's decision is part of a process by which a voluntary organization makes, or at least tries to make, a morally significant choice. There are two others.

First, mechanical engineers are, according to their code of ethics, supposed to "consider environmental impact in the performance of their professional duties" (ASME, 2009, Fundamental Canon 8). Changing the default setting looks like a good way to do that: save trees, save on the pollution necessary to turn trees into paper, and save on the pollution necessary to get the paper from forest to printer. The change in default should cost the engineer's employer virtually nothing; the new printer will require new software anyway. If customers do not mind, the change should be a painless improvement in the printer. Engineers are supposed to incorporate improvements into their designs whenever possible at reasonable cost. This certainly seems to be an improvement. The engineer's decision, if approved, will then probably change the state of the art in her company—and perhaps among printer manufacturers generally. In this respect (as in some others), an individual engineer never acts as a mere employed individual but also as one engineer setting standards for the rest (i.e., as part of a group).

Second, the engineer's decision will, if approved, impose (a little) sustainable development on anyone who unthinkingly uses the printer. Only those who take the trouble to change the setting each time they use the printer will be able to print in a less sustainable way (i.e., on one side only). Given that most office printers have several users, the number of people the engineer's decision directly affects could be several times 10,000, few of that number being engineers or employees of the engineer's employer. That is a significant *social* effect.

The effect, being social, may appear macro-ethical rather than meso-ethical. It is not. The effect would be achieved entirely without change in law, regulation, or governmental policy. What is not the work of law, regulation, or governmental policy is not macro-ethical (in any interesting sense). The term "macro-ethical" is not an indication of mere scale of effect but of the primary agent (political society rather than civil society or individuals).

Or, at least, the term *should* not be an indication of mere scale of effect. The price of reducing the micro-macro distinction to one of mere scale is that many ordinary engineering decisions, including the one concerned with the printer default, would become macro-ethical; much of engineering ethics, as now taught, would also concern macro-ethics; and much of the micro-macro criticism of engineering ethics would be trivially false.

The effect of the new default is also plainly not micro-ethical. The engineer in question could not achieve that social effect as a mere individual, say, as an inventor working alone in his basement (though he could conceive the improvement acting alone). He could only achieve that social effect as part of civil society, for example, as an engineer working for the company in question—or (with a different set of facts), in some other engineering role. In such a context, the engineer's ethical problem is neither a micro-problem (what a mere individual should do) nor a macro-problem (what a citizen, official, or other agent of government should do), but a meso-problem (what an engineer as such, a member of civil society, should do). That is true even when engineers try to change government policy. As long as they are acting *as engineers*, whether as individual engineers or as agents of some association or interest group, they are acting as members of civil society (whatever effect they have). The same is true when a voluntary organization of engineers acts in its corporate capacity. Though political society may treat it as an individual (a corporate person), its members must treat it as *their* organization—at least while *they* are acting as engineers. (By "acting as engineers," I mean claiming whatever respect, authority, or power comes from being recognized as engineers rather than, say, as Quakers, lawyers, taxpayers, or citizens.)

Consider, then, what Herkert (2005, 374) has to say about issues that are micro or macro:

> Microethical issues in engineering include such matters as designing safe products and not accepting bribes or participating in kickback schemes. Macroethics in engineering includes the social responsibilities of engineers and the engineering profession concerning such issues as sustainable development and product liability.

One lesson we might draw from the printer example is that Herkert (2005) is simply wrong about sustainable development's status as macro-ethical. Issues of sustainable development can occur in engineering exactly as issues of safety do, be subject to similar professional standards, and seem to require the same sort of design work. There is nothing inherently macro about sustainable development. Another possible lesson, the one Herkert himself prefers, is that he could have been clearer about what he meant (private communication, April 13, 2009). For Herkert, it seems, the decision becomes macro when engineering societies, or the profession as a whole, rather than individual engineers or groups of engineers working for a single employer, must address it. I still disagree. The engineers in question, that is, whether an engineering society or the profession as a whole, are still supposed to be acting according to their professional standards (which, by definition, are meso-ethical). Professional societies are part of the profession, not above or beyond it; and even the profession as a whole is bound by the same professional standards as an individual member (unless the code says otherwise). I shall return to this point later.

The decision to use two-sided printing as the default setting—whether categorized as micro, macro, or meso—is an *ethical* decision in the special-standards sense. The engineer in question got her job (we may suppose) in part by claiming (truthfully) to be a mechanical engineer. To claim to be a mechanical engineer rather than, say, merely someone good at designing mechanical devices, is to claim to be a member in good standing of a certain profession, in other words, to be a mechanical engineer reliably working as mechanical engineers are supposed to work. To work as mechanical engineers are supposed to work is in part to work as the profession's code of ethics requires. To get and keep a job by giving the impression that one will work in a certain way gives one a prima facie moral obligation to work in that way (an obligation arising from implied promise or justified reliance). To get and keep a job by claiming to be a member of a certain profession also puts the profession's reputation at risk, giving one another source of moral obligation (one arising from fairness, i.e., the standard requiring one not to claim the benefits of a voluntary morally permissible practice while declining the burdens that make those benefits possible).

The special standards of engineering ethics are, therefore, as morally binding as obligations arising from membership in a chess club—even though, like the moral obligations arising from club membership, engineering's special standards are morally binding only on some moral agents, engineers, not on all moral agents. Engineering ethics is at least *in part* meso-ethics.

So, Ladd (1980, 156) seems to have jumbled together propositions the status of which are quite different:

> Any association, including a professional association, can, of course, adopt a code of conduct for its members and lay down disciplinary procedures and sanctions to enforce conformity with its rules. But to call such a disciplinary code a code of ethics is at once pretentious and sanctimonious. Even worse, it is to make a false and misleading claim, namely, that the profession in question has the authority or special competence to create an ethics, that it is able

authoritatively to set forth what the principles of ethics are, and that it has its own brand of ethics that it can impose on its members and on society.

Ladd has jumbled together a profession having "its own brand of ethics" (which I just demonstrated it can have) with "setting forth what the principles of ethics are" (i.e., with philosophical ethics) and with "imposing" its own brand of ethics on society (something quite different from either, an act of government). What I claim is that engineering ethics is for engineers and no one else. Engineers no more set forth the principles of ethics or impose new standards *on society* when they adopt a code of *engineering* ethics than I do when I privately make a promise to you. They impose the standards on themselves and no one else—though, of course, their following those standards may affect others (and, indeed, are designed to benefit others as well as themselves).

3. ANOTHER OBJECTION

The printer example may seem too easy a decision to count as an ethical *problem*. If so, it may also fail to show that problems of sustainable development can arise as ordinary problems of engineering ethics. I therefore think it worth noting some ways in which the decision to recommend two-sided printing as the default setting might be difficult enough to count as a problem.

Sticking with an organization's customs tends to be risk-free in most organizations; recommending a change, a gamble. If the change in default setting is accepted and works, the individual engineer may gain in authority, pay, and promotion. But if it is rejected or does not work out, the engineer may lose in all those ways. There is no guarantee the change will be accepted and work. Consumers may reject the printer in part at least because of the change. The history of technology has many instances of "sure things" that failed—like the Edsel, "new Coke," or Microsoft Vista. In addition, the actual contribution to sustainable development of the new default setting may not be what it now seems likely to be. If most users of the present equivalent of the printer in question already recycle paper or routinely change from the default setting to two-sides, the new default may simply be a convenience for users, while failing as a contribution to sustainable development. As engineers know, the world often does not work as it "should."

I chose the printer example because its simplicity made it easy to discuss. It is far from the only example of a question of sustainable development arising as part of ordinary engineering. Here's another, one obviously belonging to a substantial category of harder decisions:

The sales department has asked you, a mechanical engineer in charge of a design team, to design a "self-opening wastebasket for the kitchen." Your employer already makes kitchen wastebaskets that are open at the top, that have swinging covers, and that have step-on levers to raise the cover. You ask why these are not good enough. The sales department responds that some consumers object to the open basket in the kitchen because they do not like looking at rotting food. They object to the swinging lid because it sometimes catches the hand as the user is withdrawing it. The step-on levered lid, while avoiding these problems, tends to fail because the lever sticks or breaks. Your employer does not make a self-opening wastebasket, though some of its competitors do—with mixed success. Your team considers the problem and determines that the wastebasket should have an electric eye to send a signal to a small motor when a hand is close to the basket's lid. The motor will raise the lid; gravity, close it. There will be an on-off switch to allow, for example, for turning off the electric eye at night. Given the amount of water in a typical kitchen, a plug-in basket would risk electric shock or electrocution. The electric eye and motor must rely on batteries (probably four class A). Batteries are, however, not good for the environment. Many end up dumped where they can leak toxic chemicals into the ground water. All batteries depend on a manufacturing process that tends to be hard on the environment. The self-opening wastebasket is plainly not a sustainable technology. Should you propose telling the sales department to forget it?

4. IS MACRO-ETHICS ETHICS?

I have so far shown that micro-ethics and meso-ethics are both ethics—in the ethics-as-morality sense (though meso-ethics is a special form of it). Now I want to argue that macro-ethics, *as applied to engineers or their organizations*, is not ethics in this sense. Because defenders of macro-ethics sometimes admit as much, this point may seem trivial. It is not, as I will now show.

One way in which defenders admit as much is that they sometimes *propose* macro-standards rather than report them. For example, citing Langdon Winner, Herkert (2005, 375), asserts, "Our moral obligations must ... include a willingness to engage others in the difficult work of defining what the crucial choices are that confront technological society and how intelligently to confront them." The use of "must" here at least implies that "our" moral obligations do not now include the obligation in question (even though they should). The argument that accompanies this assertion seems to make that clear.

Here, Herkert (and Winner) show one *dis*advantage of emphasizing macro-ethics. Many codes of engineering ethics now include a provision imposing (something like) the obligation in question, for example, "Engineers shall endeavor to extend public knowledge, and to prevent misunderstandings of the achievements of engineering" (ASME, 2009, Criteria 7.a). The obligation in question is an actual ethical obligation of engineers to inform the public, rather than a merely desired one, but its source is meso rather than macro. (Neither government nor any other organ of political society imposed this obligation on engineers; they took it on themselves.) Herkert's emphasis on macro-ethics seems to have blinded him to the resources that meso-ethics already provides to make the claim on engineers that he actually wants to make on them, that is, that they should "engage others in the difficult work of defining what the crucial choices are that confront technological society and how intelligently to confront them."

The second way in which advocates of macro-ethics admit, in effect, that macro-ethics is not ethics-as-morality (ordinary or special) is that much they describe as macro-ethics is in fact information, not ethics of any kind. Here again Herkert is instructive. He praises Lynch and Kline (2001) for advocating an approach:

> grounded in the history and sociology of engineering, [that] is to provide increased attention to "culturally embedded engineering practice," that is, institutional and political aspects of engineering such as "contracting, regulation, and technology transfer." Knowledge of such non-technical, but nonetheless "ordinary" engineering practice, they argue, would provide engineers with the insight to anticipate safety problems before they escalated into technological disasters. (Herkert, 2005, 377)

As I tried to make clear in chapter 10, I agree that such knowledge should be included in any engineering ethics course (and have long included it in mine). But what is recommended, however desirable, is simply information about practice, not anything ethical in *any* of our four senses. Herkert seems to have confused macro-ethics with knowledge-of-society-relevant-to-ethical-decisions.

Is there any other uses of "macro-ethics" in the advocates' repertoire that does concern ethics in any of our four senses? Yes, and Herkert provides a few examples. Here's a typical one:

> Political scientist E. J. Woodhouse is another scholar who notes that engineering ethicists have traditionally overlooked macroethical issues. Chief among these overlooked areas, he argues, is the problem of overconsumption. (Herkert, 2005, 377)

I agree that overconsumption (using more resources than necessary) is a problem that has not received much attention from engineering ethicists, but I deny that it is (in any nontrivial way) at once a *macro*-ethical problem (i.e., concerned with political decisions) *and* the proper subject of *engineering ethics*. I have just presented two ordinary engineering ethics problems, the printer default and the self-opening wastebasket, that are in fact about overconsumption (as well as about sustainable development). There is no reason why engineering ethics textbooks and courses could not include more like them—except the sacrifices necessary to make room (about which I have said enough in chapter 10). But, like engineering ethics problems generally, these are, or so I just argued, meso-ethical, not macro-ethical. Their existence provides no support for the claim that courses in engineering ethics should include more *macro*-ethical problems.

But (Herkert might respond), those two problems in fact illustrate what is wrong with contemporary engineering ethics. The two are presented as problems for one or a few engineers, working for a business, not as problems for the engineering profession as a whole or society as a whole. That response is in part right, but mostly wrong in an important way.

The response is right insofar as problems of what society as a whole should do, do not, on my account, automatically belong in engineering ethics—because they do not pose engineering problems but problems for social decision, that is, decisions engineers as such do not make. That, however, is not a weakness of engineering ethics as such. Such problems belong in political philosophy, social philosophy, philosophy of technology, technology assessment, or the like. The problems are legitimate but not every legitimate problem belongs everywhere. For example, ethical problems of ordinary health-care administration, though legitimate ethical problems, typically do not belong in an engineering ethics course. They are, as such, not problems about what engineers, as engineers, should do. This is a fundamental point about a reasonable division of intellectual labor, not about which questions are important.

That brings me to the way in which Herkert's (possible) response is mostly wrong. There are problems closely related to these excluded problems that could be, probably should be, and may well be a routine part of engineering. Consider the self-opening wastebasket again. Suppose the design team recommended dropping the project and the sales department rejected that recommendation. The engineers might then proceed with designing the basket in the environmentally destructive way they sketched—but they need not. They might instead consider going higher in the organization to reverse the sales department's decision. They might also consider going outside the organization to a professional society, such as ASME, or some international association such as ISO, asking it to adopt standards to prevent such wasteful technology. The engineers might even consider going to one or another governmental body, such as the EPA or Congress, to seek restraining regulation. All this they could do as individual *engineers* or as part of an *engineering* organization, not as mere individuals or citizens, because engineers are, as their code of ethics says, supposed to "consider environmental impact and sustainable development in the performance of their professional duties" (ASME, 2009, Fundamental Canon 8). Indeed, presenting themselves as engineers, rather than as mere individuals or citizens, is likely to be more effective. If they do present themselves as engineers when they appeal upward, what began as a local problem of a few engineers may become a problem about the role of engineers or an organization of engineers in society. There is nothing in engineering ethics as now conceived to rule that out. In fact, current standards of engineering ethics seem (with a minimum of interpretation) to rule it in. The problem, though meso-ethical for engineers, would, of course, be macro-ethical for citizens, EPA administrators, members of congress, and the like. They would have to act as citizens or officials, not as engineers (except for the few who are engineers).

I see nothing novel in this move from the first particular decision to later decisions of policy at the organizational, professional, or governmental level. It is, in fact, a routine part of any engineering ethics course I teach. I give my students a seven-step decision procedure which makes that clear. The last step is:

7. *Make final choice (after reviewing steps 1–6), act, and then ask*: What could make it less likely you would have to make such a decision again?
   - What precautions can you take as an individual (announce policy on question, change job, etc.)?
   - What can you do to have more support next time (e.g., seek future allies on this issue)?
   - What can you do to change the organization (e.g., suggest policy change at next dept. meeting)?
   - What can you do to change the larger society (e.g. work for new statute or EPA regulation)?

This is the 2008 (improved) version of the procedure. The latest published version appeared more than a decade ago in: Davis (2011, 59). That others who teach engineering ethics have adopted this method (or something like it) suggests that what I do is a widespread practice in such courses. Herkert's response to this criticism confirms the point: "This is pretty close to my position, except I don't think the involvement of engineers needs to begin with a dispute over a design" (Email, April 13, 2009). I agree with Herkert that the engineer's involvement need not begin with a dispute over a design but might instead begin when an engineer orders parts for repair of machinery she oversees, volunteers engineering services to the Environmental Defense

Fund, or runs for the U.S. Senate citing among her qualifications her status as an engineer. My disagreement with Herkert here is largely about theory, not about what we would like to see engineers doing.

The problem with the macro-ethics critique of engineering ethics is, then, that it systematically confuses two sorts of problem, one concerned with social policy as such (macro) and the other concerned with the part engineers as such (even when organized) should take in helping to make social policy (meso). Behind that confusion may be a picture of social institutions, especially engineering societies and government, as acting more or less independently of the individuals composing them. Such a picture may be useful for some purposes, such as political science or social theory, but not engineering ethics. Insofar as social institutions operate independently of the human beings constituting them, they belong to the realm of necessity; they operate according to scientific laws, responding to various social "forces" including law and public opinion, but not to ethical standards as such. For that reason, I think it appropriate for a course in engineering ethics to consider ways in which various professional organizations and employers could be made more responsive to the ethical concerns of engineers—but only if that consideration includes ways in which engineers, whether ordinary engineers or officers of engineering societies, could help to achieve that responsiveness.

Those who argue that engineers should engage macro-ethical problems more tend to overlook how much routine engineering already engages those very problems as meso-ethical problems—and how much more effective engineers can be when they speak as engineers (rather than as individuals, citizens, or government officials). Consider, for example, the enormous array of technical standards ASME, IEEE, and other engineering societies have developed for design, manufacture, distribution, maintenance, and disposal of various forms of technology. In the end, if sustainable development is to become a vital practice, it will have to be transformed from an abstract idea into thousands, perhaps millions, of technical standards. Government may author some of those standards. But if the future resembles the past, most will be the work of the engineering profession itself—of individual engineers and of the organizations they establish, populate, and administer.

5. CONCLUDING REMARKS

There is an irony in the argument I have been making. For almost four decades, I have tried to convince those interested in engineering ethics that the subject is not *micro*-ethics, that is, ordinary moral problems in which engineers happen to be the individuals involved. Engineering ethics concerns moral problems only engineers have, the problems that arise in a certain kind of (meso) institution or organization, a profession. Those who advocate macro-ethics are often trying to make much the same point for certain problems. When that is what they are trying to do, all I have to say against them is that they could do it better by just saying that engineering ethics is a kind of professional ethics—rather than "individual" or "personal" ethics (see, e.g., Hudspith, 1991.)

Often, however, those advocating macro-ethics seem to be making a different point. They want to change the subject of engineering ethics from professional ethics to social policy. These advocates are my primary target here. They are as confused about engineering ethics as an economist would be about his subject if he wanted to devote a substantial part of a course in micro-economics (say, Theory of Auctions) to questions of taxation or unemployment. There is no reason why problems of sustainable development cannot be a routine part of an ordinary course in engineering ethics. But, to be a course in engineering ethics, the problems discussed will have to be problems engineers have to resolve as engineers, not problems they have to resolve as individuals, citizens, or public officials. They have to be problems of the sort I have discussed here. Of course, nothing here is meant to rule out other courses, courses not purporting to be engineering ethics, in which questions of social policy, constitutional reform, consumer movements, or the like are the primary concern.

Ethics is about certain decisions and the standards that should guide them. Engineering ethics is about the decisions of engineers as such, whether individual engineers or organizations of them, not about the decisions of anyone else. Time permitting, we will have little trouble including more problems of sustainable development in courses in engineering ethics so long as we remember that and work accordingly.

NOTES

Earlier versions of this chapter were presented at: the Department of Philosophy and the History of Technology, the Royal Institute of Technology, Stockholm, Sweden, June 1, 2009; the Humanities Colloquium, Illinois Institute of Technology, November 6, 2009; and the Fourth International Conference on Applied Ethics, Hokkaido University, Sapporo, Japan, November 15, 2009. Thanks to those present and to Chris DiTeresi, Joe Herkert, Robert Ladenson, Shunzo Majima, Aarne Vesilind, Vivian Weil, and two anonymous reviewers for *the Journal of Applied Ethics and Philosophy*, for their helpful comments on one or another draft. Originally published as: "Engineers and Sustainability: An Inquiry into the elusive distinction between Macro-, Micro-, and Meso-Ethics," *Journal of Applied Ethics and Philosophy* 2 (2010): 12–20.

1. For a good description of the micro-macro distinction in economics, see, for example, O'Sullivan and Sheffrin, 2003, 10–13. See Brummer (1985) for some important disanalogies with the micro-macro distinction in practical ethics.

2. See also the extended explanation of that claim, Ladd, 1980, 156. Having eliminated the space for professional ethics, Ladd's micro-macro distinction cannot divide it. A similar list, omitting Devon, appears in Herkert, 2003, 163–167. No reason is given for the omission.

REFERENCES

American Society of Mechanical Engineers (ASME), "Code of Ethics of Engineers (2009)," http://ethics.iit.edu/ecodes/node/4406 (accessed February 9, 2018).

Brummer, James, "Business Ethics: Micro and Macro," *Journal of Business Ethics* 4 (1985): 81–91.

Davis, Michael, "Thinking like an Engineer: The Place of a Code of Ethics in the Practice of a Profession," *Philosophy and Public Affairs* 20 (Spring 1991): 150–167.

Davis, Michael, "Developing and Using Cases to Teach Practical Ethics," *Teaching Philosophy* 20 (December 1997): 353–385, at 374–375.

Davis, Michael, "The Usefulness of Moral Theory in Teaching Practical Ethics: A Response to Gert and Harris," *Teaching Ethics* 12 (Fall 2011): 51–60.

Harris, Charles E., Jr., Michael S. Pritchard, and Michael J. Rabins, *Engineering Ethics: Concepts and Cases*, 4th ed. (Wadsworth: Belmont, CA, 2009).

Herkert, Joseph R. "Sustainable Development, Engineering and Multinational Corporations: Ethical and Public Policy Implications." *Science and Engineering Ethics* 4 (1998): 333–346.

Herkert, Joseph R. "Future Directions in Engineering Ethics Research: Micro-Ethics, Macro-Ethics, and the Role of Professional Societies," *Science and Engineering Ethics* 7 (2001): 403–414.

Herkert, Joseph R., "Professional Societies, Microethics, and Macroethics: Product Liability as an Ethical Issue in Engineering Design," *International Journal of Engineering Education* 19 (2003): 163–167.

Herkert, Joseph R. "Ways of Thinking about and Teaching Ethical Problem Solving: Microethics and Macroethics in Engineering," *Science and Engineering Ethics* 11 (2005): 373–385.

Hudspith, Robert, "Broadening the Scope of Engineering Ethics: From Micro-Ethics to Macro-Ethics," *Bulletin of Science, Technology & Society* 11 (1991): 208–221.

Kates, Robert W., Thomas M. Parris, and Anthony A. Leiserowitz, "What is Sustainable Development?" *Environment* 47 (April 2005): 9–21.

Ladd, John, "Morality and the Ideal of Rationality in Formal Organizations," *The Monist* 54 (1970): 488–516.

Ladd, John, "The quest for a code of professional ethics: an intellectual and moral confusion," in Rosemarie Chalk, Mark S. Frankel, and S.B. Chafer, editors, *AAAS*

Professional Ethics Project: Professional Ethics Activities in the Scientific and Engineering Societies (AAAS: Washington, DC, 1980), 154–159.

O'Sullivan, Arthur and Steven M. Sheffrin, *Macroeconomics: Principles and Tools,* 3rd ed. (Prentice Hall: Upper Saddle River, NJ, 2003).

Son, Wha-Chul, "Philosophy of Technology and Macro-ethics in Engineering," *Science and Engineering Ethics* 14 (2008): 405–415.

Van Gorp, Anke, *Ethical issues in engineering design: Safety and sustainability* (Dissertation: Technological University Delft, 2005).

Chapter 13

Doing "the Minimum"

The popular textbook *Engineering Ethics* (Harris et al., 2000, 101) distinguishes three conceptions of professional responsibility: "minimalist"; "reasonable care"; and "good works." The minimalist conception "holds that engineers have a duty to conform to the standard operating procedures of their profession and to fulfill the basic duties of their job as defined by the terms of their employment." Nothing more. In contrast, the reasonable care conception (the authors say) "moves beyond the minimalist view's concern to 'stay out of trouble', requiring the engineer to exercise reasonable care in the performance of engineering tasks" (Harris et al., 2000, 103). I shall argue that the "minimalist conception" (so defined) is deeply confused, if not incoherent, because it necessarily includes not only reasonable care but also at least some of what *Engineering Ethics* classifies as "good works." I conclude with a suggestion for what the text might say instead of what it does say. Its subject is a certain "minimalist *attitude*" one finds in business (though, in truth, it might better be called "sub-minimalist").

The question of what is "the minimum" an engineer owes an employer is important. It is a recurring question in business ethics as well as professional ethics, not so much in the classroom as in practice. Those who seek "the minimum" tend to fall substantially below it.

1. THE MINIMUM

What is "the minimum?" The first thing to notice in the first passage that I quoted is that there are at least two different minimums expressly referred to: (1) "the standard operating procedures of their profession"; and (2) "the basic duties of their job as defined in the terms of their employment." To these two, I would add at least three other standards of conduct: (3) the requirements of ordinary morality; (4) the requirements of the criminal law; and (5) the requirements of the civil law, including various forms of governmental regulation.

These five standards of conduct are, of course, not independent. For example, for engineers at least, the standard operating procedures of their profession include ordinary morality, ordinary criminal law, and whatever their contract of employment says (consistent with the other standards). That is an important point because it already suggests how much might be implicit in the minimum as *Engineering Ethics* describes it. But I mention such interdependence only to put it aside. My concern is something a bit harder to get at, something requiring us to peel away two other potential sources of confusion.

One of these other potential sources of confusion is the term "standard operating procedures of their profession." If we understand the engineers' code of ethics to be part of the standard operating procedures of their profession, we will find ourselves far from the minimum that *Engineering Ethics* describes. Consider, for example, the Preamble and Fundamental Canons of NSPE's "Code of Ethics for Engineers." The Preamble enjoins engineers to "perform under a standard of professional behavior which requires adherence to the highest principles of ethical conduct on behalf of the public, clients, employers, and the profession." "Highest principles of ethical conduct" sounds much more like a maximum than a minimum.

Of course, this language is preamble, that is, the opening guide to interpretation of the specific standards of conduct to follow, not an independent standard of conduct. But even the code's Fundamental Canons require engineers, "in fulfillment of their professional duties, [to] hold paramount the safety, health, and welfare of the public ... [and so on]." If the phrase "standard operating procedures of their profession" includes this code of ethics, as I have been arguing throughout this book that it does, then the "minimum" for engineers is a high standard, maybe not as high as "good works," but certainly higher than "[ordinary] reasonable care." So, in what follows, I shall assume that "the minimum" ignores the profession's code of ethics and interprets "standard operating procedures of their profession" to refer only to technical standards.

2. CONTRACT AND REASONABLE CARE

Engineering's technical standards are (more or less) mere specifications of what engineers understand to serve the safety, health, or welfare of the public, their employers, and their clients. If the technical standards are not interpreted with that in mind, they are likely to be misinterpreted and at least some of those misinterpretations will be sufficiently far from what engineers expect of one another to count as incompetent. For professions generally, engineering included, the ethical is an integral part of the technical.

The conclusion to draw from this is that minimalism is inconsistent with professional responsibility (at least as I have been interpreting it). But that inconsistency is not my concern now. Though important, that inconsistency is merely one potential source of confusion. The contrast between the "minimalist conception" and "the reasonable care conception" of doing one's job would, I believe, be incoherent even if engineers had no code of ethics. So, let us suppose, contrary to fact, that engineers are not professionals bound by a code of ethics. We are now ready to consider a second potential source of confusion, the contract of employment.

The contract of employment—what *Engineering Ethics* seems to mean when it refers to "the basic duties of their job as defined by the terms of their employment"—is a moral as well as a legal document. It represents an exchange of promises as well as an exchange of legal commitments. Ordinary morality may, then, require an employee to do things even if the law does not, for example, to give reasonable notice before quitting even if the law does not enforce that term of the contract. Some of the promises or legal commitments are explicit, but many are implicit. For example, every contract of employment makes the person employed an agent of the employer. All agents (whether engineers or not) are, unless expressly exempted, supposed to avoid conflicts of interest, to act as faithful trustees of the property, secrets, and reputation of their employer, and to exercise the skill and judgment necessary to do the work for which they are paid (Reuschlein and Gregory, 1990, 123–128). The law of agency seems to demand a good deal more of every employee than what *Engineering Ethics* describes as "the minimum."

That brings us to the chief point I want to make here. Agency is one part of the legal substructure of employment. Tort, the area of law from which *Engineering Ethics* draws the term "reasonable care," is another. According to tort law, *everyone* has a duty of reasonable care with respect to everyone else. Legal negligence is a failure to exercise reasonable care that results in harm. If you fail to exercise reasonable care and no one is hurt, there can be no tort liability. You are, though negligent in fact, not legally negligent. Legal negligence, that is, liability for negligence, presupposes a harm that money damages can repair at least in part. If the harmful failure of reasonable care is neither conscious nor great, it is mere negligence. If, however, that failure is conscious, the (harmful) negligence is "recklessness." If the departure from reasonable care, though not conscious, is nonetheless great, the negligence is "gross negligence."

The law has always had trouble keeping recklessness and gross negligence separate; the inference from gross negligence to recklessness is generally hard to resist. So, for example, if workers throw huge tiles from a roof of a tall building onto a crowded street, we find it hard to believe that they were not aware of how gross a departure from reasonable care they were engaged in. We move instantly from the premise of gross negligence to the conclusion that they were reckless.

That is one reason the law has trouble with the distinction between recklessness and gross negligence. There is another. Either can be the basis for punitive damages in tort and for criminal prosecution. So, for example, a driver who causes another's death while moving west at high speed in the eastbound lane of a busy interstate, will not be able to defend against a charge of "reckless homicide" by claiming that he was not aware that he was going the wrong way. Given the design of interstates, his mistake is conceivable only supposing a gross departure from the standard of care expected of drivers. That departure, though not recklessness, is its legal equivalent (Prossser, 1971, 145–166, 180–186, 209, 211–215).

This description of tort law is, I believe, uncontroversial, but is not reflected in the description of minimalism in *Engineering Ethics*. Consider this claim: "Liability insurance is already an expense, and those whose aim is simply to minimize overall costs might calculate that a less than full commitment to standards of reasonable care is worth the risk" (Harris et al., 2000, 194). Those who would make such a calculation should recognize that they are thinking about conduct that falls below the minimum standard set by law in *two* (related) ways. First, their conduct will violate the duty of reasonable care. There would be no need for the insurance in question if it did not (or, at least, did not appear to) cover the cost of liability for conduct below the legal minimum. Because the conduct falls below the legal minimum, what they are risking, besides harm to others, is being caught. Second, if they are ordinary employees, they are gambling with their employer's property and reputation—without permission. In other words, unless their employer has given them authority to engage in such gambles (making them "special employees"), they are not acting as faithful agents; they are not doing what they were hired to do. They are breaching the legal (and moral) obligations of their employment contract. They have fallen below the legal minimum in that respect too.

But that is not all. The engineers in question seem to have misunderstood their legal situation. Even if what they are thinking of doing would be ordinary negligence if they did it unknowingly, once they decide to "run the risk" of being caught, they are no longer engaged in ordinary (legal) negligence. Conscious of the risk they are running, they are reckless. Such reckless conduct may be criminal even if the negligent form is not. And even if not criminal, it will add a new risk to consider, punitive damages. Punitive damages are not limited in the way ordinary damages are (Prosser, 1971, 9–14).

The only obvious principle governing punitive damages is that they tend to be so high that the defendant, knowing in advance how high they would be, would have seen running the risk as clearly a bad bet. The insurance that protects the defendant from losses due to negligence may or may not cover punitive damages. The insurance may fail to cover the punitive damages either because such damages are explicitly excluded from coverage, or because the face value of the policy is not enough to cover those damages. The defendant may have no choice but to pay the remaining punitive damages out of its assets. Few employees have authority to "bet the company" in this way, but even those who do, fall below the minimum standard of conduct that the law requires of everyone. From the law's perspective, the conduct of such employees falls below the minimum (reasonable care).

There is, I think, nothing in what I have said so far that should be controversial. But, given what I have said so far, the true minimum for *any* employee is, if anything, substantially higher—and certainly no lower—than reasonable care. How then is it possible for *Engineering Ethics* to suppose that reasonable care is a standard higher than the minimalist? The answer, I think, is that the subject in view is not a coherent conception but a set of attitudes that are often powerful factors in organizational decision-making even when incoherent.

3. THE BLAME GAME

Much of the discussion of the minimalist conception is, in fact, concerned with what anyone interested in organizations will recognize as "the blame game." For example: there has been a spill of caustic chemicals because someone at a chemical plant forgot to close a valve. What is to be done? Of course, the valve should be closed. But what then? Should Rick, the worker who forgot to close the valve, be fired for the mistake? Should Carl, the young engineer supervising Rick, be blamed? After all, he noticed (on his first day on the job) that the manual valve was an accident waiting to happen but allowed Rick's shrug and description of company policy to stop him from pursuing his concern. Should Kevin, Carl's supervisor, be blamed because he helped maintain an environment encouraging Carl not to pursue his concern? Should someone above Kevin be blamed for Rick's impression of company policy?

Part of the blame game is to define one's "responsibilities" (domain of tasks) so narrowly that one cannot be blamed for what happened—so that he can truthfully say, "That was not my responsibility." The blame must then go elsewhere. This defining is often described as "legalistic," but it has as much to do with logic or hermeneutics as it does with law. It is generally an instance of what I have elsewhere called "malicious," "negligent," or "stupid" obedience (Davis, 1999).

Engineering Ethics points out that avoiding blame in this way interferes with fixing the underlying problem. Anyone who treats fixing the problem as her responsibility opens herself to questions about why she did not do something about the problem before. In taking responsibility for the future, she invites blame for the past. She risks her reputation, her future in the company, and perhaps her present job.

Engineers have, I think, long appreciated the destructiveness of the blame game. That is why most engineering codes include some provision more or less forbidding engineers to play. For example, the NSPE Code (III.1.a) tells engineers to "admit and accept their own errors when proven wrong and to refrain from distorting or altering the facts in an attempt to justify their decisions."

But even if engineering codes lacked such provisions, there would be a problem about engineers—or, indeed, any other employees—playing the blame game. Employees are supposed to be faithful trustees of their employer's property. Fighting over blame in a way that interferes with fixing the underlying problem amounts to putting the job-related interests of the employee ahead of those of his employer, something a faithful agent or trustee would not do. So, even in its natural habitat (the blame game), the minimalist conception is sub-minimum—and so, incoherent (a minimum that is below the minimum). Engineers who appreciate this incoherence should be better able to cut the game short when it threatens to interfere with what they should be doing.

NOTE

Originally published as: "Case Study in Engineering Ethics—'Doing the Minimum'," *Science and Engineering Ethics* 7 (April 2001): 283.

REFERENCES

Davis, Michael, "Professional Responsibility: Just Following the Rules?" *Business and Professional Ethics Journal* 18 (Spring 1999): 65–87.

Harris, Charles E., Michael S. Pritchard, and Michael J. Rabins, *Engineering Ethics: Concepts and Cases* (Wadsworth, Belmont, California, 2000).

LaFave, Wayne R. and Austin W. Scott, Jr., *Criminal Law* (West Publishing Co., St. Paul, MN, 1972).

Prosser, William L., *The Law of Torts*, 4th ed. (West Publishing Co., St. Paul, MN, 1971).

Reuschlein, Harold Gill and William A. Gregory, *The Law of Agency and Partnership*, 2nd. West Group: St. Paul, MN, 1990).

Chapter 14

Re-inventing the Wheel
"Global Engineering Ethics"

Judging both by its call and what went on at the workshop entitled "Engineering Ethics for a Globalized World," there seems to be a widespread belief in the following four propositions: (1) that there are no international standards of engineering practice (or, at least, none suitable for a globalized world); (2) that engineers need international standards of professionalism, including registration or licensing; (3) that engineers need (and do not have) an international code of engineering ethics; and (4) that there is no international curriculum for engineering ethics.[1]

What I shall do here is challenge all four of these proposition. I shall do that not because I doubt the workshop's usefulness (or the usefulness of the volume that came out of it), but because I believe re-thinking all four propositions would support the argument being made there. The problem of cultural relativity, insofar as it exists, is not primarily a problem within engineering, that is, a problem to be solved by radically revising engineering's standards of practice or education to fit a new situation, "the globalized world." It is, instead, primarily a tension between engineering and every culture in which engineers must operate—national, religious, corporate, and so on—a tension to be treasured rather than escaped. The chief revisions in engineering standards now needed are those clarifying engineering's independence of other cultures. That, anyway, is what I shall now argue.

1. CULTURE

By culture, I continue to mean those distinctive ways of doing certain things (with the attitudes, beliefs, and the like that typically accompany them). So, for example, classical music is one (musical) culture; jazz, another. General Motors has one (business) culture; Microsoft, another. And so on.

We may distinguish at least three (overlapping) kinds of cultural difference relevant to our subject: manners, nomenclature, and technology. By manners, I mean those ways of doing something that are indifferent until adopted as *the* way (or one of the ways) of doing that thing. For example, bowing is one way to greet; hugging, another; shaking hands, a third; and air-kissing, a fourth. Every human association should have at least one way to greet but (within a broad range) which of these ways it adopts does not matter. What are good manners in one setting may be bad manners in another. For example, in a society where bowing is the only form of greeting, a bear hug may seem threatening rather than friendly. I shall ignore differences in manners hereafter. The old saying seems to apply, "When in Rome, do as the Romans do."

By nomenclature, I mean what words, names, or terms are to be used. Nomenclature is more than manners insofar as nomenclature itself matters. So, for example, giving measurements in metric rather than English is, though a matter of nomenclature, not indifferent (hence, not a mere difference in manners). The metric system has advantages that the English system lacks. Not only is it more likely to be understood everywhere (an extrinsic advantage), it is easier to use (an intrinsic advantage).

By technology, I mean not artifacts as such but artifacts when embedded in a social network that designs, builds, distributes, maintains, uses, and disposes of them, along with the standards governing that network. Technological culture can vary quite a bit, even within a given locale. So, for example, until recently, the Chicago Transit Authority had one system for collecting fares (certain plastic cards, card readers, and so on); Chicago's Metropolitan Transit Authority, another; the national passenger trains running through Chicago (AMTRAK), yet another. There were three railroad cultures in one city (affecting everything from track configuration to retirement income).

2. CULTURE AND GLOBALISM

Given this way of understanding culture, engineering is undoubtedly itself a culture, that is, engineers have their own distinctive nomenclature and technology, what I have been calling "the discipline of engineering." This culture is (in large part at least) what distinguishes engineers from other technologists, such as architects, computer scientists, and synthetic chemists.

Engineering's culture is international. That internationality is what allows an engineer in Japan or North America to read without difficulty (once translated into English) the design, documentation, or schedule prepared by an engineer in Europe or China. It is what allows an Egyptian or Brazilian engineer who speaks English to teach in an Irish school of engineering or work in an Australian engineering firm.

Given the scale of international trade today, it is not surprising that engineering is an international discipline. What is surprising is how long engineering has been such a discipline. For example, when the U.S. Military Academy at West Point began teaching engineering soon after its founding in 1802, it used French textbooks—and soon had a French military engineer on its faculty (Davis, 1998, 18–19).

There is, of course, a distinction to be made between "international" and "global" culture. To be international, a culture need only cross one international border, say, the U.S.-Canadian border. To be global, the culture must be the same (more or less) everywhere on earth, from the most developed country to the least. What I now want to argue is that engineering is global in this sense—and useful largely because it is. Here I pick up a theme first touched on in Chapter 2.

Consider again why a country without engineers might want them—want them enough to hire them from outside or establish an engineering school of its own with much the same curriculum as engineering schools elsewhere. The answer cannot be that, without engineers, there can be no technology. No country is so undeveloped as not to have its own craftsmen, inventors, tinkers, and the like who can repair old artifacts or build new ones. Indeed, such local technologists are more likely than engineers to create artifacts appropriate to the locale. What engineers offer, what they have offered for at least two centuries, is technology made according to international standards—so that, for example, a part cast in Thailand will fit into an engine that, though assembled in Brazil, is composed of parts made in Germany, Mexico, and Canada.

A country that does not want to participate in international trade in this way is free to develop its own curriculum to train "mechanics," "techtons," "technical managers," or whatever it chooses to call its own brand of technologist. But, insofar as its schools train local technologists to standards different from those engineers are trained to, engineers (strictly so called) will have difficulty with the work of the local technologists. The country will be shunning the international culture in which engineers work. They will have no engineers—nor the efficiencies of scale, the precision of modern tools, interchangeability of engineered part, and so on.

3. A ULYSSES CONTRACT

When a would-be employer makes being an engineer a condition of employment (or just employs an applicant in part because she truthfully declared herself to be an engineer), that engineer is obliged to bring engineering's international culture into that workplace. Indeed, to employ an engineer is to enter (something like) a "Ulysses contract" concerning engineering culture.

Recall Ulysses' ingenious solution to a problem he faced while trying to return home after the Trojan War. He wanted to hear the Sirens' song. The only way to hear it was to sail close to their rocks. Yet, anyone hearing the Sirens became temporarily mad. Since Siren-induced madness generally caused sailors to wreck their ship upon the rocks, Ulysses did not want his sailors to hear the Sirens or to obey him while he heard them. He solved the problem by having his sailors block their ears with wax, tie him to the mast, and let him listen, gesture, rant, and struggle until the ship was again in safe water. Though he remained captain of the ship, Ulysses gave up the right to give new orders until he no longer heard the Sirens' song. A Ulysses contract is an arrangement by which someone freely gives up certain rights, especially the right to do as he judges best, to protect himself from his own poor judgment.

To employ an engineer is (in part) to employ someone who should *dis*obey orders that an ordinary employee should obey. For example, to employ an engineer is to employ someone who is (in part) supposed to insist on standard safety factors even when the employer orders him to use a less demanding standard (and even when doing as ordered is both legal and likely to benefit the employer—in the short run at least). For the ordinary employee, loyalty means giving the employer the benefit of the doubt on such questions; for an engineer, it does not. Employers nonetheless employ engineers for certain work instead of other plausible candidates, such as industrial designers, chemists, or technical managers. They do that for at least one of three reasons:

First, experience has taught employers that they should not depend on themselves or their ordinary employees in such matters (whatever the employees' rank in the organization). Employers have found that an engineer's ways of doing such things is more reliable than their own or that of other employees.

Second, employers have reason to believe that suppliers, customers, consumers, or the public will trust them more if they are known to do as engineers recommend than if they are not so known. They employ engineers to have the benefit of the good reputation that engineers have earned by adhering to engineering culture.

A third reason employers have to employ engineers is that a law requires engineers rather than others to make certain decisions because experience has taught lawmakers that engineers are better trusted with such decisions than others are. An engineer who puts engineering standards before her employer's wish, rule, or apparent welfare is not being disloyal to her employer, paternalistic, or otherwise overstepping her bounds—no more than Ulysses' crew were when they ignored his Siren-induced gestures. The engineer is protecting the employer from itself, giving it what it bargained for, the benefit of engineering's discipline.

4. GLOBAL STANDARDS

It follows from what I have just said that engineering has a global culture, its own ways of doing certain things. It does not follow, however, that engineering has exactly the right ways of doing such things. Since new global standards are adopted every year—by ASME, IEEE, ISO, and so on—that conclusion is not troubling. The important question is not whether engineering needs new standards (of course, it does) but which new standards it needs. My answer is that there is no need for global standards of professionalism because they already exist, no need for a global code of engineering ethics because that too already exists, and no need for a global curriculum for engineering ethics because even that already exists. I shall now defend this answer.

By professionalism, I mean practice according to the appropriate professional standards. A "true professional" (in this sense) is a member of a profession who acts as members of that profession are supposed to act. Professionalism has nothing to do with licensure or registration. It is not surprising, then, that Canadian engineers, who must be licensed, do not seem to show more professionalism than Dutch engineers, who live under a government that has no system for licensing or registering engineers. The claim that professionalism requires licensure or registration relies on a mistaken understanding of professions, one I criticized at length in Chapter 1.

In earlier chapters, I defined "ethics" (in its special-standards sense) as any morally permissible standard of conduct that all members of a group (at their most reasonable) want all others to follow even if the others' following the standard would mean having to do the same. Given this definition, engineering ethics includes not only engineering's code of ethics (so called) but also (as I argued earlier) its technical standards—for example, those concerned with details of safety, quality, and documentation—*provided* the standards are morally permissible and what all engineers at their most reasonable want all others to follow even if their following them would mean having to do the same.

Is that proviso ever satisfied? That is an empirical question about which it would be good to have more research. But I do think that, absent clear evidence to the contrary, there is enough reason to believe not only that (most) engineering standards are morally permissible but also that they are what engineers would (at their most reasonable) endorse. If there is any standard that is clearly morally wrong, let it be pointed out. I am sure most engineers would willingly see it repealed. Meanwhile, engineers have good reason to want their present standards followed. Those standards are what distinguish engineers from other technologists. They are what make it possible for engineers to make a living as engineers.

Do engineers nonetheless lack a global code of engineering ethics? The answer may seem to be, "Obviously, yes." After all, there are many codes of engineering ethics: ASME has one code; the European Federation of National Engineering Associations, another; and so on.[2] These codes seem to differ in geographical origin, style, language, and substance. None seems designed for use in less-developed countries.

But that "Obviously, yes" rests on at least two mistakes. First, it assumes that no code of engineering ethics is global when *most* are, that is, they apply to "engineers" as such, not merely to members of an association (as the IEEE code does), to a geographical division of engineers (as the Asian Declaration of Engineering Ethics does), or to engineers working in certain places (say, developing countries). So, most codes of engineering ethics at least implicitly claim to apply globally.

Second, that "Obviously, yes" assumes that existing codes, whatever their global pretentions, differ so much from each other that they cannot jointly guide. They cancel each other out. Yet, the codes do not differ much, if at all, in substance. Even differences that at first seem large generally disappear upon inquiry. For example, engineers following a code of ethics without a provision on sustainable development seem to interpret the environmental or public-welfare provision in their code to include it. I have therefore suggested in Chapter 7 that we think of the many formal codes much as we think of the many dictionaries of American English. Though they differ, they (more or less accurately) report the same underlying reality (what engineers typically do). One code may omit what another includes because of a different purpose, format, publication date, or the like, not a difference concerning the underlying reality. The variety in formal codes is consistent with (more or less total) agreement on the "unwritten code."[3]

So, there is a global code of engineering ethics. But is it adequate for use in less-developed countries? After looking through proposed alternatives, I think the answer is plainly yes. Consider, for example, the best alternative I have seen, Harris's "Guidelines." One Guideline suggests: "Respect the cultural norms and laws of host countries, insofar as this is compatible with the other Guidelines" (Harris, 2004, 516). As I understand the Ulysses contract into which the host country (or, rather, the specific employer) enters by employing an engineer, the engineer is under an obligation *not* to respect the culture of the host country *insofar as that geographical culture is inconsistent with engineering's culture* (something about which Harris' Guidelines are silent). Indeed, the engineer is there (in part) to do what a fully acculturated local would not do. This "absolutist" way of thinking about the relation between host-country culture and engineering culture may seem less surprising if we consider one of Harris's own examples (nepotism; Harris, 2004, 517).

An engineer not from the host country is told to hire several assistants to help oversee a building project. Ordinarily, she would hire the engineers who seem most likely to do a good job. But her employer suggests one of his relatives, noting (correctly) that looking after family is part of "host-country culture." An engineer faced with the same question in Chicago (her "home country") would know what to do. She would say,

> You can hire whom you want. You're the boss. But, if you want me to take responsibility for the hire, you must let me choose those I think will do the best job. One thing I must consider is that, if I have to fire an assistant for incompetence or laziness, it will be harder if he is one of your relatives. Hiring your relative could bias my judgment in a way that might affect the public health, safety, or welfare. I won't risk such a bias.

Now, contrary to what Harris's Guidelines suggest, I think an engineer should say precisely the same in a host country, however distant from his home country, family-centered, and underdeveloped. The engineer was given the job of selecting her own assistants because she is best placed to choose well. She is unlikely to choose well if she is required to use criteria that interfere with choosing well. If the employer wants someone to do what any other employee would, he should not have assigned that responsibility to an engineer.[4]

5. CURRICULUM

I therefore think that a good "global ethics curriculum" for engineering would look much like the ethics curriculum we have now. There could be a few improvements, of course, but these would be useful at home as well as abroad. I shall mention three improvements.

The first is *explicitly* rejecting the claim that engineering ethics as understood in the United States must end at its borders—or, at least, at the borders of the "developed world." This relativist claim should be rejected because it relies on the unstated premise that geographical culture takes precedence over technological culture. I have yet to see an argument for that claim. In addition to the arguments already made against that relativist claim, I offer this one:

The Asian Declaration of Engineering Ethics (2004) was adopted by the chief engineering societies of China, Japan, and Korea. It is unusual in being explicitly limited to a geographical area (Asia) and in being the work in part of "developing countries" (China and Korea). It seems to have been conceived as a way to document differences between Asian and non-Asian standards. Yet (as noted in Chapter 7), the only significant difference between the Declaration's standards and American or European standards seems to be the last: "[Asian engineers shall] . . . Promote mutual understanding and solidarity among Asian engineers and contribute to the amicable relationships among Asian countries."[5] Since all engineers should, I think, promote mutual understanding and solidarity among engineers everywhere and contribute to amicable relations among all countries, the Asian Declaration's last provision is no more than a special case of what is (or should be) a more general obligation, one that could be included in any code of engineering ethics. Even that last provision, though unusual, does not reveal a significant difference in geographical culture within engineering.

The second improvement in curriculum I would like to see is increased use of examples drawn from outside the United States. I would like that in part because I am tired of the small store of U.S. examples that most engineering ethics texts now share. During Fall 2012 and again during Fall 2014, I used a text by two Dutch philosophers—written in English—and enjoyed discussing the European examples (van de Poel and Royakkers, 2011).[6] My students had no trouble appreciating the examples. They helped my students understand engineering as a global profession (while also teaching them how to practice in the United States). The addition of more examples from less-developed countries would do the same.

The third improvement in curriculum I would like to see is more explicit discussion of the two senses in which engineers seem to use "state of the art." In one, a piece of technology is state of the art if it is "the most advanced" (i.e., has all the "bells and whistles" money can buy). In another sense, a piece of technology is state of the art if it gives the best fit between budget, conditions, purpose, and engineering standards. State of the art in the first sense is an esthetic criterion, not properly an engineering criterion at all. Insofar as engineering is supposed to improve the material condition of society, only technology that is state of the art in the second sense is good engineering. State of the art in the first sense is generally a waste of resources, a breach of engineering ethics. So, for example, a complex system of pumps for keeping a high-rise's basement dry might be state of the art in the first sense, that is, exceed U.S., European, or Japanese standards in impressive ways, but not be state of the art in the second sense because the system will not keep a basement dry in the less-developed country in which it will be installed, a country where electrical power is off several hours a day almost every day. State of the art in the second sense means giving a country without reliable electric power a technology appropriate to it—whether that means a backup of batteries or a large diesel generator or, instead, a high rise without a basement, or no high rise at all. The idea of state of the art as the best engineering solution is, of course, an idea that applies everywhere, not just in less-developed countries.

My hope is that the discussion of "global engineering ethics," re-conceived in this way, will generate ideas as important for engineering in developed countries as in less-developed ones, smaller ideas perhaps, but also more useful ("appropriate technology").

NOTES

Originally published as: "In Praise of Emotion in Engineering," in *Philosophy and Engineering: Exploring Boundaries, Expanding Connections*, edited by Diane P. Michelfelder, Byron Newberry, Qin Zhu (Springer, 2017): 181–194.

1. Email invitation from Colleen Murphy, organizer of the conference, May 23, 2012, especially, "Specific themes that will be discussed during the workshop include the prospects for (1) international standards of engineering practice; (2) international standards of professionalism and registration; (3) an international code of ethics: and (4) an international engineering ethics curriculum."

2. For a large selection of such codes of engineering ethics, see: http://ethicscodescollection.org/search?query=engineer (accessed August 18, 2020). To be charitable, I am ignoring the Model Code adopted by the World Federation of Engineering Organizations in 1990, though it was meant to guide the writing of codes of engineering ethics in "all nations." http://ethics.iit.edu/ecodes/node/3301 (accessed April 12, 2017). Thanks to Jun Fudano for reminding me of this document.

3. The unwritten code may be amended either by an informal shift of custom or by a formal decision of suitably important engineering associations.

4. Harris' solution was to accept one relative, but only one, assuming (I suppose) that saying yes once would make it easier to say no next time. That, I think, is a mistake. He has set a precedent. Next time, his boss can say, "You did it once: why not again?" Harris' solution amounts to advising the engineer to set foot on a slippery slope.

5. http://ethics.iit.edu/ecodes/node/5076 (accessed March 1, 2021).

6. Yes, van de Poel was one of the three Dutch philosophers with whom, almost two decades earlier, I had discussed what differences, if any, existed between "American-style engineering ethics" and their own course in engineering ethics. Their 2011 textbook did nothing to clarify those differences. It was a good text but not radically different from the American texts I had used before or have used since.

REFERENCES

Davis, Michael, *Thinking like an Engineer* (Oxford University Press: New York, 1998).

Harris, Charles E. "Internationalizing Professional Codes in Engineering," *Science and Engineering Ethics* 10 (2004) 10: 503–521.

van de Poel, Ibo and Lambèr Royakkers, *Ethics, Technology, and Engineering: An Introduction* (Wiley-Blackwell: Chichester, UK, 2011).

Chapter 15

In Praise of Emotion in Engineering

Mr. Spock: Interesting. You Earth people glorify organized violence for 40 centuries, but you imprison those who employ it privately.
Dr. McCoy: And, of course, your people found an answer?
Mr. Spock: We [Vulcans] disposed of emotion, Doctor. Where there is no emotion, there is no motive for violence.

—*Star Trek*, "Dagger of the Mind," November 3, 1966

1. SPOCK AS ENGINEER

Spock is probably an engineer (in the sense we have been understanding that term). In addition to his high rank in a graduating class of Starfleet Academy (the future's Annapolis), there are at least two other reasons to think he is an engineer. First, though he is nominally the USS Enterprise's "Science Officer," much of what he does looks like engineering rather than science. For example, he invents useful devices to order. Second, he is the opposite of the "mad scientist." He is accurate, cool, laconic, prompt, orderly, and practical. He presents himself as an agent of reason in a world that emotion might otherwise overthrow. He embodies an ideal to which many of my engineering students, including many of the women, feel attracted. Indeed, most practicing engineers I know have stories in which they present themselves in just this way, for example, when they must explain why the heat pump that Marketing promised a customer cannot be built: the specifications violate the first law of thermodynamics. The engineer must tell his superior (something like), "Whatever you would like, no amount of team-building, incentivization, negotiation, budget, skill-upgrading, motivational training, reaching out to consultants, or even table-pounding can make these specifications a reality."

Yet, there are at least two reasons to doubt that Spock is the proper ideal for engineers. First, Spock is only half human, biology somehow allowing for a Vulcan father. Spock is an outsider among humans as well as among Vulcans. Of course, the popular view seems to be that engineers are also a bit less than human ("nerds," "dweebs," "geeks," or the like)—or a bit more ("technical wizards," "demi-gods of the future," or supercomputers like HAL in *2001: A Space Odyssey*). Nonetheless, all the engineers we have (or are likely to get any time soon) are entirely human; most will marry, have children, and drive to work; they will worry, hope, love, and otherwise have an emotional life much like the rest of us.

Second, there is the question whether even full-blooded Vulcans could have (as Spock put it) "disposed of emotion." How we answer that question must, of course, depend (at least in part) on how we understand "emotion." Much the same dependence between understanding and answer exists when we ask about the place that emotion should have in engineering. What I shall argue here is that, on the most defensible analysis of emotion, emotion is unavoidable in engineering—not as a necessary evil but, at least sometimes, as a positive good. The problem for engineers, Vulcan as well as human, is not to do without emotions but (as Aristotle might have said) to have the right emotions—at the right time, to the right degree, in the right way, and directed toward the right object. Not only is this true of emotion in the most defensible sense but even in some popular but less defensible senses. Not only is this true of human engineers in general, whatever their national culture, but even of nonhuman engineers (such as Vulcans) whatever their species' culture.

This chapter has four parts. The first, the philosophical, provides an analysis of emotion in enough detail for our purpose, sketching a defense of that analysis along the way. The second and third parts show how that analysis helps us understand the relation between emotion and engineering (adding another defense to this analysis). The fourth, the pedagogical, briefly considers what that analysis suggests about the content of a course in engineering ethics—anywhere in the world.

In making this argument, I may seem to be entering a debate older than philosophy, one concerned with the danger that emotion poses to the good life. Among emotions, anger seems to have been the most condemned. For example, the *Iliad*, one of the earliest surviving works of human literature, is about the anger of Achilles, how it injured Achilles, among others, and almost wrecked the siege of Troy. Many ancients, especially the Stoics, anticipated Spock by millennia. Cicero held that "in anger nothing right nor judicious can be done" (Cicero 1887, *De Officiis*, bk. I, sec. 38). Horace explained why: "Anger is a short madness." Seneca warned that humanity "is born for mutual assistance; anger for mutual ruin" (Seneca 1889, 54). And so on. While what I say here is undoubtedly relevant to such claims, I may, I think, answer the practical question I have posed (about the place that emotion should have in engineering) without worrying about what the ancients had to say. They were concerned with the place of emotion in life generally; our concern is the place of emotion in the work of engineers in particular.[1]

I shall say nothing here about what is now often called "emotional intelligence," that is, the ability to monitor one's own and others' emotions, to distinguish one emotion from another, and to use that information to guide one's thinking and action (Mayer, DiPaolo, and Salovey 1990, 189). My subject is having emotions, not knowing about them. How important having emotions is to having emotional intelligence, though an important question, is another that I shall not address here. Much the same is true for the importance of emotional intelligence in controlling emotion. I do not want to attempt too much in one chapter.

For the same reason, I shall also try to say as little as possible about psychological states that are generally not considered emotions (inklings, sensations, imaginings, calculations, and the like). Whatever their interest to philosophers of mind, they are beyond the scope of this book.

2. ANALYZING EMOTION

What then should we mean by "emotion?" If we define emotion as "a strong feeling, such as anger, fear, joy, love, or revulsion" (as many dictionaries do), Spock may be right. We can imagine something like a human life without *strong* feelings—and so, perhaps, without violence.[2] There are nonetheless at least four objections to this popular way of defining emotion. The first concerns measurement. Even assuming that we had an "emotion meter" (whatever that may be), we would still have to decide how strong a feeling like anger or fear must be before it is strong enough to count as "strong" for the purpose of counting it as an emotion. Presumably, a feeling strong enough to "preempt good judgment" or otherwise "overcome reason" would be strong enough to count as an emotion in this sense. (This sort of strong emotion seems to be what used to be called "a passion.") But using that overcoming-reason criterion would define reasonable emotions out of existence (making "emotion" a mere synonym for "passion").[3] That certainly seems a mistake. We think that some emotions, such as horror upon seeing a young child cruelly killed, are, though very strong feelings, quite reasonable, indeed, appropriate, and their absence a sign of a damaged psyche.[4] Of course, it is not good for even such emotions to overcome reason. But that is a point logically distinct from whether the feeling in question is strong or weak. We should not try to decide by definition what seems to be an empirical question, for example, whether even a weak feeling can preempt good judgment or whether even a very strong feeling can be reasonable.

What would constitute an emotion on the overcoming-reason way of measuring strength would, of course, also depend in part on how we defined "reason." Defining "reason" is itself a long-standing problem in philosophy, a problem we should avoid here insofar as possible, but one we cannot avoid entirely. For our purposes then, I think, it is enough to say this: Reason is not mere logic (the avoidance of inconsistency) or mere instrumental thinking (the capacity to choose means appropriate to one's ends, whatever ends one has). Few emotions are unreasonable in the logical or instrumental sense. We should, then, adopt a richer definition of "reason," one that (as much as possible) keeps open the question whether strong emotions can be reasonable. The following definition seems both to do that and to stay close to ordinary usage: *reason is the capacity that reasonable agents have because (and only insofar as) they are reasonable agents*. Defining reason in terms of "reasonable agents" would be (more or less) circular without a definition of "reasonable agent" that does not refer to "reason" (or some similar concept, such as "wisdom"). The following (partial) definition does that: An agent is reasonable insofar as she has: (a) certain "common sense" beliefs (such as that people must breathe to live); (b) certain evaluations (such as preferring, all else equal, pleasure to displeasure, life to death, and opportunity to the lack of it); (c) certain abilities (such as the ability to plan taking into account her beliefs and evaluations); and (d) certain ways of conducting herself (such as acting on her plans). A substantial loss of reason (in this sense)—whether in dimension (a), (b), (c), or (d)—is (as a matter of fact) a form of insanity or incompetence.

This analysis of reason, though incomplete, is nonetheless rich enough to lead to interesting conclusions, for example, that Hume was wrong to claim, "Tis not contrary to reason to prefer the destruction of the whole world to the scratching of my finger" (Hume (1739), Bk. II, Pt. III, Sec. III). That the loss of the world is worse than a scratch to a finger, even one's own, is one of those evaluations that reasonable agents share (in part because loss of the world includes the loss of one's finger and, indeed, the loss of one's life). One would have to be crazy (or in the clutches of the logical or instrumental definition of "reason") to accept Hume's claim.[5] Appealing conclusions such as this constitute significant support for (something like) this analysis of "reason." Another appeal of this analysis is that it does not rule out the distinction between strong emotions that are reasonable and strong emotions that are not.

A second objection to the strong-feeling way of defining emotion is that avoiding emotion (so defined) is not obviously desirable. A world that altogether avoided strong emotions would have mild pleasure but no joy, "love" that is hardly more than tepid affection, resentment but never anger, and so on. Life in a world without emotion (so defined) seems deeply impoverished, too impoverished even for "Vulcans." After all, Spock's father, though entirely Vulcan, must have felt strongly about Spock's human mother, since he flaunted Vulcan prejudice to marry her. There would be no Spock without a certain strong emotion, the love of Spock's father for Spock's mother.

A third objection to defining "emotion" as strong feeling is that so doing seems to exclude many gentler feelings commonly counted as emotions, for example, anticipation, contentment, liking, pity, regret, surprise, and trust. The strong-feeling definition seems designed to catch the pejorative use of "emotion"—as in "Don't be so emotional"—but to ignore many emotions that have an important place in life—Vulcan as well as human.[6]

A fourth objection to the strong-feeling definition is that it fails to connect emotion to action. Yet, even on Spock's understanding of emotion, emotions have a connection with action. According to Spock, after all, disposing of emotion is the way to end violence. Ending violence by ending emotion is possible only if emotions are (the only) causes of or reasons for violent action. But, not all feelings are causes of or reasons for action. Some feelings, such as the fright one suffers during a nightmare, are simply feelings, however strong (since nightmares typically do not lead to action).

I therefore suggest that we adopt the following definition of emotion instead: *emotion is any feeling that is a reason to act or refrain from acting.* By feeling, I mean (roughly) any conscious mental state that includes both (a) a mental representation (e.g., "this gasket leaks") and (b) a positive or negative response to that representation (e.g., attraction or distaste).

Given this pair of definitions, some pleasures and some pains are emotions, while others are not. For example, the pain I feel upon seeing my son hurt is an emotion, but the pain I feel immediately after accidentally hitting my finger with a hammer is not. The first includes a mental representation ("my son is hurt"); the second does not, producing instead an automatic response (a shriek, the bruised finger quickly moving toward my mouth, and so on). The pain of the bruised finger is not, strictly speaking, even a feeling but (we might say) a physiological shock or eruption (until I calm down enough to realize what has happened). Because emotion is a kind of feeling, there can be no unconscious emotions (unless feelings can also be unconscious). The unconscious, insofar as it motivates, must be the domain of other kinds of motive.

Emotions are reasons for acting in at least two senses. First, we can *explain* an action by pointing to an emotion, for example, "Taylor is protecting Aaron because she likes him." Liking Aaron moves her to protect him, whether she realizes it or not. Liking Aaron explains Taylor's protecting him. It is the reason she acted. Explanations, though typically offered by someone other than the agent, may also be offered by the agent (though the agent must then view herself as others do). Second, emotions may be reasons for acting in the sense of providing a *justification* for an action, for example, "I'm protecting Aaron because I like him." Taylor's liking of Aaron justifies the trouble she has taken to protect him (if justification is required). It is her reason for protecting him. Justifications, though typically something an agent offers, may be offered by another (provided that other understands the action from the agent's perspective—from the "inside," as it were).

Even a good explanation does not show that the act in question is reasonable to do, only reasonable to expect. In contrast, justifications have a closer connection with reason, that is, with what all reasonable people would (at their most reasonable) do, encourage, or at least allow. A good justification succeeds in making such a connection; a bad one tries to do it but somehow fails. So, for example, while love may justify marriage, it cannot (all else equal) justify murder. "I murdered for love" is a justification only in the way counterfeit money is money. It cannot (all else equal) be a good justification, only something that improperly seeks to pass for one.

An emotion may, or may not, arise from the corresponding disposition. So, for example, fearlessness is the emotion corresponding to the disposition to be fearless. Yet, those who are fearless at a given time (i.e., those who act from a conscious indifference to expected harm) may not have the corresponding disposition (a sustained tendency to act in a certain way). One can be fearless once in a lifetime—because, say, one is drunk—and fearful the rest of it. The absence of fear on that one occasion may be an outcome independent of a disposition to be fearless. Nothing in our concepts entails that a particular emotion can only arise from a particular disposition.

An emotion also does not necessarily have a corresponding virtue (as, say, the feeling of compassion corresponds to the virtue of compassion). An emotion may fail to correspond to a virtue in at least one of two ways. First, some emotions fail to correspond to a virtue because the emotions in question are not good. For example, the emotion of jealousy (the unjustified feeling that someone is better off than you combined with the desire to harm him enough to put him in his place) is never justified. Indeed, the disposition to be jealous is a vice. Similarly, boredom, even when justified, corresponds to no virtue, since there is nothing virtuous about being bored. To be a virtue, a disposition must dispose one to good acts (of a certain kind) rather than to bad or indifferent ones or none at all.

Second, an emotion may fail to correspond to a virtue because the virtue in question requires a feeling that does not correspond to it. For example, while courage is a virtue, it does not correspond to the feeling of courageousness. Courage is reasonable conduct when one is aware of significant danger and inclined to avoid it. The courageous person must feel fear (i.e., be inclined to avoid a significant harm in view) even when he chooses to risk the harm. Anyone who does not feel fear cannot be courageous, only fearless. What counts as "feeling courageous" is typically a kind of fearlessness or even foolhardiness. Fear is the emotion that courage requires, the same emotion that its opposite, the vice of cowardliness, also requires. The virtue of courage has no emotion to which it corresponds in the way feeling compassion corresponds to the virtue of compassion.[7]

One last point. An emotion is a reason to act (or refrain from acting) but not necessarily a decisive reason. One may have an emotion appropriate in a situation and yet not act on it or even be justified in so acting. One may, for example, feel hostility and yet show kindness—and be justified in so doing—when, say, one has incurred a great debt of gratitude to someone whom one dislikes because of past slights. In such a case, one emotion (gratitude) may (or may not) trump another (hostility). One may also act from motives that are not emotions, such as habit, prudence, or convenience. Emotions are not the only reasons to act (in either the explanation or justification sense of "reason").

Given this analysis of emotion, an engineer—even a Vulcan engineer—seems unlikely to avoid emotion. So, for example, we want our engineers, even our Vulcan engineers, to be courageous and courage requires an emotion (though fear rather than the feeling of courageousness). We do not want our engineers to be merely fearless (as an emotionless Vulcan would have to be). Being indifferent to danger, the fearless tend to take risks the courageous do not, risks that no one should take.

What if someone objected that Vulcans might get by with the appropriate attitudes without the corresponding emotions? For example, could Vulcans not have a fearful attitude without ever feeling fear? The fearful attitude would, we may assume, truly be theirs, not a mere stance or pretense, even though they did not fear. If so, then the attitude must exist in part at least as a disposition to have the corresponding emotion on appropriate occasions. Why even speak of a "fearful attitude" if it can never be manifested in the corresponding emotion, that is, in moments when one fears? Crucial to having a fearful attitude is the tendency to feel fear on certain occasions. An attitude of fearfulness without the possibility of the corresponding emotion is no attitude at all. Attitude, then, is no substitute for emotion.

3. THE EMOTIONAL LIFE OF A GOOD ENGINEER

Having established that even Vulcan engineers cannot do without emotions, we must now consider what part emotions *should* play in engineering. To avoid seeming to beg off the question with which we began, let us focus on the strong-feeling kind of emotion (which, of course, our analysis recognizes as a kind of emotion, though not as the only kind). What part, if any, should strong feelings such as anger, fear, joy, love, and revulsion have in the professional work of ordinary engineers? Consider this case:

> Your employer, Extravagant Electronics, asked you to design a "reclamation facility" for waste produced by its Chicago plant. Though the facility will be located in the Republic of Cameroon, a central African country of about 20 million people, you, a civil engineer, designed it to meet the same standards it would have to meet if located in the US. The US standards are, in part, meant (a) to protect workers from suffering injury from contact with heavy metals and other toxins present in the waste to be recycled and (b) to prevent heavy metals and other toxins, whether processed or dumped raw, from entering the air, ground water, or water table in the neighborhood of the facility. When you present the design to senior management at Extravagant, they object to the cost, ask what Cameroon requires, and—upon hearing "practically nothing"—suggest meeting only the local standards. "After all," they add, "Cameroon needs the jobs and following US standards will make processing the waste there more expensive than processing it in the US, depriving the Cameroonians of jobs they need. The low wages there are more than enough to compensate for the cost of shipping the waste so far but not if we meet US standards there." You point out that doing as management asks would mean at least 30 otherwise unanticipated deaths *annually* among the workers and neighbors of the facility during the facility's projected useful life and perhaps for several decades after that. Most of those deaths would be from poisoning of one sort or another. There would also be considerable environmental damage locally, much of it irreversible. Management responds, "That's their problem, not ours, as long as we satisfy US and Cameroon law and provide Cameroon's government with whatever information it requires to assess the risks."

When I put this case to working engineers, the initial response is typically a frown (a sign of dissatisfaction with management's position). Some engineers will go on to give more explicit signs of dissatisfaction. One might say, "It's their money; so, I'll do as asked, if I can't change their mind; but it's not work I can be proud of." Another might say, "What they're asking for is not engineering but murder. I'd refuse to do as asked." Still another might say, "I'd revise the specifications to include the cost-savings needed to have a facility that can both meet US standards and be profitable in Cameroon. I would then see what I could come up with." All three responses, even the first, seem to fall within the bounds of acceptable conduct, while a happy-to-do-whatever-you-want response does not. A good engineer is an engineer who cares about doing good engineering; the more he cares about that, the better an engineer he is (all else equal). An engineer who cares only about doing whatever his employer asks is not a good engineer. Few, if any, engineers, would view what management asks here as good engineering.

If what I just said is right, then we have found one place for strong emotions in engineering. A good engineer has a strong (positive) feeling about how engineering should be done, a feeling justifiably capable of affecting how she does her work. Indeed, we might say that caring a lot about doing engineering well (an attitude) is part of what *constitutes* a good engineer. An engineer who cares little or nothing about engineering is not a good engineer—even if she desperately wants to keep her job, do what management wants, and so on. Though there may be empirical evidence for this connection between caring and being a good engineer, the connection is, I think, primarily conceptual. While we may be able to imagine an engineer who does not care about engineering but who, from other motives, does reasonably good engineering, we—or, at least, the engineers among us—may nonetheless hesitate to call her "a good engineer." She seems, at best, the functional equivalent of a good engineer—one whose good luck is unlikely to last long. Good engineers typically have a strong negative response to bad engineering, not only to the bad engineering of others, however distant in space or time, but also to their own bad engineering. They typically try to do a good job (and generally succeed).[8] That attitude of caring must, of course, be realized in moments when the engineers actually care, that is, have the emotion of caring.

Such caring might or might not be an emotion strong enough for our purpose. Can I give at least one example of an undoubtedly strong emotion that undoubtedly seems part of what constitutes a good engineer? Fear seems as good an example as any of a strong emotion (since, unlike care, it appears on most lists of strong emotions). A good engineer will fear certain consequences of her work, especially the loss of life or substantial damage to the environment. By "fear," I mean a strong negative response to an anticipated harm. A weak negative response is a mere concern or cautiousness; an overwhelming response, fright or terror. An engineer who fears every harm that her work may produce is unreasonable. There is no engineering without harm; indeed, there is little any engineer can do without significant risk of harm. The proper emotional response to the prospect of some harms—the minor or socially tolerated ones—is concern or caution, not fear. Fear is the proper response to the larger harms, especially if they are relatively probable, poorly understood, and likely to fall upon people unable to protect themselves. A strong negative feeling in response to the prospect of producing such harm is fear—by definition.[9]

Some emotions, such as caring and fear, are part of what constitutes a good engineer. Are there any emotions that, though not constitutive of a good engineer, are still good for engineers to have *on occasion*? That is not an easy question to answer convincingly because there is a tendency to pack into the concept of a good engineer everything that might be good for an engineer to have. Before we can confidently point to an emotion that is good for engineers to have on occasion but is not part of what constitutes a good engineer, we need a criterion for distinguishing the constitutive from the merely good or useful to have. I lack such a criterion. I shall nonetheless offer an example of an emotion that seems to me to be good for engineers to have on occasion but not to be part of what constitutes a good engineer. Not much turns on the example. If I am wrong about it and the emotion in question is constitutive of a good engineer rather than just good for an engineer to have on occasion, I may resume the hunt for such an example (though not in this book). In the meantime, I will still have added to the list of emotions appropriate for engineers to have, strengthening this chapter's claim that emotions have a significant place in engineering, no matter where one is practicing engineering and even if the culture one is practicing in is as alien as that of Spock's Vulcan.[10]

The emotion I offer as one good for engineers to have on occasion but not constitutive of a good engineer is (the much-maligned emotion) anger. By "anger," I mean the feeling that someone has been wronged, slighted, or otherwise improperly treated, together with the impulse to strike back. One may be angry on another's behalf as well as on one's own. Humans show anger in many ways short of violence, for example, by speaking louder than usual, using strong language, baring their teeth, going red in the face, saying "That makes me angry," or making threats. We can, *I think*, imagine an engineer who, though obviously a good engineer, never feels anger concerning professional work (someone like Spock, perhaps). Our ability to imagine such an engineer is one reason to think anger (unlike fear) is a good example of an emotion that, while good for engineers to have on occasion, is not part of what constitutes a good engineer.

Though we can imagine a good engineer who never feels anger concerning professional work, anger would nonetheless be a reasonable response for an engineer on some occasions, for example, when management rejected her design for the waste facility in Cameroon in the way described above. Indeed, it seems right to interpret the second engineer's response, "What they're asking for is not engineering but murder," as a clear expression of anger. It is reasonable for an engineer to be angry under such circumstances insofar as (a) she is in fact being wronged (her professional judgment is being improperly discounted) and (b) it is important for management to appreciate the resulting impulse to strike back. It is important for management to appreciate that impulse if, not being anticipated, it is a cost not included in management's original calculation. The engineer's anger is, then, a reason for management to revise its decision apart from any damage the engineer's striking back would actually do. The anger indicates a problem both serious and (for now anyway) hard to measure. It would then be reasonable for management to take into account not only the fact that the engineer is angry but how angry she is, how she is likely to strike back, and how much support she may find among other engineers, other employees, and even the world outside Extravagant Electronics.

More important, I think, is that the engineer's anger would in fact help management appreciate the weight that the engineer's judgment itself deserves. All else equal, the more serious the affront to her standards of engineering, the angrier the engineer should be ("should" here including both explanation and justification). The more serious the affront, the less likely, all else equal, that management's reasons for overriding the engineer's judgment are adequate. Much that engineers are hired to do is not calculation, reporting facts, or other (more or less) algorithmic activities. Much of engineering is a matter of judgment, something hard for non-engineers to evaluate (at least in the short term). The emotion that engineers express in response is one way, an important way, for engineers to communicate what is at stake in management's decision to override an engineer's judgment. (For more on engineering judgment, see Davis, 2012.)

Anger has its own costs, of course. The most obvious is that one typical response to anger is anger, with one display of anger leading to another, until someone—as we say—"loses his temper" and there is a break in relations (the "ruin" that concerned Seneca). One way to avoid such a break in relations is to adopt the first engineer's approach, that is, to try to change management's mind by calm argument and, failing that, shrug and do as asked. That way avoids the risks of escalating anger but also abandons some of anger's power to communicate. Another way to avoid such a break in relations is to try to find a creative way out of what now seems a true dilemma (as the third engineer said he would do). That is doubtless the best approach for an engineer to take, especially initially. If it works, there is no longer a problem. If, however, it does not work, the first and second approaches (among others) are still open. A show of anger may then be a reasonable response, indeed, given what is at stake—considerable loss of human life and damage to the environment—a response more reasonable than a mere shrug. Instead of responding angrily, management may rethink its decision.

4. EMOTIONS AND ENGINEERING ETHICS

From what I have said in the last section, it should already be clear that emotions have a place not only in engineering generally but also in engineering ethics in particular, indeed, at least three places. First, some emotions, primarily those that are in part constitutive of a good engineer, help engineers appreciate what they are doing. So, for example, if—after checking the facts—contemplating a design's outcome causes an engineer dread, worry, or even mere discomfort, the engineer should certainly try to revise the design to remove the cause of such negative emotions. The revision may come at any step in the design process, but plainly earlier is better than later, since fixing a problem tends to become harder as the design process nears its end. Similarly, if an engineer enjoys contemplating a certain design, that joy is itself a reason to continue with the design. Insofar as engineering judgment tracks engineering's ethical standards (as well as its technical standards), emotions contribute to an engineer's ethical sensitivity (as well as to her technical sensitivity).

Second, the strength of an emotion may provide a measure of the weight that the considerations provoking that emotion deserve. All else equal, the stronger the emotion, the greater weight the provoking considerations should have in the engineer's deliberations. Insofar as the considerations in question are ethical (e.g., loss of human life or damage to the environment), the strength of the emotion should also provide a good measure of ethical importance.

Third, emotion has a place in the communication of engineering judgment. My examples so far probably suggest that emotion only has a place in communication with non-engineers. Actually, emotion has a similar place in many communications with other engineers. While one engineer can sometimes see through another engineer's judgment to the underlying evidence that supports it, perhaps more often she cannot. Even another engineer may lack the experience, ability, time, or special training necessary to absorb the relevant evidence fully. ("Without walking the ground, you can't imagine how inadequate the infrastructure in Cameroon is for dealing with heavy-metal waste.") When an engineer cannot see through another engineer's judgment to weigh the underlying evidence, how the engineer presents the judgment may matter a good deal. Strong words are evidence for the judgment; diffidence is evidence against; and so on.

There is, then, an ethical reason for engineers, as engineers, not only to have emotions but to show the emotions they have. Hiding one's professionally proper emotions is much like making deceptive statements about one's professional judgment. It tends to mislead those relying on the engineer. Instead, engineers should take care to give a true impression of the emotions they have as engineers. They should not, for example, try (out of modesty) to hide the pleasure they feel in a certain design or (to avoid hard feelings) tone down the fear that the location of a certain waste facility awakens in them.

Controlling emotion is, however, different from not revealing emotion. The control of emotion—assuring that we have the right emotion at the right time, to the right degree, in the right way, and directed toward the right object—is part of being a reasonable person. How then is an engineer to control his emotions? The best way to control emotion is probably to confront the emotion in question with the relevant facts available or easily obtained long enough for the facts to sink in, produce reflection, and mature into a plan. A reasonable emotion is one that can survive vivid contact with all the relevant facts over an extended period, especially when one is sober, well-rested, in good health, and otherwise at one's most reasonable.[11] Of course, "the relevant facts" include not only governmental regulations, the laws of nature, and the details of ordinary life, but also engineering standards, including engineering's code of ethics.

5. PEDAGOGICAL CONCLUSIONS

Given what I have said so far, it seems to me that teachers of engineering ethics (like other teachers of engineering) should take time in class to help students appreciate that engineering has an emotional side, for example, that they are likely to enjoy doing engineering well and to find doing engineering badly depressing. More important, though, students should be helped to see how engineering's technical standards—everything from safety factors to routines for documentation—contribute to engineers deserving an important place in many social decisions. The technical standards contribute to making the products of engineering safer, less wasteful, more reliable, more useful, and more easily disposed of than the corresponding products of engineers not following those standards (or of non-engineers following different standards or none at all).

Most important, students should be helped to understand how they internalize engineering standards, technical as well as ethical, so that they can use their "gut" to help them identify ethical issues. They should, of course, not be allowed to let their "gut" *automatically* decide what they should do. We all know what fills the gut. A churning gut—or the rough equivalent, whether called "scruples," "conscience," or "a bad feeling about this"—is nonetheless a good reason to think again, gather more information, and so on. The "gut" is like one of those "pretty good" sensors that engineers use with considerably less than total trust—but use nonetheless when they lack anything better. (Compare Mayer, DiPaolo, and Salovey, 1990.)

Having learned how to use their emotions to help them decide what, as engineers, they should do, students should be given practice turning their reasonable emotions into plausible arguments. Part of being a good engineer is being able to win others, non-engineers as well as engineers, over to her recommendations (when the recommendations deserve it). An engineer who cannot do that will seldom achieve much, however knowledgeable, skilled, and creative she is. So, for example, suppose a student is assigned the role of the lead engineer in the Extravagant Electronics case discussed earlier. She is told that her manager has just dismissed her expression of concern about safety with the (accurate) comment, "You can't prove it's unsafe" or "You don't *know* it's unsafe." The student should be able to come up with some such response as this:

> A proof is a set of statements that together should win all reasonable people over to its conclusion. Unfortunately, much of engineering is not susceptible to proof in this sense. Information is too scant to be decisive and there is not time to get enough. The problem is only partly defined; it changes as related technology, government policy, culture, and the environment change. The problem also changes, or at least becomes clearer, as we dismiss some solutions and invent others. And so on. In short, much of engineering deals with what are called "wicked problems" (Rittel and Webber, 1973). Engineers have learned to deal with such problems in a number of ways. One of these is to rely on engineering judgment when decisive proof is lacking. If one engineer is uncomfortable with the safety of the artifact he is working on, that discomfort should itself be treated as evidence that it is unsafe (reason enough to gather more information). If most engineers working on the artifact, especially the more experienced, share the same discomfort, then, though not proof, that discomfort is itself reason enough to act as if the artifact is unsafe. Of course, as a manager, you always have the *power* to overrule your engineers, even on a question of safety. But, being human, you certainly may be wrong and, if you are and disaster results, you will not have the defense of having taken the best advice in making that disastrous decision. Indeed, you will have to admit to substituting your individual judgment for the collective judgment of engineers who are the experts on such questions, experts that your decision deeply angered. Do you want to put yourself in that position?

This sort of argument strikes me as more likely to convince an engineer's manager to go along with the engineers in question than a Spock-like appeal to "logic" or "the facts." Engineers should not deny themselves the use of such *arguments from judgment*. Logic settles few debated questions in engineering, and the facts often do not speak for themselves.

NOTES

Thanks to Justin Hess, Kelly Laas, Dan McLaughlin, Diane Michelfelder; participants in the Philosophy Colloquium, Illinois Institute of Technology, January 24, 2014; and participants at the annual meeting of the Forum for Philosophy, Engineering, and Technology, Blacksburg, VA, May 28, 2014, for comments on one or another draft of this chapter. An earlier (and substantially shorter) version of this chapter appears in Sethy (2015, 1–11); and in something close to its present form in Michelfelder et al. (2017, 181–194).

1. Thanks to Jack Snapper for reminding me of that old debate.

2. By "violence," I suppose Spock means (something like): using force in a way that violates a moral rule. Murder, mayhem, kidnapping, false imprisonment, and the like are all acts of violence. Justified self-defense is not (however much force proves necessary for the defense). Being justified, the force in question is not violence.

3. This collapse of emotion into passion even occurs now and then in the scientific literature. For example, one introductory text in psychology defined "emotion" as "a disorganized response, largely visceral, resulting from the lack of an effective adjustment" (Schaffer, Gilmer, and Schoen, 1940, 505, quoted in Mayer, DiPaolo, and Salovey 1990, 185).

4. For more on rational emotions, see, for example, Wallace (1993) or Davis (1986).

5. The same would, of course, be true of the claim (common among economists) that rationality (reasonableness) consists in pursuing one's own self-interest (whatever interests the self happens to have). Part of being reasonable is having certain interests other than self-interest, indeed, that caring for some others more than oneself can be more reasonable than caring only for oneself. The economist's "rational man" is often unreasonable, for example, when, already rich, he prefers a small percentage increase in his own wealth even though million starve to death as a result.

6. Literally, the expression "so emotional" leaves open the possibility that less emotion would be okay even if it is still much more than no emotion. Idiomatically, however, the expression seems to carry the message that no emotional response would be best—as do such similar expressions as "Cool it," "Get a grip on yourself," and "Think with your head, not your heart." "Don't be so emotional" is, then, not the equivalent of "Don't be too emotional."

7. It may be of interest to note that, like courage, the other cardinal virtues of the ancient Greeks—wisdom, moderation, and justice—also seem to lack a corresponding emotion. While emotion seems important to understanding virtues, virtues seem to be largely irrelevant for understanding emotion.

8. For empirical evidence for the connection between caring and doing a good job, see, for example: Gaudine and Thorne, 2001; or Roeser, 2012.

9. Note that I have not argued that engineers should have a fearful attitude even though I have argued that fearing certain outcomes is part of being a good engineer. This at least suggests that the list of emotional *attitudes* constitutive of a good engineer may differ substantially from the list of *emotions* constitutive of a good engineer.

10. When I say "appropriate to have," I mean to leave open the question whether the emotion should, on that occasion, be a decisive reason for some act. Sometimes it may be appropriate to have an emotion without it being appropriate to act on it. For example, loving admiration seems the appropriate emotion to feel when I come into a courtroom and see my spouse, a lawyer, eloquently addressing the court, but it is not, all else equal, a reason good enough to justify kissing her right then.

11. Compare Brandt, 1979, 148. Of course, this full-information approach has serious theoretical problems. For more on those problems, together with attempts to fix them, see Carson, 2000, 222–239. I offer my version of the full-information approach

merely as a practical ideal, a useful way to try to assure the reasonableness of one's emotions. I do not mean to prejudge any theoretical question.

REFERENCES

Brandt, Richard, *A Theory of the Good and the Right* (Oxford University Press: Oxford, 1979).

Carson, Thomas, *Value and the Good Life* (University of Notre Dame Press: Notre Dame, IN, 2000).

Cicero, Marcus Tullius, *Ethical Writings of Cicero: De Officiis, De Senectute, De Amicitia, and Scipio's Dream*, translated by Andrew (Peabody. Little, Brown, and Company: Boston, 1887), http://oll.libertyfund.org/titles/542 (accessed July 25, 2015).

Davis, Michael, "Interested Vegetables, Reasonable Emotions, and Moral Status," *Philosophical Research Archives* 11 (March 1986): 531–550.

Davis, Michael, "A Plea for Judgment," *Science and Engineering Ethics* 18 (2012): 789–808.

Gaudine, Alice and Linda Thorne, "Emotion and Ethical Decision-Making in Organizations," *Journal of Business Ethics* 31 (2001): 175–187.

Hume, David, *A Treatise of Human Nature*. Adam Black and William Tait: Edinburgh, 1826. http://files.libertyfund.org/files/1482/0221-02_Bk.pdf (accessed January 14, 2015).

Seneca, Lucius Annaeus, *Minor Dialogues Together with the Dialogue on Clemency*, translated by Aubry Stewart. G. Bell: London, 1989). https://archive.org/details/minordialoguesto00seneuoft (accessed July 25, 2015).

Sunderland, Mary E., "Taking Emotion Seriously: Meeting Students Where They Are," *Science and Engineering Ethics* 20 (2013): 183–195.

Mayer, J. D., M. T. DiPaolo, and P. Salovey, "Perceiving affective content in ambiguous visual stimuli: A component of emotional intelligence," *Journal of Personality Assessment*, 54 (1990): 772–781.

Michelfelder, Diane P., Byron Newberry, and Qin Zhu, ed., *Philosophy and Engineering: Exploring Boundaries, Expanding Connections* (Springer, 2017).

Rittel, Horst W. and Melvin M. Webber, "Dilemmas in a General Theory of Planning," *Policy Science* 4 (1973): 155–169.

Roeser, Sabine, "Emotional Engineers: Toward Morally Responsible Design," *Science and Engineering Ethics* 18 (2012):103–115.

Schaffer, Laurance F., B. von Haller Gilmer, and Max Schoen, *Psychology* (Harper&Brothers: New York, 1940).

Sethy, Satya Sundar, ed. *Contemporary Ethics Issues in Engineering* (IGI_Global, 2015).

Wallace, Kathleen, "Reconstructing Judgment: Emotion and Moral Judgment," *Hypatin* 8 (Summer 1993): 61–83.

Part IV

ENGINEERING'S GLOBALISM

Chapter 16

The Whistle Not Blown

VW, Diesels, and Engineers

This chapter is a "case study," that is, a collection of facts organized into a story (the case) analyzed to yield one or more lessons (the study). Collecting facts is always a problem. There is no end of facts. Even a small event in the distant past may yield a surprise or two if one looks carefully enough. But the problem of collecting facts is especially severe when they change almost daily as a story "unfolds" in the news. One must either stop collecting on some arbitrarily chosen day or go on collecting indefinitely. I stopped collecting facts for this study on October 3, 2016 (the day on which I first passed this chapter to the editor of the volume in which it first appeared). There is undoubtedly much to be learned from the facts uncovered since then, but this chapter leaves to others the collecting and analyzing of those newer facts. The story told here is good enough for the use I make of it: showing the importance of whistleblowing to the "global profession" of engineering whatever the status of whistleblowing in the local culture.

The facts this chapter collects concern what appears to be illegal conduct by employees of Volkswagen (VW), including senior officers and board members. There are several ways to analyze these facts. One can analyze them as a failure of management such as might be included in a course in business ethics; or as a failure of government supervision suitable for political scientists to study; or even as a failure of the news media that the public may want to ponder; but this chapter analyzes the facts as a failure of engineers, something appropriate here both because my chief interest in these facts arises from my interest in engineering ethics and because that interest corresponds to the focus of this book.

This chapter is about whistleblowing—or, more exactly, about its absence when it seems there should have been some. Since "whistleblowing" is another important word having several meanings, I should say that, for the purpose of this chapter, one "blows the whistle" when one belongs to a legitimate organization (as would a VW employee) and goes out of the organization's normal channels to report serious moral wrongdoing in the organization for a morally permissible reason. Whistleblowing is a certain sort of going out of channels. Some whistleblowing is "internal," that is, the channels used are inside the organization even though outside of normal channels, Some whistleblowing is "external," that is, the channel used is outside the organization.

1. THE SCANDAL BEGINS

On September 9, 2016, James Robert Liang, a VW employee for more than thirty years, pled guilty in the U.S. District Court for the Eastern District of Michigan to having a significant role in VW's conspiracy to mislead U.S. regulators concerning pollution-related emissions of diesel engines that VW had sold in the United States. Liang seems to be a technically adept mechanical engineer much of whose career was in Europe. He is credited as the inventor on at least one European patent related to motor technology. His indictment was unlikely to be the last. Facing up to five years in prison and a fine of $250,000, Liang agreed to cooperate with prosecutors (Tabuchi and Ewing, 2016a). Prosecutors do not seek "cooperation" unless they expect it to lead to indictments of "higher-ups."

We still know relatively little about the scandal sometimes called "Dieselgate." For example, we do not know what part Liang actually had in it. All that we could be sure of in 2016 was that U.S. prosecutors pursued him because he did enough to be indicted and, being resident in the United States at the time, could be arrested, questioned, and threatened with long imprisonment if he did not cooperate. The sovereignty of various European states, especially Germany, protects most of his co-conspirators from similar treatment in the United States.

That protection is important because—from what we did know, or at least thought we knew—the center of the conspiracy was in Wolfsburg, VW's corporate headquarters in the German state of Lower Saxony, not in the United States. Software similar to that giving false readings for pollution tests on about a half million VW diesels sold in the United States between 2007 and 2015 did the same for more than eight and a half million VW diesels sold in Europe during the same period (and another two million sold elsewhere; Tabuchi and Ewing, 2016b). After carrying out its own investigation in 2015, VW claimed that the software in question was the work of a small group of its employees ("a handful of rogue engineers"; Smith and Parloff, 2016). That claim is probably true (except for the "rogue"). The software in question seems relatively simple, a few hundred lines of code in a system having a hundred million lines, something a few engineers could have written and inserted.[1] However, the question in any corporate scandal of this scale is not so much who did the "dirty deed" itself as who ordered it, who aided it, who supervised it, who lied to authorities rather than reveal it, who knew but did nothing about it, and who took pains to remain ignorant, typically a much larger group. Of course, even that much larger group is probably no more than a few hundred employees out of VW's approximately 600,000 worldwide (Volkswagen, 2016).

But that much larger group, even though still relatively small, is important. By the time I was writing this chapter, it had already cost VW much of its good reputation worldwide as well as $2 billion to be spent on cleaner-automobile projects, $2.7 billion for a U.S. government fund to compensate for the environmental damage the diesels may have caused, and $10 billion to buy back affected autos in the United States at their pre-scandal value (and otherwise compensate owners).[2]

That group of "rogues" seems to have included at least four high-ranking *engineers* at Wolfsburg, all of whom have now left VW, probably because of the scandal: Martin Winterkorn, VW's CEO when Liang was indicted; Wolfgang Hatz, then head of engine and transmission development at VW and Audi; Ulrich Hackenberg, then head of development for Audi; and Heinz-Jakob Neusser, then head of development for the VW brand.[3] Unlike most scandals involving engineers, Dieselgate seems to have no heroic engineer inside the organization.[4] What I want to do here is consider why that might be and what might be done about it. What part, if any, did national culture have in the scandal?

2. THE "FACTS"

What we think we know about Dieselgate may change substantially should any of those accused of wrongdoing go to trial, or the prosecutors open their files as part of a plea agreement, or someone leak VW files about the case. Until then, this is what we think we know:

1. For reasons of convenience and cost, regulators had been evaluating automobile pollution primarily by using a laboratory test rather than more realistic road tests. The regulators had known for some time that automobile performance on their test rigs can be several times better than on the road. But for autos other than VW's, the discrepancy between test-rig performance and road performance was generally the consequence of the testing protocol itself, both the rigidity necessary to apply the same

standard to a large number of quite different autos and a failure of the test to keep up with the computerization of what used to be simple mechanical devices. Autos were much "smarter" in 2016 than they were even a decade earlier; the tests were not (Hakim, 2016).

2. All modern autos must have a test-rig mode as well as an on-the-road mode because the test rig requires the front wheels to move while the back wheels do not. Without the test-rig mode, the computerized traction-control system would interpret the rear wheels not rotating while the front wheels are rotating as a skid and try to correct, giving results having nothing to do with ordinary driving. The test-rig mode is necessary to keep the auto's computerized traction-control system from interfering with the tests.

3. The test-rig mode does not require any special hardware, only special software. The software switches from standard mode to test-rig mode when the traction control senses that the front wheels can rotate freely while the back wheels cannot move at all. What distinguished VW's arrangement for testing diesel autos from others in the industry was the way the test-rig mode preempted the standard mode. The test-rig mode is not supposed to change any setting affecting pollution-control devices. VW's standard mode for diesels ("the cheat code") differed from the test-rig mode in the fuel pressure, injection timing, exhaust-gas recirculation, and (in models with AdBlue) the amount of urea fluid sprayed into the exhaust. All these differences are, of course, relevant to pollution control. The standard mode typically delivered the better mileage, responsiveness, and power on the road that drivers expected in their diesels because, unlike the test-rig mode, the standard mode permitted a much larger amount of nitrogen oxide to exit the exhaust pipe, up to 40 times more than the U.S. limit (Nitrogen oxide is a smog-forming pollutant linked to lung cancer. Atiyeh, 2016; Robinson, 2015). Anyone who has ever driven a car when it lost its muffler will have some idea of the advantage VW's standard mode had over its test-rig mode.

4. On November 2, 2015, the U.S. Environmental Protection Agency enlarged Dieselgate by issuing notice of similar violations for the Porsche Cayenne and five Audi models as well as VW's Touareg (a joint venture between Audi, Porsche, and VW). Though Audi and Porsche are VW brands, their engines were developed by Audi in Ingolstadt, about 300 miles south of Wolfsburg. Thus, at least two groups of engineers were simultaneously breaking the law in much the same way for seven years, with little in common except the senior executives in Wolfsburg to whom both groups reported. Since it seems unlikely that two groups of engineers widely separated and working on engines belonging to different "families" would both "go rogue" in the same way at the same time (when auto

engineers in other companies did not), the obvious conclusion is that senior executives in Wolfsburg oversaw Dieselgate. If there were any rogue engineers, some of them at least were in senior management.

5. Bosch provided the software for sensing the test-rig mode in 2007, warning against its misuse (just the sort of warning lawyers would insert in a contract to protect their client). VW promptly used Bosch's test-rig mode in software that had the pollution-spewing standard mode. Why? When Martin Winterkorn became VW's CEO in 2007, he announced that he would make VW the world leader among auto-makers in volume, profit, and quality. One of his early actions was to order VW's engineers to deliver a clean diesel that could be sold worldwide (Ewing, 2015). The engineers soon realized they could not immediately deliver such an engine without increasing the cost of a VW auto by at least $300, requiring owners to refill their urea reservoir inconveniently often, and otherwise making the diesel less marketable than desired. The engineers may have adopted the "cheat code" as a stopgap while they continued to look for a better way to control diesel pollutants.[5] But months became years as their search proved unsuccessful. Did Winterkorn know about the cheat code? He has denied any knowledge of it (Robinson, 2015).

6. That denial is implausible for at least three reasons. First, VW has a long history of strong management. Winterkorn was apparently brought in as CEO because he was that sort of manager. He was known to "rule by fear." He was not someone to be secretly disobeyed. Second, Winterkorn was not only a strong manager but a strong micro-manager, someone who paid attention to details. The effectiveness of pollution controls was more than a detail. Responsive low-pollution diesel engines were a significant part of Winterkorn's strategy to make VW the world's largest automaker. Subordinates were therefore unlikely to adopt the cheat code on their own. Adopting it was just too important a decision. (Smith and Parloff, 2016) Third, Winterkorn was an engineer, though a metallurgical rather than a mechanical engineer. He had enough experience at VW to understand its diesel technology, including the pollutant emission controls. If Winterkorn could honestly deny knowledge of the cheat code, it seems likely that he could do so only because he took pains to avoid such knowledge.

7. The cheat code seems like software that a few computer scientists or engineers could write, test, and insert, but not software that they could write, test, and insert without others, especially VW's management and engineers, knowing. VW's documentation system should have tracked all changes in software; hence, some managers should have known about the cheat code, especially what it was designed to do. Some engineers would also have to know because they had to write the specifications for the software, calibrate VW's test-rigs, run the physical tests, and so on. The

skill of VW's engineers is evident in the deception succeeding as long as it did (Smith and Parloff, 2016).
8. Like Watergate, Dieselgate is actually two scandals. VW engineers seem to be prominent in both. The first involves the original deception just described; the second, a cover-up. For almost a year after the California Air Resources Board first asked VW to explain why its diesels did so much better in rig tests than in road tests, VW engineers tried to explain away the difference, for example, by pointing to inadequacies in the road tests. Then one day they admitted what they had denied for almost a year: Starting in 2008, VW had installed undisclosed software in its diesel engines that triggered a "second calibration intended to run only during certification testing [on test rigs]." On September 3, 2015 (almost exactly a year before Liang's guilty plea), a VW official formally signed a document so stating (Smith and Parloff, 2016).
9. Some VW engineers in Germany may have told a manager outside their department of the cheat code—but none ever told anyone *outside* VW even after the managers, upon being told, seemed to do nothing.[6] There may, then, have been some internal whistleblowing at Wolfsburg, perhaps reaching senior managers. Unfortunately, those senior managers seem to have been the very people who had approved the deception. Winterkorn was "Chairman of the Board of Management of Volkswagen AG," the body to which Group Auditing and other internal watchdogs reported. There was also (something like) internal whistleblowing in the United States once the cover-up reached it. But there was no *external* whistleblowing in Germany or the United States, unless we count as whistleblowing answering truthfully the official questions of VW's lawyers or a government's prosecutor (Ewing, 2016b).

3. THE ENGINEER AS EMPLOYEE

Any event as complex as Dieselgate is likely to have several causes, not just one, but one cause seems to stand out: VW's (German) "corporate culture" (VW's distinctive way of doing certain things). Let me explain.

If we accept the *Code of Conduct* issued under the signature of "Prof. Dr. M. Winterkorn, Dr. H. Neumann, and Bernd Osterloh" in 2010 as a (rough) statement of VW's culture during Winterkorn's term of office, we can see much that is good.[7] The *Code* applies not only to all employees of every company in the VW Group, their suppliers, and dealers anywhere in the world but also to members of its executive bodies. The *Code* sets the baseline for all, but any organization subject to it may set a stricter standard. Among the *Code*'s "General Conduct Requirements" are the following chapters:

The Whistle Not Blown 225

Responsibility for the Reputation of the VW Group; Responsibility for Basic Social Rights and Principles; Equal Opportunity and Mutual Respect; Avoiding Conflict of Interest and Corruption; Privacy and Data Security; Secrecy; Handling Insider Information; Occupational Safety and Health Protection; Environmental Protection; and Protection and Proper Use of VW Group Property. There is an "ombudsman," an "anti-corruption officer," and other channels for reporting a violation of the *Code*. Yet, on close examination, it is clear that the *Code* had nothing useful to say about Dieselgate.

The jurisdiction of the ombudsman, the anti-corruption officer, and the like was only over conflict of interest, secondary employment, corruption, and similar "white-collar crime":

> Each of our employees is obligated to seek help or advice upon suspicion or legal uncertainty about the existence of corruption or white-collar crime. Advice and assistance are provided by the superior, the responsible internal departments (e.g., Auditing, Legal, Compliance, Group Security, or Human Resources), the anticorruption officer, or the ombudsmen. In addition, every employee can also turn to the Works Council [the labor union]. (Volkswagen Group, *Code of Conduct* (05/2010), 11)

Though this language (especially the term "white collar crime") might seem to justify blowing the whistle internally in a case like Dieselgate, the context forbids that interpretation. The language just quoted appears in the section entitled "Conflict of Interest and Corruption." The context thus limits the (internal) whistleblowing obligation to conflict of interest, corruption (presumably schemes of self-enrichment), and the like. That need not be so. VW America (VWGoA) does not so limit the corresponding provision. In the United States, the corresponding provision has its own section ("Reporting Code Violation, Corruption and Conflicts of Interest"). It is also much more explicit about when an employee should blow the whistle. It says (in part):

> Any employee who has knowledge of, or information concerning a past or present violation or possible violation of any law, regulation, policy or provision of this Code, or has knowledge or information indicating that such a violation may occur in the future, must promptly report such information to the Company's Compliance Officer, the Office of General Counsel, or the Ethics Hotline. In addition, all employees, contractors, suppliers or business partners may also turn to the VWGoA Ethics Hotline upon discovering indications of corruption or unethical or illegal practices.[8]

The text of the U.S. version of VW's whistleblowing provision, like its placement, makes it clear that it would cover the kind of wrongdoing involved in

Dieselgate—since, whatever else Dieselgate did, it violated a law or regulation. But, like its German counterpart, VWGoA's code makes no mention of *external* whistleblowing.

The other provisions of VW's (German) code would also not be of much help to employees in Wolfsburg or Ingolstadt aware of Dieselgate. The obvious place for them to look for guidance in the Code is "Environmental Protection." What would they find? Here is the whole section:

> We develop, produce, and distribute automobiles around the world to preserve individual mobility. We bear responsibility for continuous improvement of the environmental tolerability of our products and for the lowering of demands on natural resources while taking economic considerations into account. We therefore make ecologically efficient advanced technologies available throughout the world and implement them over the entire lifecycle of our products. At all of our locations, we are a partner to society and politics with respect to the configuration of social and ecologically sustainable positive development. Each of our employees make[s] appropriate and economical use of natural resources and ensure[s] that their activities have only as limited an influence on the environment as possible. (VW Group, *Code*, 19)

The language in this section takes the form of statements of fact, not of obligations as in "Conflict of Interest and Corruption." The focus is on "we," that is, VW as a corporation, not on what individuals should do. The sentiments are noble (something one might expect in a public relations brochure). But the only responsibility VW recognizes for itself is "continuous improvement of the environmental tolerability of our products and for the lowering of demands on natural resources while taking economic considerations into account." Dieselgate was arguably trying to do that, that is, lower demands on natural sources *while* also taking economics considerations into account. The economic considerations won. The only reference to individuals in this section declares that each employee "make[s] appropriate and economical use of natural resources and ensure[s] that their activities have only as limited an influence on the environment as possible." The Code provides no measure of what is appropriate in a case like Dieselgate—as VWGoA's code does by requiring that the employees at least act within legal bounds.

The VW Group (the German part of VW) thus seems to have given little thought to ethics apart from conflict of interest and corruption. It certainly did not provide employees with internal channels for blowing the whistle on Dieselgate's selfless wrongdoing, much less make any provision for dealing with wrongdoing by senior management.

The effect of the VW Group's code of ethics is, of course, an empirical question, one I am in no position to resolve. My claim here is that, even if the code of ethics did guide the conduct of everyone at VW, Dieselgate might well have occurred just as it did. Whatever its effect, the Code may be a good guide to the way VW employees thought about what they should be doing.

4. THE ENGINEER AS MEMBER OF A PROFESSION

The VW *Code* (like VWGoA's) is for all employees, suppliers, and so on, not just for VW engineers. We might suppose, then, that VW's German engineers do not need a clear whistleblowing provision in their employer's code because European codes of engineering ethics, like their American counterparts, already have such a provision. We should not suppose that. Consider FEANI's Code of Conduct. There is no whistleblowing provision. The only provision that seems at all relevant to blowing the whistle on Dieselgate is the last, "Social Responsibility":
The Engineer shall

- respect the personal rights of his superiors, colleagues and subordinates by taking due account of their requirements and aspirations, provided they conform to the laws and ethics of their professions,
- be conscious of nature, environment, safety and health and work to the benefit and welfare of mankind,
- provide the general public with clear information, only in his field of competence, to enable a proper understanding of technical matters of public interest,
- treat with the utmost respect the traditional and cultural values of the countries in which he exercises his profession (FEANI, 2016).

We certainly can argue that the engineers at VW violated at least one of this section's four clauses. While VW's engineers might claim that they "work[ed] to the benefit and welfare of mankind," were "conscious of nature, environment, safety and health [of mankind]," and treated "with the utmost respect the traditional and cultural values of [Germany when exercising] his profession', they could not claim to "provide the general public with clear information" about VW's diesels, a technical matter within their competence. Instead, they lied to U.S. regulators for at least a year. In that respect at least (assuming U.S. regulators count as part of "the general public"), they failed to blow the whistle on VW *externally* when they should have.

But, another provision of FEANI's code, one near the beginning, seems to forbid external whistleblowing (while allowing for informing the public with the consent of the employer): "He shall consider himself bound in conscience by any business confidentiality agreement into which he has freely entered."

Employees of large employers, such as the VW Group, typically must sign such an agreement as a condition of employment.[9] Even if they did not, VW's Code of Conduct (presumably a part of every employment contract) declares:

> Each of our employees is obligated to maintain secrecy regarding the business or trade secrets with which they are entrusted within the scope of the performance of their duties or have otherwise become known. Silence must be maintained regarding work and matters within the Company that are significant to the VW Group or its business partners and that have not been made known publicly, such as, for example, product developments, plans, and testing. ("Secrets")

There is no exception for engineers.

FEANI's code is meant to be "additional to and does not take the place of any Code of Ethics to which the registrant might be subject in his own country." Since 2002, there has been such a code for German engineers: "The Fundamentals of Engineering Ethics" of the Association of Engineers in Germany (VDI).[10] At least five of its provisions seem relevant here. The first (1.3) requires engineers to "honour [laws and regulations of their countries] insofar as they do not contradict universal ethical principles." Of course, honoring laws or regulations is not necessarily the same as obeying them. But, even if it were, this provision would only tell engineers in Germany to obey German law, not that of other countries, such as the US, for which their products are destined. In any case, the point of 1.3 seems to be to deny that positive law automatically takes precedence over "universal ethical principles" (such as "Don't lie," "Don't cheat," and "Keep your promises").

Another provision of the VDI code (2.2) nudges German engineers in the direction of sustainable development rather than (Dieselgate's) "short-term profitability":

> The fundamental orientation in designing new technological solutions is to maintain today and for future generations, the options of acting in freedom and responsibility. Engineers thus avoid actions which may compel them to accept given constraints (e.g. the arbitrary pressures of crises or the forces of short-term profitability).

Section 2.4 provides more than a nudge in the same direction. It specifies that, in cases of conflicting values, engineers should give priority:

- to the values of humanity over the dynamics of nature—to issues of human rights over technology implementation and exploitation,
- to public welfare over private interests, and
- to safety and security over functionality and profitability of their technical solutions.

Having set these priorities, 2.4 then hedges a bit: "Engineers, however, are careful not to adopt such criteria or indicators in any dogmatic manner." But 2.4 ends with what seems to be an invitation to blow the whistle *externally* when appropriate: "[Engineers] seek public dialogue in order to find acceptable balance and consensus concerning these conflicting values."

Section 3.3 seems to reassert the priorities of 2.4: "national laws have priority over professional regulations, such professional regulations have priority over individual contracts." Engineers are, it seems, not to be bound by the employment contract when obeying any part of it, even a pledge of secrecy, would violate national law or professional obligations.[11] That is important because another provision of the VDI code seems to *allow* external whistleblowing:

> 3.4 There may be cases when engineers are involved into [sic] professional conflicts which they cannot resolve co-operatively with their employers or customers. These engineers may apply to the appropriate professional institutions which are prepared to follow up such ethical conflicts. As a last resort, engineers may consider to [sic] directly inform the public about such conflicts or to refuse co-operation altogether. To prevent such escalating developments from taking place, engineers support the founding of these supporting professional institutions, in particular within the VDI.

Presumably, applying to "appropriate *professional* institutions" is a kind of external whistleblowing. The last resort, "consider to directly inform the public" does not quite allow informing the public, but it at least comes close.

To summarize: According to the DVI code, engineers (of whatever country) are *allowed* to blow the whistle on an employer who is violating human rights or harming the public welfare. They may blow the whistle by going to a "professional institution" and, as a last resort, perhaps by going public. Engineers are under no obligation to do either. While the engineers involved in Dieselgate did act contrary to their professional obligations—both because what they did was illegal and because it violated universal ethical principles, such as "Don't lie" and "Don't cheat"—all they were required to do was say no (i.e., to refuse to do the illegal or unethical acts in question). More could be required. Consider, for example, the equivalent provision of a US code of ethics for engineers:

> Engineers having knowledge of any alleged violation of this Code shall report thereon to appropriate professional bodies and, when relevant, also to public authorities, and cooperate with the proper authorities in furnishing such information or assistance as may be required. (NSPE, 2007, II.1.f.)

5. CONCLUSION

The VW Group seems to have been ill-prepared to deal with wrongdoing by its senior management. Its internal whistleblowing arrangements covered only a narrow range of possible wrongdoing (conflict of interest and corruption), far from the full range of wrongdoing its own code of conduct covered. Arrangements for internal whistleblowing seem to have assumed that senior management would never do anything in violation of VW's *Code of Conduct*. There was, for example, no provision allowing the ombudsman, the legal department, or the auditors to turn a complaint over to an outside investigator who, upon establishing that senior management was implicated in wrongdoing, could report directly to stockholders or the appropriate governmental agency.

Nowhere in VW's *Code of Conduct* is there any recognition that engineers, or members of any other profession, might have their own code of ethics, much less that VW might want to encourage them to follow that code even when, indeed, especially when, the profession's ethics collided with management's plans.

Last, Dieselgate seems to show that both the code of engineering ethics of FEANI and the code of ethics of VDI need much stronger provisions for external whistleblowing. Engineers having knowledge of a violation of their code of ethics should have an explicit obligation to report it to appropriate professional bodies and, when that is not enough to set things right, also to public authorities. Europeans should give much more attention to whistleblowing (internal as well as external) than they have.

Whether changes in any code of engineering ethics, or in VW's *Code of Conduct*, would have any effect must, of course, depend in part on what engineering schools, engineering associations, and VW then do to inform engineers of their ethical responsibilities and to support engineers when they act responsibly. A code of ethics forgotten in a drawer is unlikely to have much effect on conduct even if it says everything it should. But such a code integrated into the everyday operations of a corporation is.

It is clear from our comparison of codes of engineering ethics here that European codes did not—as of 2016—require whistleblowing in the way the corresponding American code did. Is that a cultural difference or just an accident of history? How might we go about answering that question? If it is a cultural difference, should it be "respected"? How?

NOTES

This chapter was prepared in part under a grant from the MacArthur Foundation (Ethics Codes Collection), https://www.macfound.org/grantees/7562/(accessed October 12, 2016). My thanks to the Foundation, and to Keith Miller, Donald

Gotterbarn, and Alexander Davis-Jones for help thinking through the technical aspects of this case. The VW sites cited here have all disappeared since this chapter was first prepared (2016). Originally published as: "A Whistle Not Blown: WV, Diesels, and Engineers," *Next Generation Ethics: Engineering a Better Society*, edited by Ali Abbas (Cambridge University Press, 2019): 217–229.

1. Computer scientists I have consulted think one or two computer scientists must have been involved. The engineers I consulted think the engineers could have written the software on their own.

2. Tabuchi and Ewing (2016b). For comparison, VW's annual profit in the last year before the scandal (2014) was $12.3 billion (Boston, 2015). So, the 2016 U.S. settlement alone wiped out the annual profits of VW worldwide (in a good year).

3. Hiroko Tabuchi and Jack Ewing, 2016c. For engineering credentials, see: Martin Winterkorn, https://en.wikipedia.org/wiki/Martin_Winterkorn (accessed September 22, 2016); Wolfgang Hatz, http://www.bloomberg.com/research/stocks/people/person.asp?personId=73467911&privcapId=875012 (accessed September 22, 2016); Ulrich Hackenberg, http://www.volkswagenag.com/content/vwcorp/info_center/en/news/2014/01/Stanford_University.html (accessed September 22, 2016); Heinz-Jakob Neusser, http://www.bloomberg.com/research/stocks/people/person.asp?personId=222431348&privcapId=377732 (accessed September 22, 2016).

4. The chief engineer-heroes of this story are in West Virginia. See, for example, Ewing (2016).

5. "The engineers viewed the ruse as a stopgap measure, Volkswagen has suggested, and hoped to abandon it when better technologies became available" (Smith and Parloff, 2016).

6. "A whistleblower allegedly revealed the use of a defeat device to Heinz-Jakob Neusser, a Volkswagen brand-development boss and, later, management board member, according to the *Süddeutsche Zeitung*" (Smith and Parloff, 2016). There was also at least one other internal European "whistleblower" *after* the scandal broke. Under questioning by the US law firm (Jones Day) that VW hired to carry out an internal investigation of the scandal, the employee (probably an engineer) testified:

> "[The pressure seemed to intensify inside VW. It wasn't] acceptable to admit anything is impossible ... Instead of telling management that they couldn't meet the parameters, the decision was taken to manipulate. No one had the courage to admit failure. Moreover, the engine developers felt secure because there was no way of detecting the deceit with the testing technology that existed at the time. [It was, the employee said] an act of desperation" (Smith and Parloff, 2016). The employee might have added that over time the deception might have reduced the pressure to improve pollution-control devices, allowing senior management's attention to drift to other priorities.

7. Like Winterkorn, Neumann (VW's Supervisory Board's member for Human Resources) retired soon after the scandal broke. http://www.volkswagenag.com/content/vwcorp/info_center/en/news/2015/11/Neumann.html (accessed September 26, 2016). Osterloh (Supervisory Board member for Labor, a lawyer) is still in office, http://www.volkswagenag.com/content/vwcorp/content/en/investor_relations/

corporate_governance/supervisory_board.html (accessed September 26, 2016). The version of the *Code* now posted at http://www.volkswagenag.com/content/vwcorp/content/en/the_group/compliance.html (accessed September 28, 2016) is unsigned and dated 09/2015 but seems otherwise without significant change.

8. VWGoA, 2015, 8. This is the "1st Edition" of the code, suggesting that VWGoA learned from Dieselgate.

9. While I have not been able to obtain a copy of VW's (German) confidentiality agreement, I think the English version used by VWGoA will give a general idea of what it contains. See Volkswagen, Group America, "Confidentiality Agreement", https://www.google.com/search?q=vw+employee+confidentiality+agreement&ie=utf-8&oe=utf-8 (accessed September 30, 2016).

10. For full text, see VDI, "Fundamentals of Engineering Ethics," March, 2002, https://www.vdi.de/fileadmin/vdi_de/redakteur_dateien/.../engineering_ethincs.pdf (accessed October 1, 2016).

11. Note that this provision seems to give national law priority over professional regulation. This is a priority that seems to be disappearing from codes of engineering worldwide as engineers contemplate the abuse of national law from Russia to Venezuela, from China to South Africa.

REFERENCES

Atiyeh, Clifford, "Everything You Need to Know about the VW Diesel-Emissions Scandal," *Car and Driver* (July 26, 2016), http://blog.caranddriver.com/everything-you-need-to-know-about-the-vw-diesel-emissions-scandal/ (accessed September 23, 2016).

Boston, William, "Germany's Volkswagen Posts Rise in 2014 Profit," *Wall Street Journal*, February 15, 2015, http://www.wsj.com/articles/germanys-volkswagen-posts-rise-in- 2014-profit-1425052616 (accessed September 29, 2016).

European Federation of Engineering Associations (FEANI), "Code of Conduct," http://www.tendrup.dk/feani.htm (accessed September 29, 2016).

Ewing, Jack, "VW Says Emissions Cheating Was Not a One-Time Error," *New York Times* (December 10, 2015), http://www.nytimes.com/2015/12/11/business/international/vw-emissions-scandal.html (accessed September 24, 2016).

Ewing, Jack, "Researchers Who Exposed VW Gain Little Reward From Success," *New York Times* (July 24, 2016a), http://www.nytimes.com/2016/07/25/business/vw-wvu-diesel-volkswagen-westvirginia.html?action=click&contentCollection=International%20Business&module=RelatedCoverage&region=EndOfArticle&pgtype=article (accessed September 25, 2016).

Ewing, Jack, "VW Whistle-Blower's Suit Accuses Carmaker of Deleting Data," *New York Times* (March 14, 2016b), http://www.nytimes.com/2016/03/15/business/energy- environment/vw-diesel-emissions-scandal-whistleblower.html?_r=0 (accessed September 25, 2016).

Hakim, Danny "Beyond Volkswagen, Europe's Diesels Flunked a Pollution Test," *New York Times* (February 7, 2016), http://www.nytimes.com/2016/02/08/business

/international/no-matter-the-brand-europes-diesels-flunked-a-pollution-test.html?_r=0 (accessed September 24, 2016).

National Society of Professional Engineers (NSPE), Code of Ethics for Engineers, 2007, http://ethics.iit.edu/ecodes/node/4098 (accessed October 3, 2016).

Robinson, Aaron, "Caught Black-Handed: Why Did Volkswagen Cheat?" *Car and Driver* (November 3, 2015), http://blog.caranddriver.com/caught-black-handed-why-did-volkswagen-cheat/ (accessed September 23, 2016).

Smith, Geoffrey and Roger Parloff, "Hoaxwagen: How the massive diesel fraud incinerated VW's reputation—and will hobble the company for years to come," *Fortune* (March 7, 2016), http://fortune.com/inside-volkswagen-emissions-scandal/ (accessed September 25, 2016).

Hiroko Tabuchi and Jack Ewing, "VW Engineer Pleads Guilty in US Criminal Case Over Diesel Emissions," *New York Times* (Sept. 9, 2016a), http://www.nytimes.com/2016/09/10/business/international/vw-criminal-charge-diesel.html?_r=0 (accessed September 20, 2016).

Hiroko Tabuchi and Jack Ewing, "VW's US Diesel Settlement Clears Just One Financial Hurdle," *New York Times* (June 28, 2016b), http://www.nytimes.com/2016/06/29/business/vw-diesel-emissions-us-settlement.html (accessed September 21, 2016).

Hiroko Tabuchi and Jack Ewing, "Volkswagen Scandal Reaches All the Way to the Top, Lawsuits Say," *New York Times* (July 19, 2016c), http://www.nytimes.com/2016/07/20/business/international/volkswagen-ny-attorney-general-emissions-scandal.html?module=Promotron&region=Body&action=click&pgtype=article (accessed September 21, 2016).

Volkswagen Group of America (VWGoA), *Code of Conduct*, December 2015 (accessed September 28, 2016),

Volkswagen, Human Resources, http://www.volkswagenag.com/content/vwcorp/content/en/human_resources.html (accessed September 21, 2016).

Volkswagen Group America (VWGoA), "Confidentiality Agreement," https://www.google.com/search?q=vw+employee+confidentiality+agreement&ie=utf-8&oe=utf-8 (accessed September 30, 2016).

Chapter 17

Three Nuclear Disasters and a Hurricane

Reflections on Engineering Ethics

This chapter is another case study highlighting engineering's globalism. It began with an invitation from Japan to an American a month after the disaster at the Fukushima I Nuclear Power Plant (March 11, 2011). There was to be a presentation at a conference in Japan including guests from Europe, North America, Australia, China, and so on (October 30, 2011). Like Dieselgate when I first prepared Chapter 16, the Fukushima disaster was still very much the province of journalism. Its outlines certainly lacked the stability of history. I therefore thought it useful to compare the Fukushima disaster with three other disasters that both resemble the Fukushima disaster in ways that seem important and better understood. The results surprised me, especially results relevant to engineers generally.

1. CAVEATS

Of course, even history, though it generally seems stable, is not entirely so, being subject to dispute here and there and to radical revision every now and then. At first, Fukushima's facts changed almost daily—if by "facts" we mean those descriptive propositions about which there is general agreement at a certain time. After a while, the changes were less frequent and more a matter of addition than correction. The general outline of the disaster seemed settled. Dispute by then concerned only details, such as how much land, if any, would have to be abandoned for some years, how many premature deaths were to be expected because of radiation released during the disaster, and so on. At some point, I had to stop worrying about the facts and sit down to report my reflections. That was on October 15, 2011. Since then, I have changed "a fact" only when a reader or auditor pointed out that it was no longer a fact. My effort has gone into thinking about a certain array of facts, not in making sure that the facts I present here are complete and accurate.

The invitation from Japan presented me with a practical problem. The newspapers, websites, and other sources available (at least in English) seldom identified anyone as an engineer. The stories focused on "workers," "managers," and machinery. I had to use what I knew about nuclear power plants in the United States to interpret the facts thus given. I had similar problems, though less severe, when interpreting the other disasters to which I chose to compare Fukushima. As often happens, the technology erased the engineers.

Interpretations are, of course, open to objection but, without interpretation, facts merely pile up, becoming in time an unmanageable heap. There is no understanding without interpretation. But interpretation of an event relying on changing facts is necessarily the sort of time-stamped enterprise philosophers are inclined to avoid—and I would have avoided it but for that invitation. There is not much that a philosopher can do about a disaster such as that at Fukushima—except help those seeking to understand it and thereby help to prevent similar disasters. I felt I owed my Japanese hosts atleast that much.

This chapter's title promises "reflections" on Fukushima, not systematic or definitive understanding. Reflections are what one gets when, focusing thought on certain facts, one captures connections that seem to jump out of the dark. Reflection is a source of hypothesis rather than proof, the beginning of a discussion rather than the end. We do not need reflection when we can derive a conclusion from what we know. Reflection is useful when we want to discover a conclusion that, though far from provable given the facts we have, invites investigation. There is no algorithm for reflection, no test of success beyond useful surprise. But it is sometimes useful to save reflections as a reminder of what insight is, or is not, possible when a philosopher steps into the stream of current events. That is what I have done here (as I also did in Chapter 16). So far, I find that I have had no reason to change the conclusions I jumped to in 2011.

2. WHY COMPARE THESE FOUR DISASTERS?

The nuclear disaster that Japan suffered at Fukushima has been compared with other major *nuclear* disasters, especially Three Mile Island (1979) and Chernobyl (1986). It is more like Chernobyl in immediate destructiveness, the only other 7 on the International Nuclear Event Scale (the upper limit of which is 7). It is more like Three Mile Island in probable long-term effects (though Fukushima's long-term effects are likely to be substantially worse than Three Mile Island's). Chernobyl seems to be an order of magnitude worse than either Three Mile Island's or Fukushima's. To date, Chernobyl seems to have directly killed thirty-one reactor staff and workers, to have caused between 200,000 and 1,000,000 premature deaths worldwide, to have forced the permanent abandonment of a city of about 50,000 (Pripyat), and to have ruined perhaps a 100,000 square km of farmland. Over 300,000 people lost their homes to contamination.[1]

In contrast, the radiation released from the Fukushima plant, though significant, will, it seems, leave little long-term contamination, except at the plant itself and in a plume perhaps fifty km beyond. At least six workers have exceeded lifetime legal limits for radiation and more than three hundred have received significant radiation doses. Estimates of future cancer deaths from accumulated radiation exposures in the population living near Fukushima have ranged from none to a non-peer-reviewed "guesstimate" of a thousand. No one died in the explosions at the plant or from subsequent radiation exposure (though the tsunami killed two workers and evacuation of hospitals in the exclusion zone may have caused as many as forty-five more deaths). The earthquake or tsunami, rather than the nuclear accident, seems to be responsible for the few employees severely injured or killed at the plant.[2]

Though certainly a nuclear disaster, Fukushima is not just another nuclear disaster. In ways important to engineering, it is much more like Katrina's destruction of New Orleans than like any other nuclear disaster. It is (primarily) a consequence of a natural—or, at least, much larger—disaster, the enormous earthquake and tsunami that wrecked much of northeast Japan on March 11, 2011, killing about 28,000 people. Fukushima has many lessons to teach, especially if we compare it with the other disasters I shall discuss. I shall focus on four: The first lesson concerns the different roles engineers have at different stages in an engineering project, especially the relative powerlessness of engineers to affect certain early large-scale trade-offs between public safety and public welfare. A second lesson may be the need to evaluate risk in ways beyond ordinary cost-benefit analysis when the risks are improbable but catastrophic. A third lesson is the importance of not leaving complex technical systems untended. Engineering systems do not work long or well without engineers. A fourth lesson may concern the way engineers should respond, and typically do respond, to engineering disasters. They should take responsibility for limiting the harm as well as for fixing the underlying problem, even if limiting the harm involves risking their lives. To see what I mean, let us consider these four disasters in greater detail, beginning with the earliest.

Note that two of these disasters are American and the other two are Japanese or "Russian" (Soviet or Ukrainian, depending on how we think about Chernobyl's "culture"). Note that the engineering ethics does not seem to be different just because the national cultures are different.

3. THREE MILE ISLAND

Three Mile Island was a "normal accident," that is, it began with normal failures of equipment and practice within a plant itself operating normally. Perrow (1984) also describes Three Mile Island as a "normal accident." While I agree that it was a "normal accident" in his sense, my use of that term is somewhat different. I mean simply that the accident was a product of what engineers normally do rather than a product of incompetence, negligence, corruption, or other unusual conduct (such as experimentation). What distinguishes Three Mile Island from most of the other accidents at nuclear power plants is the way normal failures combined to produce an abnormal result.

During the night of March 27–28, 1979, workers were engaged in routine cleaning of a blockage in one of Reactor 2's eight condensate polishers (filters for the secondary cooling loop). At 4 am, the pumps feeding the polishers stopped. We still do not know the cause of the stoppage. When a bypass valve failed to open, water ceased flowing to the secondary loop's main feed-water pumps. These then also shut down. No longer receiving water, the steam-driven generators stopped and the reactor automatically carried out an emergency shutdown. Within eight seconds, control rods were inserted into the core to halt the nuclear chain reaction. The reactor nonetheless continued to generate heat (a byproduct of the normal decay of its fuel). Because steam was no longer being used by the turbine, heat was no longer being removed from the reactor's primary water loop.[3]

Once the secondary loop's feed-water pumps stopped, three auxiliary pumps started up automatically; but because some valves were closed for routine maintenance, the system could not pump water. So, the secondary loop was no longer working. Without the secondary loop removing heat, pressure in the primary loop began to increase, automatically triggering a relief valve. The relief valve should have closed again when the excess pressure had been released; instead, it stayed open. That open valve permitted coolant water to escape from the primary system. It was the principal mechanical cause of the coolant-loss meltdown that followed.

The mechanical failures were compounded by the failure of plant operators to recognize the situation as a loss-of-coolant accident for more than two hours. (One cause of their failure seems to have been an indicator light blocked from view.) That initial failure led an operator to override the reactor's automatic emergency cooling system manually. With the release valve still open, the quench tank that collected the discharge from the release valve soon overfilled, causing the containment building's sump to fill and sound an alarm at 4:11 am (eleven minutes after the first pumps failed). That alarm, along with higher than normal temperatures on the discharge line and unusually high temperatures and pressures in the containment building, clearly indicated that there was a loss-of-coolant accident, but the operators did not respond to these indications. At 4:15 am, the quench-tank relief diaphragm ruptured and radioactive coolant began to leak into the general containment building. This coolant was pumped from the containment building sump to an auxiliary building, outside the main containment, until the sump pumps were stopped at 4:39 am.

After almost eighty minutes of slow temperature rise, the primary loop's four main pumps began to suffer damage as a mixture of steam and water passed through them. The operators then shut down the pumps, believing that natural circulation would continue the water movement, but steam in the system (itself the product of rising temperature) prevented coolant flow through the core. As the coolant stopped circulating, it increasingly turned to steam. Just over two hours after the first sign of trouble, the coolant level fell so low that the top of the reactor core was exposed to the steam. Intense heat then caused a reaction between the steam in the reactor core and the nuclear fuel-rod cladding. That reaction burned off the cladding and damaged the fuel pellets. The pellets then released more radioactivity into the reactor coolant, producing hydrogen gas accumulation which probably caused a small explosion in the containment building in the afternoon.

At 6 am (two hours after the incident began), there was a change of shift in the control room. A new arrival noticed that temperatures in the relief valve tailpipe and holding tanks were too high and used a backup valve to shut off the coolant venting through the relief valve. But, by then, about 120,000 liters of coolant had leaked from the primary loop. Not until almost 7 am (about three hours after the incident began) did contaminated water reach radiation-activated alarms. By then, the radiation in the primary coolant water was around three-hundred times higher than usual. The plant was seriously contaminated and the reactor's core had suffered a partial meltdown.

The Nuclear Regulatory Commission (NRC) made an extensive investigation of the disaster, a typical engineering response. Its report ended with recommendations for changes in controls, quality assurance, maintenance, operator training, management, and communication of important safety information. There was no finding of negligence or more serious wrongdoing having caused the disaster, no suggestion that major redesign of nuclear plants was needed, and no proposal to rethink the place of nuclear energy in the US supply of electricity.[4] The report did, however, contain much criticism of other aspects of how Three Mile Island operated.

4. CHERNOBYL

Chernobyl was not a normal accident. Its cause was an engineering *experiment* which, though successful, lacked proper approval. That is not to say that the experiment was unjustified, fundamentally improper, or indeed abnormal.

Even when not actively generating power, nuclear reactors require cooling to remove heat produced by the natural decay of nuclear fuel. Chernobyl's pressurized water reactors (different in design from Three Mile Island's) used water flowing at high pressure to remove waste heat (about 28,000 liters of water an hour). After an emergency shutdown, the core could still generate a significant amount of residual heat. If not removed, the heat could cause core damage (as it did at Three Mile Island). If the (outside) power grid failed during a shutdown, Chernobyl would have to depend on batteries to run the plant's cooling system.

For that reason Chernobyl's reactors had three backup *diesel* generators. Each generator required fifteen seconds to start up but took over a minute to attain the speed required to run one of the main coolant pumps. Chernobyl's engineers judged this one-minute power gap unacceptable. Too much can happen in a nuclear reactor in a minute when the cooling system is not working. Analysis indicated that one way to bridge the one-minute gap was to use the mechanical energy of the steam turbine and residual steam pressure to generate electricity to run the main coolant pumps while the generator was reaching the correct RPM, frequency, and voltage. But, of course, the analysis had to be confirmed experimentally. The engineers had to work out and then test a specific procedure for effectively employing residual momentum and steam pressure.

Previous experiments—in 1982, 1984, and 1985—had ended in failure. The 1986 experiment was scheduled to take place at Reactor 4 during a maintenance shutdown. The experiment focused on refinements in the switching sequences of the electrical supplies for the reactor. The experiment was to begin with an automatic emergency shutdown. Because the engineers did not anticipate any danger to the reactor, they did not formally coordinate the experiment with either the reactor's chief designer or scientific manager. Indeed, the experiment did not even have the approval of the onsite representative of the Soviet nuclear oversight agency. Only the director of the plant approved it (and even his approval did not follow standard procedures). Everyone involved seemed to be working on the premise that the experiment was "no big deal."

The experiment began just after 1:23 am on April 26, 1986. The diesel generator started and sequentially picked up loads. The turbine generator supplied the power for the four main circulating pumps as it coasted down. The experiment was all but complete forty seconds later. But, as the momentum of the turbine generator that powered the water pumps decreased, the water flow decreased, producing more and more steam bubbles in the core. The reactor was now ready to begin a destructive feedback loop: The production of steam would reduce the ability of the coolant to absorb neutrons, increasing the reactor's output of heat. The increased heat would cause yet more water to become steam, further increasing heat. During almost the entire period of the experiment, the automatic control system successfully counteracted this destructive feedback, inserting control rods into the reactor core to keep the temperature down.

If conditions had been as planned, the experiment would almost certainly have been carried out safely. The Chernobyl disaster resulted from attempts to boost reactor power—and, therefore, temperature—once the experiment had started (something inconsistent with approved procedure). The approved procedure called for Reactor 4's power output to be gradually reduced to 700–1000 MW. The minimum level established in the procedure (700 MW) was achieved about an hour before the experiment began. However, because of the natural dampening effect of the core's neutron absorber, reactor power continued to decrease, even without further operator action.

As the power dropped to approximately 500 MW during the experiment, one of the engineers conducting the experiment mistakenly inserted the control rods too far, nearly shutting down the reactor. Control-room personnel soon decided to restore the power and extracted the reactor control rods, but several minutes elapsed between the extraction and the time that the power output began to increase and stabilize at 160–200 MW. The extraction withdrew the majority of the control rods to the rods' upper limit, but the rapid reduction in the power during the initial shutdown and subsequent operation at less than 200 MW led to increased dampening of the reactor core by the accumulation of xenon-135 (an unstable fission product of uranium that absorbs neutrons at a high rate). To counteract this unwanted high absorption, the operators withdrew additional control rods from the reactor core.

Then, about the time the experiment ended, there was an emergency shutdown of the reactor. The shutdown started when someone pressed the button of the reactor's emergency protection system. (We do not know whether the button was pressed as an emergency measure, by mistake, or simply as a routine method of shutting down the reactor upon completion of the experiment.) Because of a flaw in the design of the graphite-tipped control rods, the dampening rods displaced coolant before inserting neutron-absorbing material to slow the reaction. The emergency shutdown therefore *briefly* increased the reaction rate in the lower half of the core. A few seconds after the start of the emergency shutdown, there was a massive power increase, the core overheated, and seconds later this overheating produced the first (nonnuclear) explosion. Some of the fuel rods fractured, blocking the control-rod columns and causing the control rods to become stuck at one-third insertion. Several more explosions followed, exposing the reactor's graphite moderator to air, causing it to ignite. Since the reactor lacked a containment (a thick concrete shell protecting the world from the reactor), the fire in the reactor sent a plume of highly radioactive smoke into the atmosphere, causing dangerous fallout over a huge area (as much as 500 km away)—and, eventually, less dangerous fallout over much of the world. The effort to avert a much greater disaster soon involved over 500,000 workers and cost an estimated eighteen billion rubles, crippling the Soviet economy.

Because many of the managers, engineers, and ordinary workers directly involved in the Chernobyl disaster soon died of radiation poisoning, there are many uncertainties about the exact sequence of events. Nonetheless, we can be sure that the actual disaster would not have occurred had the experiment not been carried out. The Chernobyl disaster combines the "normal failures" of operators and equipment we saw at Three Mile Island with an experiment of the sort engineers often perform. An experiment, however "small" and "routine," necessarily invites the unexpected. Chernobyl was as much an engineering disaster as Three Mile Island: both the immediate and underlying causes were ordinary engineering decisions, whether in operation or design.

5. FUKUSHIMA

The disaster at Fukushima fits neither of these patterns. The accident was not normal or the result of an engineering experiment. It was also not the result of operator negligence, incompetence, or misconduct. The disaster began with a large earthquake, one larger than any Japan had experienced in 1400 years of recorded history.[5] The quake was followed by an enormous tsunami. That double disaster would have happened even if the Fukushima nuclear power plant, one of the twenty-five largest in the world, had never existed. The nuclear disaster is a byproduct of that larger natural disaster.

At the time of the quake, 2:46 pm, Reactor 4 had been de-fueled while 5 and 6 were in cold shutdown for planned maintenance. The remaining three reactors shut down automatically in response to the quake. After the reactors shut down, the plant's own generation of electricity ceased, eliminating one source of electricity used to run cooling and control systems. One of two connections to the national electrical grid also failed. That loss of power started up thirteen on-site emergency diesel generators. These would ordinarily have provided enough power to operate the reactors' control and cooling systems until the lost connection to the national grid could be restored. Had the earthquake been the only disaster to hit the Fukushima plant on March 11, there would have been little to discuss here. The tsunami changed that.

The plant was protected by a seawall designed to withstand any tsunami of up to 5.7 meters, but the great wave that struck forty-one minutes after the quake was 15-meters high. It crashed over the wall and flooded the entire plant, including generators and electrical switchgear in reactor basements. It also broke the remaining connection with the national electrical grid. All conventional power for cooling was lost. Only one backup remained: emergency batteries, able to run some of the monitoring and control systems for up to eight hours. Replacement batteries and mobile generators were soon dispatched to Fukushima, but collapsed bridges, debris-strewn roads, and similar obstacles delayed them. The first replacements did not arrive until 9:00 pm (six hours after the first call went in and seven hours after the quake).

The arrival of the replacement batteries and mobile generators did not end the crisis, however. They had to be installed. The normal connection points were in flooded basements. There was also difficulty finding suitable cables. Work to connect batteries and generators was still continuing twenty-four hours after the quake when there was an explosion in Reactor 1's building. The side walls of the upper level were blown away, the roof collapsed, and debris covered much of the floor and machinery.

The roof of the building was designed to provide ordinary weather protection, not to withstand an explosion or to act as containment for the reactor. In the Fukushima reactors, the primary containment surrounded the reactor's pressure vessel. The top floor of the reactor buildings had no reactors, only water-filled pools for storing new fuel ready to be craned into the reactor and used fuel ready for disposal.

This first explosion was probably caused when hydrogen collected under the roof. Exposed to air, the fuel rods in the containment would have become very hot and reacted with steam, oxidizing the cladding and releasing hydrogen. The hydrogen would have leaked upward. Safety devices normally burn such hydrogen before it reaches explosive concentrations, but these safety devices seem to have failed when the electrical power did.

Reactor 1's containment survived the explosion. There were no large leaks of radioactive material, although there was an increase in radiation following the explosion. The explosion injured four workers. But that was only the beginning of the nuclear disaster. Hydrogen gas was also collecting at the other five reactors. Over the next few days, hydrogen explosions destroyed the upper cladding of the buildings for Reactor 3 and 4 and the containment inside Reactor 2. Several fires broke out at Reactor 4. In addition, spent-fuel rods stored in the spent-fuel pools of Reactors 1–4 began to overheat as the water level dropped. Fear of radiation leaks led to evacuation of all nonessential persons within a 20-km radius of the plant.

In short, the Fukushima plant was overwhelmed by forces from outside well beyond what it was designed for. Without heroic efforts by plant staff, some of whom may have died over the next few years because of exposure to radiation, the Fukushima disaster might have become at least as bad as Chernobyl. Even with those heroic efforts, several weeks passed before the plant could be said to be more or less under control. One generator at Reactor 6 was restarted on March 17 (six days after the quake) allowing some cooling at Reactor 5 and 6, the least damaged. Connection to the power grid was restored to parts of the plant on March 20, but machinery for Reactors 1-4—damaged by flooding, fires, and explosions—could not be restarted for several months. Only in early October (half a year after the quake) did coolant in all the reactors reach safe temperatures.

The Fukushima plant could have been designed to withstand the natural disaster that occurred. A breakwater three times higher than the actual breakwater could have protected the plant against the tsunami (assuming the breakwater survived the quake); the plant might have been located far enough away from the ocean to be safe from even so large a tsunami; the basements of generator buildings might have been made waterproof; and so on. Even some less expensive arrangements might have considerably improved what happened. For example, storing more batteries on site would have allowed the cooling and control systems to function longer without repair or resupply, days or weeks instead of hours. But most of these changes would have been expensive enough significantly to raise the price of the electricity that the plant produced. Typically, engineers, though consulted, do not make such big decisions. Government regulators, senior management, or public opinion typically decide, for example, whether to protect against a 500-year, 1,000-year, or 10,000-year quake. Engineers acting more or less on their own might have been able to make smaller improvements, such as storing the necessary cables alongside the backup batteries and storing some backup batteries above ground level.

6. KATRINA

When it struck New Orleans on August 29, 2005, Katrina was a category 3 hurricane, a large storm but no larger than storms that strike the Gulf Coast almost every year. (The top of the hurricane scale is 5.) Katrina was nonetheless unusually destructive because it moved so slowly that anything in its path was subject to heavy rains and high winds for many hours. The heavy rain and high winds were, however, only part of what caused so much destruction in New Orleans.[6]

Even on an ordinary day, New Orleans is a city that must work to prevent flooding. One of the world's largest rivers, the Mississippi, flows through it. From Jackson Park, the jewel of the tourist-drawing French Quarter, one of the *highest* points in the city, one can see the mighty river rushing by about 2 meters *above* the street. On any day of the year, the Mississippi would flood most of the city were it not for the levees that hold it back. Nor is the Mississippi the only watery threat. Though the oldest parts of the city are as much as 10 meters above sea level, a majority of the city is below sea level, and the sea, the Gulf of Mexico, reaches New Orleans at its back, through Lake Pontchartrain, and underground, through the water table. (While the water under New Orleans is fresh, it is as high as it is in part because the Gulf's salt water is not lower.)

Mostly developed since 1900, the newer parts of the city are, like much of the Netherlands, dry only because water is constantly pumped out. Every year, there is more for the pumps to do. Sea level is rising about a third of a meter a century; some parts of the city have subsided by half a meter or so because the weight of buildings is compressing the soil or because pumping water from the ground allows the soil to compress. Were it not for huge screw pumps working day and night, New Orleans would today be a version of what it was when the French first settled there in 1718, a crescent-shaped string of small islands in a huge swamp. Like Venice in Italy and St. Petersburg in Russia, New Orleans is much more artificial, and therefore much more vulnerable to natural forces, than most cities.

Engineers did not found New Orleans, but the city has long survived only because of engineering. The floods that the city suffers from time to time are due in part to the engineering not being good enough. That is as true of Katrina as of earlier disasters, for example, the1965 disaster named for hurricane Betsy (about as destructive as Katrina).

Katrina flooded New Orleans because the levee system failed catastrophically. Much of the disaster, however, occurred hours *after* the storm had moved inland and water poured through holes in levees and filled much of the city—much as a bathtub fills once the drain is closed and the faucet is turned on. There was no attempt to repair the levees immediately—for example, with volunteers and national guard working together to plug the gaps with sand bags. Indeed, for many days, there were no officials in New Orleans even to report damage. Everyone who could be evacuated had been. By August 31 (two days after Katrina struck), 80% of New Orleans, a city almost emptied of inhabitants, was under water, with some parts under water almost 5 meters deep. The water lingered for weeks.

On March 26, 2007, a year and a half after Katrina passed through New Orleans, the Interagency Performance Evaluation Task Force (IPET) issued its (draft) *Final Report*. IPET was an independent team of more than one-hundred-fifty international and national experts from more than fifty different government organizations, universities, and private companies. The US Army Corps of Engineers commissioned IPET a few weeks after Katrina hit New Orleans. It was to analyze how the levee system performed. Though many questions of detail remained unsettled, this nine-volume report, was, the last word on both the causes of the Katrina disaster and means of preventing similar disasters at the time I wrote this case study. The "final" *Final Report* did not appear until 2011.

IPET reports a "system" that grew up piecemeal, only in part under the control of the Corps of Engineers, the governmental agency officially in charge of waterways. In some places, the system failed because a levee or other barrier to water was not high enough, often because of unanticipated subsidence rather than original design error. In other places, the system failed because, though high enough, the barriers were not designed for the forces to which they were in fact subject (an unusually slow-moving storm). Design of floodwalls along three canals was "particularly inadequate." A series of incremental decisions between the original plan and the structures actually constructed "systematically increased the inherent risk in the system without recognition or acknowledgment" (IPET 2006, I-2). Many of the failures in the system would not have occurred had implementation of plans for reconstruction not been delayed for almost twenty-five years by inadequate funding, new laws governing the environment, and similar difficulties well beyond the control of engineers. For some important "decisions," there was no decision-maker at all. The decisions were a mere byproduct of poor communication, poor information, poor coordination, an unconscious preference for new construction over repair of old, or some combination of these.

The most important lesson IPET drew from its analysis is unsurprising: The way to avoid similar disasters is to use larger safety factors ("conservative design assumptions") and good materials ("higher quality, less erodible"; IPET 2006, I-3).

The flood control system now replacing the one that Katrina overwhelmed is considerably more expensive than the old one. For example, the Corps has been replacing the 5-meter pilings holding canal walls in place with pilings that go down *fifteen and a half* meters (more than three times as deep). The Corps agreed that the use of I-walls along the canals (without or even with the support of an earthen levee) was a mistake. It is replacing the canal's I-walls with heavily-braced T-walls locked down by 21-meter H-piles angled out in two directions. The use of simple sand or gravel levees was also judged a mistake. The Corps is now "armoring" all levees where they seem vulnerable to overtopping, that is, covering them with something water will not soak through or quickly wear away. These are expensive changes in design that the government was unwilling to pay for without a major disaster and may yet lose interest in paying for before the work is complete or the next time there is a need for maintenance.

7. CONCLUSIONS

We can, I think, distinguish four sorts of engineering decision in these four case studies: planning, designing, management, and operations.[7]

By *planning*, I mean such decisions as whether to build a nuclear power plant at all, where to put it, and the upper limit of its budget. For such decisions, engineers are most important for vetoing certain options, for example, a location because the risk of earthquake makes safe construction too expensive. Engineers are also important for suggesting alternatives, for example, conservation or a wind-powered system rather than a nuclear plant. Engineers are not (or, at least, should not be) mere "problem solvers." One important function they have is helping to define problems—or re-define them when it becomes clear that the client or employer has not asked all the right questions. Planning includes planning for accidents, natural disasters, and other nonroutine events.

For any large undertaking such as a nuclear power plant or flood control system, engineers are generally only one party in a complex social decision in which the other parties include employer, government officials, experts of various sorts (such as geologists or lawyers), bankers, financiers, and civil society (or "the public"). Perhaps the most important contribution engineers can make to planning is developing minimum standards for evaluating and responding to specific risks and benefits of the technology in question.

By *designing*, I mean the actual drafting of specifications, floor plans, and so on necessary to construct or modify the technological artifact in question. Once planning has set limits, engineers are generally free to work within those limits, for example, to design a nuclear plant that will fail slowly rather than quickly or cool rather than heat up if left alone. Only when a planning limit is too strict do engineers have a reason to restart the planning process, for example, by suggesting that the budget be raised to provide an adequate margin of safety.

By *management*, I mean overseeing the operations of a plant, including choosing, training, and directing operators. Much management, such as setting wages, is not technical—and is therefore not the domain of engineers. But, for nuclear power plants or flood control systems, the managers will typically be engineers. For engineers, part of technical management is remaining alert to possible improvements in staff, procedures, and equipment. So, for example, a manager who noticed that operators at Three Mile Island often missed readings on an important gauge because equipment blocked their view of it should recommend, or order, that the control board or control room be redesigned to improve the view.

By *operations*, I mean actually doing what is necessary for the plant or other technical artifact to work as designed or to respond well to unusual conditions as they occur. While engineers do not, in general, operate plants, they do constitute most of the operators in a nuclear power plant. So, for example, at Chernobyl, engineers pushed the buttons that moved dampening rods into the core. While operators can be reprimanded, and their acts reversed, they are, while acting as operators, completely in control of their machines.

One of the features we should note from our discussion of the three nuclear disasters is how quickly things can go wrong. What goes wrong in a nuclear plant does not, of course, go seriously wrong for just one reason. Because engineers typically design nuclear plants with a large safety factor (think of all the electrical power backups at Fukushima), several systems must fail before anything goes seriously wrong. But, given the complexity of a nuclear plant, it is reasonable to expect at least one system to fail now and then because, even with proper maintenance and inspection, technical systems sometimes fail unexpectedly. That being so, it is also reasonable to expect (given the laws of statistics) that all of the independent systems will fail together sooner or later. One of the "systems" that may fail at any given time is the human operator—whether because of distraction, fatigue, poor training, misjudgment, interference from someone higher up, or the like.

How likely is a catastrophic failure of a nuclear power plant at any moment? Not very likely. Perhaps only 10^{-5}. But over many years and many reactors even such a small risk adds up. One author recently calculated that there are:

> 450 nuclear power plants in the world. There have been 4 meltdowns in history, one each at Chernobyl and Three Mile Island and two so far at Fukushima, as partial meltdowns count as meltdowns. That is a ~1% failure rate. (Lindsay, 2011)

This calculation means nothing unless the four meltdowns are statistically significant, that is, a good predictor of what will happen over, say, the next hundred years (rather than a chance concurrence of events—like winning the lottery three days in a row). Still, it is an empirical reminder that even a low-probability event will, given a large enough population or enough time, become highly probable.

If we look at our four disasters, two—Three Mile Island and Chernobyl—seem unrelated to any ordinary planning or design failure. Of course, with a higher budget, the Three Mile Island plant might have had more working backups for its cooling system; Chernobyl might have had a concrete containment for its reactor or a better way of controlling core temperature. But that will always, or at least almost always, be true. Engineering is about making things "safe enough" rather than "absolutely safe."

How safe is "safe enough" is at least as much a social decision as an engineering decision. But it is an engineering decision in part. For small risks, engineers may well make the final decision. Even concerning the largest risks, engineers will be consulted and their opinion given considerable weight. No decision-maker wants to overrule the engineers on a matter of safety only to have the decision (more or less figuratively) blow up in her face.

Engineers generally evaluate risk by multiplying the harm's (net) disvalue by the harm's probability. This method of risk analysis works reasonably well for small harms. But it does not seem to work at all well for the largest harms—those that, even if highly improbable, would be intolerable if realized—such as destruction of the earth or even the sort of devastation that Chernobyl produced. For intolerable harms, engineers should, I think, adopt something like the following principle of prudence in planning: *If we (society at its most reasonable) would reject any plausible benefit in exchange for suffering that harm, we (that part of society making the decision) should, all else equal, rule out any design that risks that harm* (however small the probability—so long as it the probability is finite). Since this principle applies when we know both the harm in question and its probability, this is (technically) not a precautionary principle (though its spirit is much the same). It is, in this respect, more like advice frequently given to gamblers betting in games of chance with known odds ("Don't bet more than you can afford to lose" or "never bet the farm"). Precautionary principles are about dealing with uncertainty. (See, e.g., Andorno, 2004.)

The principle I am proposing is only about dealing with known probabilities. Yet, it is, or at least should be, an important principle in engineering. Failure is part of engineering. While engineers have a very low tolerance for failure of any kind, even in subsystems that are not "safety sensitive," I have yet to hear of any complicated system (even one as simple as a mechanical pencil) for which engineers have not calculated a failure rate (often, to be sure, a tiny failure rate, such as 3.4 defects per million—the famous Six Sigma). No product of engineering is (strictly speaking) "failure proof" (all things considered).

Most, perhaps all, nuclear power plants now in operation seem to have been built in violation of the planning principle suggested above (at least when the calculation of probability takes into account that humans will operate the plant). The analogy with gambling may not be altogether fair, however. For such activities as gambling, we always have the option of doing something much safer, such as going to the theater or buying government bonds. But for nuclear energy, our choices today are more difficult. Fossil-fuel plants together (though not individually) threaten us with a world too hot to live in. Hydro-electric dams are often not available as an alternative to nuclear power. When they are available, there is always a risk that they will fail, flooding lowlands when they do. Failing hydro-electric dams may have killed many more people than nuclear power-plant accidents have (depending on how deaths are calculated). Just one dam failure, that of the dam at Banqiao, China, in 1975, seems to have killed at least 26,000 people directly—and another 145,000 through resulting disease and famine (Wiki, "Banqiao"). Three Mile Island itself is only a hundred miles or so from Johnstown, Pennsylvania, the site of the "Johnstown Flood," which killed more than 2,200 people in 1889, the result of a dam failing (Wiki, "Johnston Flood"). In contrast, no one died at Three Mile Island and statistical deaths worldwide to be expected from the radiation that escaped is much smaller. Nowhere has wind and geothermal energy yet met the demand for electricity in an industrial country. And so on. Even with the radical conservation that Japan has instituted since the Fukushima disaster, there is, it seems, still a demand for electricity beyond what is available without some method of generating power that violates the principle of prudence in planning. For the time at least, we may face a choice among dangerous friends. We can only minimize the risk of catastrophic disaster, not avoid it.

There are two features that neither Fukushima nor Katrina share with Three Mile Island and Chernobyl: operator error and normal equipment failure. Equipment did fail at Fukushima and New Orleans—the diesel generators failed at Fukushima as did the screw pumps at New Orleans—but both these failed because of flooding, itself produced by a natural disaster (or, at least, overwhelming external events). There was nothing like the release valves at Three Mile Island that inexplicably failed to close.

Insofar, as there were managers or operators involved in the Fukushima or Katrina disaster, they seem to have prevented an even worse outcome.

What all four disasters have in common are failures of engineering design, that is, designs that could have been better. So, for example, the canals in New Orleans could have been designed with T-walls rather than I-walls; Fukushima could have had a breakwater high enough to block a ten-thousand-year tsunami; Chernobyl could have had a better design for its dampening rods; and Three Mile Island could have had a control board that took more account of human factors such as sight lines. And, of course, after these disasters, engineering designs made some of those improvements. Engineers generally learn from their failures. But such failures are, all else equal, present at every disaster. They do not help us to see what, if anything, is special about Fukushima.

For me, what is special about Fukushima compared with New Orleans is precisely what makes Fukushima like Three Mile Island and Chernobyl. The engineers, and their supporting staff, stayed with the machinery—monitoring, trying to prevent things from going further wrong, and even making repairs.

How many of (what the media called) "workers" at Fukushima were engineers? I have been unable to determine that either from news sources or from contacts in Japan. But my one visit to a nuclear plant in the United States suggests that most of those working at Fukushima would have been engineers (say, 90%)—with the remainder divided about evenly between scientists and technicians. (I hope someone will correct me if I am wrong about that.)

The engineers at Fukushima were not as successful as the engineers at Three Mile Island and Chernobyl in limiting the disaster. Both those disasters were limited to one reactor. At Fukushima, the disaster spread to four of the six reactors—and might have spread to the other two as well but for the restarting of a diesel generator at Reactor 6 to provide power for cooling the fuel in the holding pools of Reactors 5 and 6. Workers also removed roofing from Reactors 5 and 6 to allow hydrogen to escape, thus preventing explosions similar to those that had damaged the other four units.

This aspect of what happened at Fukushima is a reminder that part of what makes engineering so reliable is that engineers design with the (usually) justified expectation that other engineers will always be present to look after what they design. The works of engineering, even of civil or mechanical engineering, do not last long without continual maintenance, including continual adjustments as experience identifies unanticipated problems or unanticipated opportunities for positive improvement. The engineering experiment at Chernobyl, despite its disastrous outcome, was part of normal engineering. The engineers were trying to reduce the risks arising from the backup system's slow startup. Even nuclear plants that are identical when commissioned, slowly differentiate as they operate, because the engineers managing a plant will continually make improvements. Those engineers should, of course, let engineers at similar plants know about the changes, thus advancing the state of the art, but other engineers may not be able to make the necessary changes immediately because of budget or schedule, or make the necessary changes at all because changes that they have made already bar the improvement in question. Engineers may also find an alternative way to achieve the same end. For these reasons (and perhaps others), nuclear plants, however, alike at birth, tend to grow into noticeably different individuals, much as biological plants do. That is another reason it is important to have engineers on site. Those who know the idiosyncrasies are more likely to be useful in an emergency.

Some people, especially philosophers, seem to think of those who stayed on at Fukushima—those who, for example, worked in the dark in cold waist-high radioactive water to restart the generators—as engaged in "supererogatory" conduct, that is, as engaged in conduct "above and beyond" what morality requires. The engineers I have discussed this with (American as well as Japanese) seem to view the conduct of Fukushima's workers as heroic *but required* (supposing the "workers" in question to be engineers). An engineer who left when needed would have acted unprofessionally; he would have failed as an engineer even if he left to save his life or look after his family. Engineering sometimes requires heroism (a significantly higher standard than proposed in Alpern, 1983, or Harris, Prichard, and Rabins, 2000)—or so the engineers I have talked with about this seem to think—whether Japanese or American. That shared higher standard is more evidence that engineering is a global profession.

NOTES

This chapter has benefited from discussion of it at: a workshop for philosophy graduate students at the Technical University-Delft, The Netherlands, May 11, 2011 ("The engineer, public safety, and economic constraints"); a seminar for the Department of Philosophy and Ethics, the Technical University-Eindhoven, The Netherlands, May 13, 2011 ("The Fukushima Nuclear Disaster: Reflections"); a talk for the Department of Philosophy and Religion, University of North Texas, Denton, October 13, 2011 ("The Fukushima Nuclear Disaster: Some Issues of Engineering Ethics"); a plenary session of the Sixth International Conference on Applied Ethics, Hokkaido University, Sapporo, Japan, October 30, 2011; and the Annual Meeting of the Association for Practical and Professional Ethics, Cincinnati, Ohio, March 3, 2012; as well as from the comments of several reviewers of the journal in which it was originally published as "Three Nuclear Disasters and a Hurricane: some reflections on engineering ethics," *Journal of Applied Ethics and Philosophy* 4 (September 2012): 1–10.

1. All information about Chernobyl here and below is drawn from Wiki, "Chernobyl," a source valuable both because it is easily accessed and regularly updated—something not true of scholarly articles or books or of most news media.
2. The discussion of Fukushima below relies not only on Wiki, "Fukushima Daiichi" but also on Wiki, "Fukushima I." Though I shall hereafter refer to "Fukushima", it is in fact Fukushima I (Fukushima Dai-ichi) that I shall be referring to. There is also a Fukushima II (Fukushima Dai-ni). For details about that plant, see Wiki, "Fukushima II."
3. Except where otherwise indicated, the discussion of Three Mile Island in this chapter relies for its facts on Wiki, "Three Mile Island."
4. Rogovin 1980, 89–93, report focused mainly on changes in emphasis and procedures at the NRC; Kemeny 1979, 61–73, report focused on "attitudes and practices."
5. See http://en.wikipedia.org/wiki/List_of_earthquakes_in_Japan (accessed April 25, 2011).
6. Except as otherwise indicated, all information in this section comes from Davis (2007).
7. In a different context, I would include "disposal" in this list. I do not include it here only because none of these disasters concerns disposal as such, though Fukushima's problems were due, in part, to fuel rods awaiting disposal.

REFERENCES

Alpern, Ken, "Moral Responsibility for Engineers," *Business and Professional Ethics Journal* 2 (Winter 1983): 39–48.
Andorno, Roberto, "The Precautionary Principle: A New Legal Standard for a Technological Age," *Journal of International Biotechnology Law* 1 (2004): 11–19.
Davis, Michael, "Imaginary Cases in Ethics: A Critique," *International Journal of Applied Philosophy* 26 (Spring 2012): 1–17.
Davis, Michael, "Perils of Katrina: Using that Current Event to Teach Engineering Ethics," *IEEE Technology and Society Magazine* 26 (December 2007): 16–22.
Davis, Michael, "Do the Professional Ethics of Chemists and Engineers Differ?" *HYLE* 8 (Spring 2002): 21–34.
Davis, Michael, *Thinking like an Engineer* (Oxford University Press: New York, 1998).
Harris, Charles E., Michael S. Pritchard, and Michael J. Rabins, *Engineering Ethics: Concepts and Cases* (Wadsworth: Belmont, California, 2000).
Kemeny, John G. *Report of The President's Commission on the Accident at Three Mile Island: The Need for Change: The Legacy of TMI* (Washington, DC: The Commission, October 1979) http://www.threemileisland.org/downloads/188.pdf (accessed April 18, 2012).
Lindsay, Robert, "1% Failure Rate For Nuclear Power," http://robertlindsay.word press.com/2011/03/23/1-failure-rate-for-nuclear-power (accessed April 19, 2012).
Interagency Performance Evaluation Task Force, *Performance Evaluation of the New Orleans and Southeast Louisiana Hurricane Protection System: Draft Final Report of the Interagency Performance Evaluation Task Force* (June 1, 2006), http://per manent.access.gpo.gov/lps71007/ (accessed December 5, 2011).
Perrow, Charles, *Normal Accidents: Living with High-Risk Technologies* (Basic Books, NY, 1984).
Rogovin, Mitchell, *Three Mile Island: A report to the Commissioners and to the Public, Volume I.* Nuclear Regulatory Commission, Special Inquiry Group, 1980, http://www.threemileisland.org/downloads/354.pdf (accessed April 18, 2012).
Wiki, "Banqiao Dam" http://en.wikipedia.org/wiki/Banqiao_Dam (accessed May 5, 2012).
Wiki, "Fukushima Daiichi Nuclear Disaster" http://en.wikipedia.org/wiki/Fukushima _Daiichi_nuclear_disaster (accessed April 19, 2012).
Wiki, "Fukushima Daiichi Nuclear Power Plant" http://en.wikipedia.org/wiki/Fukush ima_I_Nuclear_Power_Plant (accessed April 19, 2012)
Wiki, "Fukushima II Nuclear Power Plant" http://en.wikipedia.org/wiki/Fukushima_I I_Nuclear_Power_Plant (accessed April 18, 2012).
Wiki, "Johnstown Flood" http://en.wikipedia.org/wiki/Johnstown_Flood (accessed May 5, 2012)
Wiki, "Three Mile Island Accident" http://en.wikipedia.org/wiki/Three_Mile_Islan d_accident (accessed April 22, 2011).
Wikipedia, "Chernobyl disaster" http://en.wikipedia.org/wiki/Chernobyl_disaster (accessed December 16, 2011).

Chapter 18

Ethical Issues in the Global Arms Industry

A Role for Engineers

This chapter has four sections. The first two seek to clarify its subject, ethical issues in the global arms industry. The third sketches the global role engineers have in much of the global arms industry. The last section considers one way that engineers might use that global role to help resolve some of the industry's ethical issues. While the first section of this chapter should contain few surprises, the last three will, I hope, contain more, providing evidence of the power engineers actually have to help shape public policy worldwide.

1. DILEMMAS AND DEFENSE

Let me begin with two differences between what this subject is usually called—"ethical dilemmas in the global defense industry"—and the title of this chapter. First, I have substituted "issues" for "dilemmas." Second, I have substituted "arms" for "defense." The purpose of these changes is to avoid unnecessary disputes rather than to change the subject. Let me explain.

A "dilemma" is a situation in which a difficult choice must be made between just two equally undesirable alternatives. (If there are three, and only three, hard choices, it is a trilemma; if four, and only four, hard choices, a quadrilemma; and so on.) This is the typical dictionary definition of "dilemma," when the term is not simply wasted as a synonym for "hard choice."[1]

If the alternatives were not *equally* undesirable, the choice would be easy: choose the more desirable alternative. There would be no dilemma (though the choice might, like most good choices, still have its cost). My impression is that the main ethical issues, questions, problems, quandaries, or whatever posed by the global arms industry are not dilemmas (in this sense) but complex situations in which most of the choices on offer are hard to assess and many of the best choices have yet to be devised. Indeed, many of the issues, questions, problems, quandaries, or whatever are so ill-defined that we cannot say what the criteria for a good choice are. We are dealing with a subject requiring the work philosophers typically do. We must understand the issues before we can have anything as tidy as a dilemma. Hence, my substitution of "issue" for "dilemma." I might have used "problem," "question," "quandary," or another catch-all instead of "issue."

My substitution of "arms" for "defense" has a different rationale. I regard "defense" in such terms as "Defense Department," "defense forces," and "defensive weapons," as a misleading euphemism. In the United States, "the Defense Department" is the department of the federal government that oversees war-making, offensive as well as defensive. In most countries that call their military "Defense Forces," the military still consists of an army, air force, navy, and so on, all of which can, and sometimes do, engage in offensive warfare. The Defense Department might more accurately be called (as it once was) "the War Department"; the defense forces, "the armed forces."

Much the same is true of weapons. Few, if any, are purely defensive. Even a shield, that epitome of defense, can be used offensively, for example, to bash the face of an opponent too focused on one's sword arm. Rather than try to sort out whether a particular piece of equipment, say, an anti-aircraft missile or landmine, is an offensive or defensive weapon (or both), I have substituted "arms" for "defense."

By "arms industry," I mean all those organizations, whether commercial or not, that design, build, sell, or service weapons or related equipment for military use, provide military research, training, or advice, or otherwise aid the military. The military is that technological system—a combination of people and things—the purpose of which is to kill on a large scale. Though the military is typically an arm of government, it can also be an arm of a nongovernmental agent, such as a business corporation ("private army"), religious organization (the Knights Templar or the warrior monks of ancient Japan), or political movement (Shining Path or Khmer Rouge).

"Arms industry" (as used here) does not include the design, construction, sale, or servicing of weapons for nonmilitary use, whether use by police, hobbyists, or civilians intent on self-defense or mayhem, even if the nonmilitary weapon is indistinguishable from its military counterpart and manufactured in the same factory. So, while the manufacturer of ordinary bandages or backpackers' dinners is not, as such, part of the arms industry, the manufacturer of field dressings or combat rations is. Products of the arms industry include aircraft, ammunition, artillery, backpacks, electronic systems, light weapons, operations support, research, software, and uniforms.

While much of the arms industry is "domestic," that is, serves the "home country," a significant part is "international" or "global," that is, serves the military of other countries, rebellions outside the home country, or other foreign military forces. Our subject is the *global* arms industry, that is, that part of the arms industry that is not domestic.

The distinction between the global arms industry and the domestic arms industry is, of course, artificial in at least two respects. First, much of the domestic arms industry may seek foreign customers if they have a product that they can sell abroad (and permission of their home government for such sales). Foreign sales can reduce the unit cost of a product, help tie foreign customers to the home country, and otherwise serve domestic interests. Much of the global arms industry is, in this respect, also part of the domestic arms industry.

Second, much of what even a strictly domestic arms industry produces depends on raw materials, research, or subsystems produced outside the home country. Even a domestic arms industry must rely on international trade to provide much of what the military of the home country needs, everything from rubber to rare earths, from computer chips to Kevlar. Much of the domestic arms industry is global in this respect—and has been for at least a century.

The distinction between the global arms industry and the domestic is nonetheless worth making. The patriotism that may justify producing arms for one's own country cannot justify producing arms for others, especially those not allied with the home country. The ethical issues of the global arms industry seem to differ in systematic ways from those of the domestic arms industry.

We must now turn to the ethical issues of the global arms industry.

2. ETHICAL ISSUES

I shall continue to reserve "ethics" for the special-standards sense, using "morality" for the first sense of "ethics" and "moral theory" for the third. Given this familiar terminology, some "ethical issues" typically identified as raised by the global arms industry seem in part moral (whether or not they are also ethical, that is, whether or not they concern an existing special standard or might lead to the adoption of a new one). For example, the threat military drones pose to people's privacy is (in part) a moral issue. Every moral agent, even the agents of the global arms industry, should, all else equal, avoid contributing to the invasion of people's privacy. Other issues, such as what to do about government officials who expect some form of *quid pro quo* for their cooperation, though ethical issues for most of the global arms industry, are no longer difficult issues. Most of the global arms industry have long since adopted a special standard resolving them. So, for example, the National

Defense Industrial Association's "Statement of Defense Industry Ethics" says that:

> When contemplating any international sale to a governmental or quasi-governmental buyer, it is imperative that effective measures be undertaken to ensure full compliance, not only with the letter, but also the spirit of the Foreign Corrupt Practices Act, as amended, and the FCPA's bar against improper payments to foreign officials. (*National Defense*, 2011)

Of course, we can debate whether such special standards are wise, morally required, merely morally permitted, or even morally wrong. But that debate is likely to do little more than return us to ground that business ethics (the philosophical study) has exhausted during the last forty years.[2] I therefore propose to limit this chapter to moral issues that the global arms industry faces while ordinary businesses do not (or, at least, do not face in the same way), issues never much discussed in business ethics.[3] I have identified six. No doubt there are others.

1. Weapons versus non-weapons. Much of what the global arms industry sells are weapons (artifacts designed to kill, wound, disable, or destroy) but much is not. For example, much of what the global arms industry sells consists of clothing, field kitchens, tents, medications, and so on, artifacts harmless in themselves even in military service. And some of what the global arms industry sells is neither clearly a weapon nor clearly not a weapon, for example, body armor, observation drones, communications equipment, circuit boards, reflective paint, software, and other nonlethal elements of a "weapons system." How morally significant, then, is the distinction between weapons and non-weapons? Should the global arms industry consider the sale of non-weapons less morally objectionable than the sale of weapons or nonlethal elements of a weapons system? After all, almost every artifact embedded in the technological system we call "the military" is there to help the military do its job, which is (in part at least) to kill other human beings on a large scale, a morally dubious undertaking, especially if the regime directing the military is itself morally dubious. (A morally dubious undertaking can, of course, turn out, all things considered, to be morally justified, but the burden of proof must fall on those claiming justification.)
2. Morally dubious regimes. Some customers of the global arms industry respect human rights but most, to varying degrees, do not. Of those that do not, some may simply deny their people certain basic rights, such as self-government or decent medical care, but many actively harm those under their control by, for example, imprisoning, torturing, or killing

them for having certain political, religious, or other beliefs, for forming various kinds of peaceful voluntary associations, or for speaking a certain language, dressing in a certain way, or the like. How abusive must a regime be before the global arms industry should refuse to sell it weapons? How abusive before the global arms industry should have no dealings with it at all? How important is the argument that if "we" do not sell to them, others will?

3. Cultural differences. By "culture," I continue to mean a distinctive way of doing something, including the beliefs and evaluations that accompany the doing. There are military cultures (distinct from the civilian culture of the home country). For example, some militaries "live off the land" on which they fight, while others routinely bring most of their supplies with them and purchase the rest. Some routinely take prisoners; some do not. Some militaries force young men to serve while others take only volunteers. Some allow "children" (adolescents under eighteen) to be soldiers; some do not. How important should such cultural differences be to the global arms industry when deciding whether to take on a certain customer? Should international standards preempt noncomplying military cultures?

4. Lawful artifacts having illegal uses. Most weapons have illegal uses as well as legal ones. For example, the same rifle that is legally used to kill enemy soldiers in combat can be used illegally to kill soldiers who have given up their weapons and clearly surrendered. Something similar is true of many non-weapons. For example, the same small electric generator that can lawfully be used to power field radar can be used illegally to deliver electric shocks to a prisoner's genitals. How important should the likelihood of illegal use of military equipment be to the decision to sell the equipment to a certain customer? Should military equipment be designed, as much as possible, to prevent illegal use? Should the contract of sale expressly rule out such illegal uses?

5. Weapons likely to fall into the wrong hands. Much of what the arms industry sells can be stolen, resold, transferred to another by capture, or otherwise "diverted." How much responsibility should the global arms industry take for preventing its products falling into the wrong hands? For example, should the global arms industry refuse to sell to unstable regimes (a regime likely to lose control of its military soon) or regimes with a record of losing many of its weapons to its non-state enemies (a regime such as that currently ruling Iraq)? Should the products of the global arms industry be designed to make diversion of its products more difficult or less attractive (e.g., by making rifles requiring unusual bullets or hard to replace parts)?

6. Relatively indiscriminate weapons. Some weapons are relatively indiscriminate, even when used by a sophisticated military. For example,

landmines can as easily be set off by a civilian as by a soldier and even the U.S. military can fail to retrieve all its mines when it departs a combat zone. Landmines may go on killing and maiming civilians for decades after the end of the war that justified their use. Something similar is true of conventional bombs (though to a lesser degree). Lost "duds" can explode long after the end of hostilities, killing anyone who happens to be nearby. While some weapons are relatively indiscriminate even in sophisticated hands, some are indiscriminate only in unsophisticated hands. For example, without good record keeping, a military may lose track of the age of artillery shells. Past-date shells may explode when they should not, say, when being trucked down a rough road or even when sitting in a warehouse. How much care should the global arms industry take to make weapons as discriminating as practical in the circumstances in which they are likely to be used?

The classic indiscriminate weapons are, of course, nuclear bombs, biological devices, and deadly gases, weapons that, I believe, are not currently part of the official global arms trade. I shall therefore ignore them here. (For more on these, see Richardson, 2014)

3. ENGINEERS IN THE GLOBAL ARMS INDUSTRY

Engineers have had a significant role in the arms industry since at least the 1700s. Their role has only increased as the products of the arms industry have become more sophisticated. At the beginning of the twenty-first century, one in ten U.S. engineers works in a military-related industry, including about 39,000 electrical engineers (just under 14% of all U.S. electrical engineers) and about 6,000 aerospace engineers (just under 19% of all aerospace engineers) (U.S. Chamber of Commerce, 2007, 8; Department of Defense, 2011, 4). Engineers design weapons and other equipment the military needs, test them, sell them, and oversee their manufacture, delivery, maintenance, and even disposal. Indeed, it is hard to imagine today's arms industry without engineers, not only "bench engineers" but technical managers up to, and often including senior management.[4] So, for example, of Lockheed Martin's eight vice presidents, three are engineers.[5] There is no reason to think that engineers do not have a similar part with respect to most products of the global arms industry or, at least, most of its most distinctive products.

Suppose, for example, that a certain large African country contacts a U.S. manufacturer of modern jet fighters to buy twenty for its air force. The sale is likely to be a long process, lasting months or even years. At an early stage, the U.S. manufacturer would have to send out engineers to assess the would-be customer's airbases, maintenance practices, pilot training, local suppliers, and so on. A jet fighter requires a complex technological system to operate. The would-be customer may be surprised to learn that its runways are too short, that its fuel storage is inadequate, that its maintenance staff will have to be larger, better trained, and provided with more sophisticated tools, and so on. While some of this information is typically public, some is not, being proprietary or classified. Much of it will, in any case, be in a form that engineers are used to, and others are not. The customer will need its own engineers to talk to those of the U.S. manufacturer.

The African country need not agree to all the requirements that the U.S. manufacturer seeks to impose as part of the sale. It may suggest changes in the design of the jet fighters so that, for example, they can use fuel that the customer is already using for other aircraft. Indeed, after a full assessment, the parties may agree on a less sophisticated fighter or simply go their separate ways. But, if they do make a deal, the final specifications for the fighter, including training, support, munitions, replacement parts, and so on, should be the result (in part) of extensive negotiations between the engineers of the U.S. manufacturer and those of the African country. (The parenthetical "in part" recognizes the role that the U.S. government would normally have in such a sale.) Though the terms of such a sale are, in principle, entirely under the control of the U.S. manufacturer's senior management and the African country's senior governmental officials, in practice many of the decisions, perhaps most, will be made by engineers, some quite junior, no one else having the information, time, and skills to appreciate their import.

The involvement of engineers typically does not end with the writing of specifications or even with the signing of the sales contract. Engineers will oversee the manufacture of the planes, not only making sure that every part satisfies the specifications and the whole is constructed properly but also changing the specifications if, say, there is difficulty getting a specified part or a better part has become available. Given that there will typically be several years between the initial writing of specifications and the delivery of the last jet fighter, there may be many changes in the specifications, most quietly made by agreement among engineers. Some of these changes will, of course, be "no brainers," but a substantial number may involve painful balancing of cost, reliability, timeliness, and so on. So, for example, a new part may be cheaper and, based on experience with similar parts, as good as the old. But, since the part is new, experience with it must be short. The part may fail long before it should. Who knows? The engineers will have to rely on experience with parts analogous in one way or another to forecast the probable failure date of the new part—and decide accordingly. There may be a good deal of discussion between the manufacturer's engineers and the customer's.

The relationship between the engineers of the U.S. manufacturer and customer's should not end when the last fighter is delivered. The U.S. engineers should keep the customer's engineers informed of problems identified in similar aircraft elsewhere in the world and the solutions devised. The customer's engineers should in turn advise the U.S. engineers of any problems they identify in the jets that they purchased, anything from unusual wear on engine blades to difficulty getting ground crews to comply with required maintenance procedures. The purpose of this exchange of technical information between the manufacturer's engineers and those of the African country is not simply to maintain the fighters; it is in part to improve them where possible, not only the fighters that the African country has purchased but other fighters in that family, both those yet to be built and those already in use elsewhere in the world. In principle, this exchange of information should continue until the last fighter delivered has ceased to exist. That is normal engineering—in a "globalized" world.

While much of this exchange of information will go on long-distance, some of it may require "site visits," for example, to see the dust clouds possibly contributing to unusual engine wear or the conditions under which maintenance must actually be performed.

The relationship between a manufacturer's engineers and those of a customer can be both intimate and enduring. There is often a tension between the legal department's "arm's length" conception of how information should be shared and the engineers' conception (something more like a long hug than a handshake). For example, engineers of a manufacturer can seldom do a good job of designing a sophisticated piece of equipment without knowing how it will be used, under what conditions, and for how long. Similarly, a customer purchasing such equipment cannot be as helpful in its design as it could be unless it knows the details of manufacture, including some trade secrets and (in the case of a fighter jet) even some highly classified information.

4. HOW ENGINEERS MIGHT HELP WITH SOME ETHICAL ISSUES

Most engineers working in the global arms industry are civilians. Most who are not have nonetheless been trained in the same way as civilian engineers, work in much the same way as civilian engineers, and have little trouble communicating with civilian engineers. Engineering is (in this respect at least) a single profession. It is also a global profession. Engineers in Brazil, China, Nigeria, or India are trained much as are engineers in Germany, Japan, or the United States. Engineers also share certain standards, whether formalized in a code of ethics or not. They are committed not simply to maintaining technology but to improving it for the benefit of humanity. Their first loyalty is (or, at least, is supposed to be) not to their employer or government but to "the *public* health, safety, and welfare."

Much of what engineers share are technical standards. Some of these are governmental, such as the standards of safety issued by the Environmental Protection Agency (EPA) or Nuclear Regulatory Commission (NRC). But (as noted before) many technical standards, perhaps most, are not the work of government. Of these, some are the work of technical associations, such as ASME or IEEE. Others are the work of trade associations or other private groups, the best known of which today is probably ISO.

Whatever the official source of engineering's technical standards, they will, in large part, be the work of engineers. They will be the work of engineers because only engineers have at hand the knowledge necessary to write them. The standards are not deduced from physics, chemistry, or any other natural science; nor are they simply common sense (though generally consistent with common sense). They are instead typically a product of engineering experience. Some of that experience derives from laboratory experiments, much like the experiments of natural science. The chief difference between the experiments of natural science and those of engineering (insofar as there is any) is that engineers typically experiment on human artifacts, not natural objects. However, much of the engineering experience on which the writing of standards depends will not be experimental but "field experience," that is, experience of artifacts in use where the control necessary for (laboratory) experiment is absent, for example, when the left wing of a fighter jet falls off at twenty-thousand feet during combat training. Engineers try to learn as much as possible from every such unhappy experience. Unlike surgeons in the old joke, engineers do not bury their mistakes. Instead, they record them, study them, and try to learn from them, typically embedding what they learn in new technical standards.

The recent history of the global arms industry offers enough examples of unhappy experiences with the products of engineering, such as the many children killed or maimed by landmines in peace time, for engineers to begin to develop international standards for the global arms industry like engineering's other international standards. There are even a few signs that now is a good time to begin developing such standards. I shall briefly describe three of those signs. Whatever they may mean individually, they seem to me together to constitute movement in the right direction.

First, there is a U.S. statute, the Arms Export Control Act, as well as the International Traffic in Arms Regulations (ITAR) issued under that Act. Since 1976, these have governed what military information and artifacts may be shared with "non-US persons." U.S. persons (including organizations) can face heavy fines if they have, without authorization or the use of an exemption, provided non-U.S. persons with access to ITAR-protected military articles, services, or technical data. Until the end of the Cold War, the focus of ITAR enforcement was to prevent the Soviet Union from obtaining U.S. military technology. Since 1990, the focus has increasingly become preventing both weapons and weapons technology falling into the wrong hands, especially the hands of terrorists or rogue states (Wikipedia, 2018).

Second, in 2004, the National Defense Industry Association (NDIA), a U.S. trade association, published a "Statement of Defense Industry Ethics," making several small revisions in 2009. Most of the larger U.S. participants in the global arms industry have adopted codes of ethics including provisions like those in the Statement. The Statement seems to reject making ethics relative to geographical or even military cultures: the arms industry is to "[i]mplement effective ethics programs for company activities at home or abroad." The Statement is, however, almost silent about the health, safety, and welfare of people outside the United States. The nearest the Statement comes to providing any guidance on that issue is the requirement that members of the arms industry "[e]stablish corporate integrity as a business asset, rather than a requirement to satisfy regulators, by making ethics compliance integral to all aspects of corporate life and culture, including employee appraisals and promotions, to foster an environment where employees aspire to do the right thing." For engineers at least, "the right thing" seems to include considering, for example, the welfare of non-U.S. children whom landmines might kill or injure. Such children, being unable to protect themselves against the danger, are part of the "public" whose safety, health, and welfare engineers are (according to most codes of engineering ethics) supposed to "hold paramount" (NSPE, 2007).

More important, the Statement does not treat ethical knowledge as proprietary. Instead, it urges members of the arms industry to "[c]ontribute to the common good of our industry and promote industry ethics whenever and wherever possible (a) by sharing best practices in ethics and business conduct among NDIA members and (b) by including ethics training in NDIA sponsored events" (National Defense, 2011).

Third, and most timely, there is an initiative of the United Nations: the Arms Trade Treaty. Since its signing on December 24, 2014, eighty-nine states have ratified it (with forty more signing; UN, 2018). The Treaty is an important step in regulating the global arms industry. It certainly provides a starting point for writing global standards for engineers.

The Treaty is (1) to "establish the highest possible common international standards for regulating or improving the regulation of the international trade in conventional arms" and (2) to "prevent and eradicate the illicit trade in conventional arms and prevent their diversion." The Treaty applies to all conventional arms within the following categories: battle tanks; armored combat vehicles; large-caliber artillery systems; combat aircraft; attack helicopters;

warships; missiles and missile launchers; and small arms and light weapons (UNODA, 2014, Art. 1 and Art. 2). The Treaty seems to cover nonlethal *parts* of weapon systems, such as radar or observation drones. However, it does not seem to cover other nonlethal equipment, such as trucks, transport aircraft, body armor, or field kitchens.

How do these three documents support the use of engineering's standards, especially its international standards, to enforce standards for the global arms industry that might help to resolve the ethical issues identified in section 2 of this chapter? Let me give a simple example: if there were an international standard prohibiting engineers from participating in the sale of complete weapons or parts of weapons of any sort to a regime likely to misuse them, the standard would simply echo the Treaty. If, in addition, the engineering standards contained criteria for identifying weapons likely to be misused and the sort of regime likely to misuse them, engineers might then inform an employer considering sale of such weapons or parts of such weapons to such a regime that the sale not only violates international standards but also is inconsistent with good engineering. Engineers can have no part in such a sale. Involvement would be unprofessional.

A source of engineering's standards, such as IEEE or ISO, would have a justification for issuing such a standard. Article 7 of the Arms Trade Treaty specifically requires a signatory State considering licensing an export to "assess the potential that the conventional arms or items [in question]" may be misused in various ways, for example, to "commit or facilitate a serious violation of international humanitarian law . . . [or] human rights law." While the Treaty's authors probably thought of the decision to license as primarily governmental, there is nothing in that understanding to forbid a member of the global arms industry from deciding not to seek its government's permission, nor is there anything in that understanding to forbid engineers working in the global arms industry from appealing to their own profession's ethical standards when asked to participate in such a transaction. Their employer is (according to the NDIA Statement) supposed to ask engineers to "do the right thing." Standing by (morally justified) engineering standards is doing just that.

Of course, engineers individually are not qualified to assess the likelihood that a particular regime will misuse a particular weapon, even though they are likely to know much about how the weapon can be misused. So, any standard developed for the use of engineers would have to include the sort of information that an engineer would need to make a competent assessment. That information might come in a quite simple form, for example, in the form of a checklist asking (among other things) how this or that human rights group rates the regime, what uses the regime has made of weapons in the recent past, whether it has signed the relevant treaties, and so on. An individual engineer could then inform the appropriate superior, "We need to check out the following to be sure that this sale meets international engineering standards."

This sort of individual response may not seem much help with the ethical issues identified in section 2. After all, the engineer's superior might simply ignore the international standard and replace an engineer unwilling to participate in the sale with an engineer who is willing or, if no suitable engineer is willing, with a willing non-engineer.

While it is true that a superior might respond in that way, there is good reason to think that such a response is, all things considered, unlikely. Such a response can have substantial costs, especially when the manager most needs an engineer. There are at least three sources of that cost: First, engineers are not indefinitely interchangeable. Indeed, they are often quite specialized. The engineer first asked to participate in the sale is likely to be the most qualified ("most up to speed"). The replacement (assuming one can be found) is likely to be less qualified for that work (even if just as qualified otherwise). Therefore, the substitution may increase the risk of bad decisions as the sale progresses. Second, the risk of bad decisions is even higher if the substitute for the engineer is a non-engineer. Engineers are generally brought into sales only when they are needed, only when they are likely to have knowledge or insight non-engineers do not have. Third, overruling an engineer on a matter involving application of an engineering standard risks harm to the manager. If anything later goes wrong, the manager who overruled the engineer will be open to blame, even if he found another engineer or a "scientific expert" to replace the unwilling engineer. He was on notice that there might be a problem and he did not "do the right thing." If, on the other hand, he goes along with the engineer's recommendation, he can at least claim that he was acting on the best technical advice available.

These are all relatively short-term consequences of one manager's respecting or not respecting the engineering standard in question. There is also at least one long-term consequence worth considering if the organization makes a practice of overruling engineers on such issues. Widespread lack of respect for engineering standards may have a bad effect on the morale of the organization's engineers generally and so, on the ability of the organization to recruit and keep the most marketable engineers, not only the most marketable "bench engineers" but also the most marketable higher-ranking engineers (including senior managers).

We have, of course, been assuming that the engineer's superior is unsympathetic to the appeal to engineering standards. That is a worst-case scenario. In practice, the superior is likely to be another engineer, one for whom engineering standards carry considerable weight even if she is now acting as a manager rather than just an engineer. And the organization in which these two engineers work is likely to have its own code of ethics, compliance procedures, and the like designed (as the NDIA Statement requires) to ensure, as much as possible, that organization employees, including engineers, "do the right thing." The ethical environment of the organization is likely to be far friendlier to engineering standards than we have tacitly been assuming in dealing with this example (even if that tacit assumption was justified not so long ago; compare Harris et al., 2014).

This is, admittedly, a relatively simple example of a standard that might be adopted, one that does not look particularly technical. The standards actually adopted—for example, criteria for "safe landmines" requiring them to resist light touches, to disarm automatically after a certain period, and so on—are likely to look much more technical, making the overruling of the engineer look even more risky.

NOTES

I presented the first version of this chapter to the Philosophy Colloquium, Illinois Institute of Technology, March 6, 2015, and a second version to a plenary session of the Ethical Dilemmas in the Global Defense Industry Conference, Center for Ethics and the Rule of Law, University of Pennsylvania Law School, April 16, 2015. I would like to thank participants in one or the other event for their encouragement and several improvements in this chapter. Originally published as:

"Ethical Issues in the Global Arms Industry: A Role for Engineers," *Ethical Dilemmas in the Global Defense Industry*, edited by Claire Finkelstein and Kevin Govern (Oxford University Press, forthcoming).

1. Until the last few decades, philosophers have had a much-more-precise definition of "dilemma," that is, as an inference having the following form: P v Q, P—> R, Q—>R, therefore R. I regret the eclipse of that technical sense.

2. That is, since the Lockheed bribery scandal of the early 1970s. Wikipedia, 2018b. For recent work on bribery in global business ethics, see, for example, Cleveland, Favo, and Frecka (2009); or the initial discussion and references in Byrne (2014).

3. By "not much discussed in business ethics," I actually mean "virtually undiscussed." The following are the only discussions I have found: Maitland, 1998; Byrne, 2007; and Halpern and Snider, 2012.

4. For a fuller discussion of the importance of engineers to the arms industry, see Fichtelberg, 2006.

5. See biographies of: Patrick M. Dewar, Executive VP; Dale Bennett, VP for Mission Systems and Training; Richard F. Ambrose, VP for Space Systems, http://www.lockheedmartin.com/us/who-we-are/leadership.html (accessed December 30, 2014).

REFERENCES

Byrne, Edmund F., "Towards Enforceable Bans on Illicit Businesses: From Moral Relativism to Human Rights," *Journal of Business Ethics* (2014): 119–130.

Byrne, Edmund F., "Assessing Arms Makers Corporate Social Responsibility," *Journal of Business Ethics* 74 (2007): 201–217.

Cleveland, Margo, Christopher M. Favo, and Thomas J. Frecka, "Trends in the International Fight Against Bribery and Corruption," *Journal of Business Ethics* 90 (2009), Supplement: 199–244.

Department of Defense, "Defense-Related Employment of Skilled Labor: An Introduction to LDEPPS" (March 2011), www.economics.osd.mil/LDEPPS_Primer.pdf (accessed December 30, 2014).

Aaron Fichtelberg, "Applying the Rules of Just War Theory to Engineers in the Arms Industry," *Science and Engineering Ethics* 12 (2006): 685–700.

Halpern Barton H. and Keith F. Snider, "Products That Kill and Corporate Social Responsibility: The Case of US Defense Firms," *Armed Forces & Society* 38 (2012): 604–624.

Harris, Charles E. et al., *Engineering Ethics: Concepts and Cases* (Wadsworth: Boston, MA, 2014), pp. 228–229, "Case 14: Halting a Dangerous Project."

Lockheed Martin (2014), *Setting the Standard: Code of Ethics and Business Conduct*, http://www.lockheedmartin.com/us/who-we-are/ethics/code-of-ethics.html (accessed December 28).

Maitland, Gavin, "The Ethics of the International Arms Trade," *Business Ethics: A European Review* 7 (October 1998): 200–204.

National Defense (2011), http://www.nationaldefensemagazine.org/articles/2011/3/1/2011march-statement-of-defense-industry-ethics (accessed January 18, 2018)

Richardson, Jacque G., "The bane of 'inhumane' weapons and overkill: An overview on increasingly lethal arms and the inadequacy of regulatory controls," *Science and Engineering Ethics* 10 (2004): 667–692.

US Chamber of Commerce, *Defense Trade: Keeping America Secure and Competitive* (March 2007), www.uschamber.com/sites/default/files/legacy/issues/defense/files/defensetrade.pdf (accessed December 30, 2014).

United Nations, "ATT Status Report," 2018, https://www.un.org/disarmament/convarms/att/ (accessed January 18, 2018).

United Nations Office of Disarmament Affairs (UNODA), The Arms Trade Treaty (2014), http://www.un.org/ga/search/view_doc.asp?symbol=A/RES/67/234&Lang=E (accessed January 18, 2018)

Wikipedia, "International Traffic in Arms Regulations," http://en.wikipedia.org/wiki/International_Traffic_in_Arms_Regulations (accessed January 18, 2018a).

Wikipedia, Lockheed Bribery Scandals, http://en.wikipedia.org/wiki/Lockheed_bribery_scandals (accessed January 18, 2018b).

Chapter 19

Temporal Limits of Engineers' Planning

In 2018, Kyoto University held a conference entitled "Science, Technology, and Future Generations." Despite the title, the conference included little about the relationship between future generations and science or technology. This chapter will not directly address that relationship either. Instead, it will address a narrower question: *How far into the future is it possible for engineers as such to plan?* Since engineers do much of the planning humans now do, especially large-scale planning, such as for power grids and road systems, answering that narrower question might help regional planners, economists, politicians, and the like answer the broader question, *How is successful large-scale planning possible, when it is, whether or not engineers do it?* This chapter will conclude with a few words about what sort of planning engineers should not do.

THE QUESTION

The temporal limits of engineers' planning is an important topic for engineers. Engineers are increasingly concerned with "life-cycle planning" and "sustainable development." Life-cycle planning is a guide for treating an artifact (or process) that begins with its conception and ends when the artifact has ceased to have a distinct existence—which can, in practice, be quite far in the future. The Code of Ethics of the National Society of Professional Engineers (NSPE) gives a definition of "sustainable development" that engineers seem generally to accept:

> the challenge of meeting human needs for natural resources, industrial products, energy, food, transportation, shelter, and effective waste management while

conserving and protecting environmental quality and the natural resource base essential for future development. (NSPE 2007n.)[1]

The Code does not set a limit to the number of future generations whose environment is to be protected. Statements about the present or past are, if meaningful, either true or false; statements about the future are not. Statements about the future merely become true or false depending on what happens (or, at least, that is what this chapter will assume). There is no knowledge of the future—or, at least, not the detailed and useful knowledge available for past or present. At best, what those who plan the future have instead of knowledge of it are mere probabilities.

Yet, engineers do plan as part of their work and many, especially civil engineers, plan for decades, and a few for even longer. For example, starting four decades ago, engineers helped to design the Yucca Mountain Nuclear Waste Repository to store highly radioactive nuclear waste safely in the Nevada desert for *at least 10,000 years*.[2] The Repository was a big project. By the time it was abandoned in 2011, it had cost $9 billion. If completed, it would have had 40 miles of tunnels and stored 77,000 metric tons of nuclear waste deep underground.[3] The project was abandoned after Congress ended funding. Congress seems to have ended funding for political reasons, especially an unhappy Senator from Nevada, not for the technical or ethical reasons considered here.

The Repository is not merely of historical interest. Recently, there has been an effort in Congress to restart the project (Wikipedia 2019d; YuccaMountain .org 2019). Making plans for such large-scale, long-term storage may seem more like science fiction than like engineering, especially to engineers, but the planning that engineers do is philosophically interesting precisely because it is so often both fantastic and successful, successful much more often than the experiments of science or the "sure things" of ordinary life seem to be. So, one might reasonably ask how such planning is possible—and what its limits are. Is 100 years beyond the limits of what engineers as such can plan? Is a thousand years? Is 10,000 years? Is there a time beyond which engineers as such cannot plan? Is there a time beyond which they should not plan?

These questions belong to the philosophy of technology insofar as engineers help to create, operate, and dispose of technology (i.e., systems of skills, methods, artifacts, and processes by which humans produce the goods and services that make their life possible, comfortable, or enjoyable). But these questions also belong to the philosophy of professions insofar as engineering is a profession and one profession may have technical standards distinct not only from the public's as such but also from those of other professions. What may be within the competence of one profession may not be within the competence of another. So, for example, dentists are not competent to design punch presses nor are engineers competent to do root canals; and, of course, most people are not competent to do either.

REFINING THE QUESTION

So, the question has become how far into the future engineers as such can plan. The answer to be considered here is that engineers as such can at most plan only a little farther into the future than they can reasonably expect engineers of the right sort to be present (two or three generations). Beyond that time, engineers can still plan, perhaps even successfully, but not as engineers. Thus, this chapter will not focus on the empirical question, How far into the future can humans called "engineers" plan? No doubt, people called "engineers" can plan as far into the future as they wish. Consider the ancient tomb known as "the Great Pyramid at Giza." It was the tallest building in the world from about 2560 BC until the spire of England's Lincoln Cathedral overtopped it just after 1300 AD. Suppose the Great Pyramid's designers had been called "engineers." They *could* have planned the Pyramid to serve a certain purpose forever, or for 10,000 years, or for just a few centuries. The Great Pyramid is now more than 4500 years old. It might still be serving its original purpose had robbers not broken into it long ago, removing the artifacts it was designed to hold safe. Whatever the rate of failure for human planning on the Great Pyramid's scale, humans can, it seems, plan a building that could serve its purpose for thousands of years, as the Great Pyramid could have (but did not).

But what was it that the ancient Egyptians did when they built the Great Pyramid: engineering rather than some other sort of building? Should we even count as engineers the "builders" of the Great Pyramid, that is, those who designed the Pyramid or managed its construction? There are at least three reasons not to count those builders as engineers.

First, and least important, the Great Pyramid is a tomb. Typically, engineers do not build tombs. Building tombs seems to have been, and remains, primarily the work of architects, stonemasons, artists, and the like. Engineers are brought in only if the tomb requires construction methods that only engineers can safely or reliably use. Claiming that ancient engineers built tombs thus raises a question we have already discussed several times: how one is to distinguish between engineers strictly so called and architects, stonemasons, artists, and other tomb builders, however called, who are similar to engineers insofar as they build on a large scale, use sophisticated methods, and so on. That is an important question to which this chapter will return.

The second reason to deny that any builder of the Great Pyramid was an engineer is that engineers, engineers strictly so called, would probably not have designed a tomb in the way Egypt's ancient builders did. For example, there is no practical reason for the Great Pyramid to be as large as it is. A smaller pyramid could have served the same purpose more economically—as later pyramids did. Indeed, so could a less conspicuous underground tomb.

The point now is not that the government of ancient Egypt did something unwise or morally wrong when it built the Great Pyramid. That is a subject for political philosophy. The point now is that the engineers of ancient Egypt (assuming there were any) would seem to have done something engineers should not have done, that is, wasted their employer's resources. Whether they ("the ancient Egyptian engineers") did something they should not have done depends in part on what "engineer" means in the assertion that there were (or were not) engineers in ancient Egypt to do it.

Third, the meaning of "engineer" relied on here is not a dictionary definition or a classical "abstract" or "verbal" definition—genus and species, necessary and sufficient conditions, or the like. The meaning is, instead, a living practice, at least several human beings organized in a certain way, the profession of engineering. This sort of definition resembles what mathematicians call "recursive" or "inductive" definition. It chiefly differs from the mathematical analog in being "practical," that is, part of the practice of engineering, not merely an activity a scholar might perform on her own. The first step is that a certain number of people recognize themselves and each other as engineers, that is, as belonging to the same discipline (a distinctive way of doing certain things). The next step is to have those people examine the credentials of the next candidate for admission and accept (or reject) him (or her) as competent to do what they do. If accepted, the new member of the set joins in the evaluation of the next candidate. And so on. (For more on this, see Davis 2009.)

Like anyone else, a philosopher can advise engineers concerning who is or is not an engineer (strictly so called), but the decision belongs to the profession much as (authoritative) decisions about what the law is belong to judges, though anyone may offer an opinion. The claim made here—that the builders of the Great Pyramid were not engineers—relies on the standards that the profession of engineering uses to identify engineers today (engineers strictly so called).

Sometimes applying those standards is easy, for example, when the candidate engineer is a recent graduate of a law school with an undergraduate degree in literature. Such a candidate is, all else equal, certainly not an engineer. In contrast, the graduate of an accredited program in mechanical engineering with five years' experience working as mechanical engineers typically work certainly is. Sometimes, however, the decision is harder to make, for example, when a chemist has been in responsible charge of a chemical plant for more than a decade: should engineers treat her as one of them?

This definition does not, please note, claim to cover all those individuals, and only those individuals, called "engineers." There are engineers strictly so called who are not called "engineers," for example, naval architects. There are also people legitimately called "engineers" who are not engineers strictly so called, for example, those technicians who, like Casey Jones, drive locomotives or, like the janitor of a high-rise building, are licensed to operate the boilers that heat the building but have only a high-school education. Who is properly called "engineer" has only a loose connection with who is an engineer strictly so called.

This definition also did not define engineers by what they do (some special function). That is because engineers do many things: design, inspect, test, manage, teach, write manuals, testify in court, lead tours, and so on. They perform those functions not merely by chance but because their training prepares them to, because they regularly offer to perform those functions, and because their employers regularly accept the offer. Of course, the designs that engineers typically prepare are engineering designs, not any sort of design; the inspections they typically perform are engineering inspections, not any sort of inspection; and so on.

The definition just sketched identifies engineers (strictly so called) using a certain practice, the profession of engineering. A profession is, in part at least, a shared discipline, a certain way of doing certain things that the members of the profession, and only its members, share. The discipline is passed from one generation of engineers to the next by a curriculum that engineers typically follow and non-engineers typically do not. The reason engineers recognize naval architecture as engineering (whatever naval architecture is called) while they do not so recognize ordinary architecture (however similar ordinary architecture is to engineering in function), is that naval architecture shares a discipline with the rest of engineering that ordinary architecture does not. Engineers have only to look at the curriculum of naval architects to see that naval architecture is engineering, not architecture, and that the graduates of that curriculum will not be architects but engineers—or, at least, in need of only a few years of engineering experience to become engineers rather than mere "engineers in training" (Davis 2010a).

The reason other disciplines similar to engineering in name and function—genetic engineering, geo-engineering, climate engineering, and the like—are not engineering (despite the name) is that they do not share the appropriate discipline. In each case, the core curriculum differs from engineering's in ways engineers typically consider important.

The exact structure of engineering's discipline is a matter of history, not logic. Engineering might have had a somewhat different curriculum, one allowing industrial chemistry or software engineering to be a part of engineering (as chemical engineering and computer engineering are). Indeed, probably nothing much but decisions of the professions in question prevents some such amalgamation from occurring tomorrow. What logic forbids is that engineering should absorb industrial chemistry or software engineering without some change in at least one of the disciplines. Disciplines are as real as nations, political parties, languages, or business corporations—and, like them, distinct historical individuals. They can change, but the change involves complex social practices, not just words. The price of change ("the transaction costs") can be high. Philosophers must take that into account if they are to understand what engineering is and why it is not many things logic or other department of philosophy says it could be.

There are, of course, verbal definitions of engineering useful in practice. Perhaps the best known comes from ABET (the body, formerly known as "the Accreditation Board of Engineering and Technology," that accredits programs for educating engineers in the United States and many other countries):

> Engineering is the profession in which a knowledge of the mathematical and natural sciences gained by study, experience, and practice is applied with judgment to develop ways to utilize economically the materials and forces of nature for the benefit of mankind.[4]

However useful and authoritative this definition is in practice, it is not a good philosophical definition of engineering. It fits architecture, industrial chemistry, industrial design, and even carpentry as well.[5] Architects, for example, also use their knowledge of mathematics and natural science, with judgment, to take advantage of the materials and forces of nature, to build structures economically for human benefit. Architects just differ enough from engineers in the mathematics and natural science they learn, the sort of judgment they exercise, and the way in which they seek to benefit humanity for (most) engineers to refuse to recognize them as sharing their discipline.

The discipline of engineering is not static, of course. Even some respected engineers of the past might have trouble entering the profession if, being reborn or revived with their knowledge, skill, and judgment intact, they asked to join today. Some engineers of the past lacked important skills or ethical commitments today's engineers think important. Whether those engineers of the past should count as engineers, nonetheless, is a different question, one hard to answer in part at least because it has no practical significance. It is a question best left to historians, philosophers, and the like.

One conclusion that does follow from the definition of engineering presented here is that there are likely to be changes in engineering's discipline even in the next few years—and so, in who counts as an engineer (strictly so called). But the details of those changes are hard to predict. For example, no one today should claim to know the answer to such questions as these: Will engineering programs reduce the calculus required from two years to one? Will they allow the substitution of biology for physics? Will they require a course in statistics?

Is it possible for engineers to appear in history more than once, say, in the Buddhist monasteries of eighth-century China as well as in the French army of the seventeenth century (with no engineers in between)? To ask that question is like asking whether it is possible for George Washington to appear in history more than once. The answer depends on how "thick" the description of the individual in question is: the thicker the description of the individual, the less likely that history can repeat itself. So, for example, if the description includes many details of Washington's life (e.g., that he commanded the American army during the Battle of Monmouth or that he wore false teeth much of his life), history cannot repeat itself. The specific details are themselves historical individuals tied to a distinct time and place (the weather, the battles that came before, the weapons available at that time, and so on). Since the definition of engineering offered here is quite thick, it is unlikely that engineering so defined will appear in history more than once.

Consider again the possibility of "ancient Egyptian engineers." The engineering curriculum did not begin in ancient Egypt. The chain of teachers and students that constitutes the profession of engineering reaches from today back to the French military of the late 1600s. For a few centuries before then there were soldiers, stonemasons, inventors, and even artists sometimes called "engineers," but no engineers strictly so called, and no school to train them in an identifiable discipline.

Given the definition of engineering presented here, there is no reason to count any of the builders of ancient Egypt as engineers (strictly so called). So, the question this chapter is to address has become: What is it possible for engineers (strictly so called) to plan while remaining within their discipline? Outside their discipline, engineers are like everyone else, individuals with knowledge, skill, and ambition. They can plan anything they like, but they are much less likely to succeed than engineers typically do. What they do will not be engineering.

HOW ENGINEERS AS SUCH PLAN

What then is known about how engineers as such plan? One thing that is known is that there is a high probability that engineers will preserve continuity between current and future practice (safety factors, standards of reliability, permissible tolerances, and so on). There are at least two reasons that such continuity is highly probable.

First, the engineering curriculum will not change radically over the next century or two because, if it did, it would cease to be an engineering curriculum. Radical change would signal a new discipline. Change in engineering's curriculum must, then, remain within certain limits, however hard it may be to specify those limits exactly. This is a conceptual truth, not an empirical one. There is an analogy here with language. A language can change much over time while remaining the same language, as English has since, say, 1600. Today's native speakers of English in Scotland, New Zealand, or Jamaica can still read the works of William Shakespeare or Sir Francis Bacon, though they would not write that way (except to imitate). But too radical a change, say, from Old English (Anglo-Saxon) to Modern English, creates a new language, not just a variation on the old. Changed that much, there would be two mutually incomprehensible ways of talking or writing. Only scholars can read *Beowulf* in the original Old English. Differences in degree can become differences in kind.

Second, there is good reason to believe that engineering practice will not change much over the next century or two because continuity is (currently) part of engineering's discipline. For example, assuming that there are engineers a hundred or even two-hundred years from now, there is good reason to believe that they will generally continue to document what they do much as engineers do now, preserve those documents, and use the information the documents contain when asked to repair, improve, or replace a particular bit of engineering. There is good reason to believe that documentation will continue to have the central place in engineering that it now has even though documentation is a mere fact about engineering, not a necessary part of it. Engineers typically require a good reason to change the way they work (since, all else equal, change is an expense). Documentation reduces duplication of effort, allowing one engineer to pick up where another engineer left off (rather than having to "re-invent the wheel"). Documentation is likely to be at least as useful to future engineers as it is to today's and unlikely to hinder good engineering in any way.

The change engineers prefer is incremental, for example, ever more precision in their documents. They do not like to go far outside their "data base." When they do introduce a new artifact (or process) into the world, they typically do it in stages, moving from laboratory tests to pilots, from pilots to field tests, and from field tests to ordinary use. Even after an artifact has entered ordinary use, engineers typically continue to monitor it. They do that not only so that they can fix problems as they appear but also so that they can make improvements as they see the opportunity.

This permanent search for problems and opportunities seems to explain much of engineering's success. Engineering artifacts do not go into the world naked; instead, they go into the world armored in an enduring process that identifies problems, recommends solutions, and sees that the solutions are carried out.

Of course, the world may change faster than engineering. For example, robots may replace engineers in such jobs as inspection, scheduling routine maintenance, and the like. Such change may cause the number of engineering jobs to decrease substantially, leaving many engineers unemployed and in search of new ways to be employed. But such change could nonetheless leave the remaining engineers not much different from today's, except that the work they do would have to be more creative, the kind of engineering that robots, even robots with great intelligence, cannot do. That is not to claim that engineering will not someday disappear altogether. Disciplines come and go. Where are the alchemists, alienists, apothecaries, conveyancers, and the like of yesteryear? The only claim made here is that, while engineering continues to exist, the way engineers work will only change slowly. The next generation of engineers will resemble today's much more than they will differ.

THE USEFUL LIFE OF AN ARTIFACT (OR PROCESS)

Asked how long the artifact (or process) they are designing will last, engineers typically answer a slightly different question. They answer with the artifact's "probable useful life." The useful life of an artifact is generally determined by the technology around it, both by what the artifact must compete with and the cost of repair, especially the cost of replacement parts. So, for example, automobiles that are a hundred or even a hundred-fifty years old exist and can be operated, but their useful life is over. They are "museum pieces" or "collector's items," vehicles that no longer meet minimum road standards (such as for tailpipe emission or crash safety). They are also too expensive to repair for ordinary use (in part because many replacement parts must be handmade, the mass market for those parts having disappeared long ago).

The oldest engineering artifacts, those that have had a useful life of a hundred-fifty years or so—a few aqueducts, bridges, canals, dykes, railway lines, sewers, and so on—have had to be updated in various ways to survive alongside newer technologies. For example, the traffic lanes of the Brooklyn Bridge were originally designed for the horse-drawn carriages and rail traffic of the late nineteenth century. They had to be reconfigured for automobiles. Without that reconfiguration, the Brooklyn Bridge would today probably exist only in the way the Iron Bridge over the Severn does, that is, as part of a museum, not as an artifact in ordinary use. The longest lasting artifacts of engineering also require regular maintenance to continue their useful life. Thus, the steel cables that support the great spans of the Brooklyn Bridge must be cleaned and painted every few years to prevent rusting. Without that maintenance, the Bridge would likely have become unsafe long ago. Like most, perhaps all, works of engineering, the Brooklyn Bridge was not designed to last for long without humans to maintain it, especially engineers to carry out routine inspections and oversee the maintenance. No doubt, the history of engineering explains why the works of engineers, unlike the Great Pyramid, require so much maintenance. But, whatever the explanation, that need should not be forgotten here.

HOW FAR INTO THE FUTURE CAN ENGINEERS AS SUCH PLAN?

With that, this chapter returns to the Yucca Mountain Nuclear Waste Repository. The discipline of engineering goes back less than four centuries. In that time until the Repository, engineers have never, it seems, tried to build an artifact with a probable useful life of even a few centuries, much less one with a probable useful life of 10,000 years. Planning for 10,000 years is well beyond the experience of engineers (the profession as a whole, not just individual engineers). When working beyond their experience, engineers prefer to scale up slowly. For engineers, even scaling up from a few decades to a few centuries would be an unprecedented leap into the unknown.

That is not the only concern engineers should have about participating in the Yucca Mountain Repository. Planning for an artifact with a useful life of even a few centuries seems to require social planning of a sort foreign to engineering. For example, to ensure that there will be suitably trained engineers even two centuries from now seems to require arrangements that will guarantee the existence of similar engineers for all the years between now and then, engineers trained to manage, inspect, repair, and update such a repository, to keep appropriate records for use by the engineers who follow, and so on. Training those engineers would seem to require at least one school of engineering to teach the requisite discipline, a school the endowment and other resources of which would not run out in less than two centuries. Such a school must have enough students every year to sustain at least one "professor of repository science" (someone who can pass on the special knowledge, skill, and judgment that only an engineer familiar with the Repository or repositories like it would have—including their "tacit know-how" which cannot, or at least has not, been put into texts). The school would also require enough employers to hire enough graduates of the school to make potential students think repository engineering an attractive career.[6] Engineers have never been trained to do that sort of social planning.

The point now is not that humans cannot engage in such planning. They can. For example, a large church might have enough maintenance to do to keep a few stonemasons busy indefinitely if they divide the work into annual packets rather than doing major projects every century so when the repairs can no longer be put off. As the centuries pass, changing styles of stonework, the church would have to choose between maintaining its original style and modernizing. If the church chose to retain its original style, its masons would in time become distinct from stonemasons generally. They would have to become a specialized guild, training "old style masons" for its employer's building (and ones like it) rather than "modern masons" of the sort the rest of the world would ask for.

Note that this specialization presupposes enough work to keep the old-style masons working more or less continuously. Without enough work, the old-style masons could not keep up their "old style." They would have to find other work just to live, leaving the church no choice but to hire modern masons the next time it needed masons. Not working as the old-style masons did, the modern masons would not produce the results the old-style masons did (not at all or, at least, not as efficiently).

It is, then, probably a mistake to bet that the future will include engineers qualified to design social institutions to manage, repair, and update the Repository for two-hundred years, much less for 10,000. The Repository was not designed to require many engineers once it was filled and sealed. Every year, old engineers die or retire and new ones, those without the experience of the old ones, replace them. One or two generations of ordinary operation once the Repository is full would probably be enough to drain the pool of those with the special knowledge, skill, and judgment of the original "repository engineers"; the same for "engineers" trained to design, maintain, and improve the social institutions to prepare the next generation of repository engineers.

A better bet for engineers than trying to design institutions to support the Repository over even two-hundred years may be to design the Repository in such a way that most futures (those without engineers as well as those with them) could manage, update, and repair the Repository indefinitely. The simpler the design, the more likely that the future that happens will have the resources to manage, update, and repair the Repository. But the simpler the Repository's design, the less likely that the Repository will need engineers to plan or build it, much less to maintain it.

The technology of the most resilient Repository might well resemble the technology of the Great Pyramid more than the technology of anything that engineers have built. After all, the Repository is, in principle, a tomb, the simplest form of which would be a mere hole in the ground filled with canisters of nuclear waste and topped with earth, the holes laid out in rows like graves at a military cemetery. The chief risk to the useful life of such a simple repository is probably the modern equivalent of the "tomb robbers" who ended the useful life of most of Egypt's pyramids, that is, those enterprising individuals or gangs who find a market for the repository's canisters or nuclear waste.

Human guards are the chief way to protect against such robbers (supported these days by cameras, motion detectors, and other "security devices"). But there will probably still be some need for engineers (or some similar discipline) to inspect the holes now and then to check for leakage that might enter the water table or air—and to figure out what to do if a hole begins to leak. However sure the scientists are about the materials holding the waste, the hydrology under the holes, the geology around the holes, and the future climate, engineers will want at least one backup for whatever seems a safe way to store waste (a safety factor). Scientists are constantly discovering that the world is more complicated than they thought. Human inspection is the ultimate backup, using qualified engineers when available and something like rent-a-cops when not.

The argument so far seems to have produced a dilemma. To plan for the distant future, even for a few centuries from now, engineers must either a) do, or at least assume, a good deal of social planning, enough to guarantee that there will be engineers competent to tend the complex artifact in question for its desired useful life (at least a few centuries), or b) design the artifact so that it is simple enough to survive any plausible change in social organization during its desired useful life, even a collapse of civilization that would leave no engineers competent to maintain, repair, or update the artifact. Neither horn of the dilemma seems to belong to the discipline of engineering in its current form—nor in a form that it is likely to take any time soon.

HOW FAR INTO THE FUTURE SHOULD ENGINEERS AS SUCH PLAN

The conclusion, then, seems to be that the future for which engineers as such can plan is quite near, perhaps no more than two or three generations away, the time for which it seems reasonable to assume that there will be both engineers of the right kind and the appropriate supporting technology.

Engineers might respond in one of two ways to this substantial limit on what they can do as engineers. First, they might just shrug and say, "If you want to plan for more than two or three generations, hire a futurist or some other technologist, not an engineer. What we do, we do well. But planning for more than a few generations is not among the things we do." Second, engineers might respond instead, "We need to revise the engineering curriculum so that engineers of the future will be competent to plan for many more generations than engineers are today. In particular, we need to include a good deal of 'social engineering' in our curriculum."

What engineers as such cannot answer is that they should just muddle through, trying to do whatever their employer asks of them. Their "paramount" responsibility is to "the public health, safety, and welfare," not to their employer (see, e.g., NSPE 2007, I.1.). Of course, as a matter of fact, some engineers might decide just to muddle through—whether to keep their job, or out of loyalty to their employer, or even because they just believe that engineers can solve any problem given them. But they cannot, as engineers strictly so called, just try to muddle through. They must adhere to their discipline. Just muddling through, they would have become engineers in name only, no more to be relied on than the scientists, managers, programmers, and the like whom they are typically brought in to augment.

For now, then, engineers have an ethical problem whenever asked to plan for more than two or three generations. Most codes of engineering ethics require engineers to perform services only in areas of their competence (see, e.g., NSPE 2007, I.2.). Engineers as such are not now competent to plan for more than two or three generations. So, it seems, today's engineers should refuse to work on a project like the Yucca Mountain Nuclear Waste Repository; or, if they do work on it, they should limit their part in the work to what both they and their employer would count as success even if the actual useful life of the resulting artifact were no more than two or three generations, for example, a repository designed to keep nuclear waste safely for two or three generations, until a better solution calls for the waste's retrieval. Engineers should leave planning for later futures to futurists or other technologists, whether such technologists exist now or not. While society has a responsibility to find a way safely to dispose of nuclear waste or store it safely for at least 10,000 years, an engineer as such only has a responsibility to do as society asks when that is within the engineer's competence (ethical as well as technical).

Most codes of engineering ethics also require engineers to "advise their clients or employers when they believe a project will not be successful" (NSPE 2007, III.1.b.). Like almost any construction designed to produce an artifact with a useful life of several centuries, much less 10,000 years, the Yucca Mountain Repository is probably something that engineers cannot, as engineers, believe could be a successful engineering project. There is no precedent for engineers producing an artifact having a useful life of even two centuries.

There are, of course, a few religious buildings, such as the Monastery of Paromeos in Egypt or the Izumo-taisha Shrine in Japan, that have been in continuous use for more than 1000 years (Wikipedia, 2019a, c).[7]

But no one has ever built a substantial structure with a useful life of even a few thousand years—and those with that longest useful life catered only to the simple needs of the dead. That being so, engineers asked to work on a project such as the Yucca Mountain Repository should tell their employer that they do not believe that it would be successful.

Indeed, they should refuse to work on such a project because it exceeds their competence. They may, of course, suggest alternatives within their competence, for example, tombs for nuclear waste likely to be safe for at least two or three generations. Any such repository should also include inspection, repair, and updating, because engineering artifacts typically have a short useful life otherwise.

Perhaps this is an appropriate place to return to the distinction between micro-ethics, macro-ethics, and meso-ethics to clarify what sort of argument is being made here. Micro-ethics is another name for ordinary morality, the standards a mere individual should bring to any decision (and the practices by which those standards should be applied, the answers that should result, and the like). Macro-ethics are the standards governments should bring to questions of social policy (e.g., whether to build the Yucca Mountain Repository). Meso-ethics are the standards that voluntary organizations, such as professions, should bring to the decisions they make—whether the decision-maker is a single member of the profession, a team, a professional society, or a whole profession. To act as an engineer (strictly so called) is never to act as a mere individual but as a member of a cooperative association (the engineering profession). What is morally right for an engineer may differ both from what is morally right for government and what it is morally right for a mere individual to do. So, for example, though both governments and individuals may properly design, establish, or operate primary schools or mental health clinics, engineers as such should not (Davis, 2010b).

PLANNING IN GENERAL

That leaves the question this chapter initially put aside: how is useful planning for the future possible (possible for anyone, not just engineers)? We may divide that question into at least three subsidiary questions—with somewhat different answers.

The first subsidiary question concerns the "near future," say, the next five or ten years. Planning for that future is, it seems, not hard because it is (more or less) for an extended present. It is probable (but not certain) that not much will change in so short a time.

The second subsidiary question concerns the "middle future," whatever is beyond the near future but not yet the "distant future," say, from ten years up to two or three generations. Planning for the middle future is harder than planning for the near future because more will probably change in unpredicted ways, with the probability (and magnitude) of unpredicted change increasing geometrically the farther into the future the planning tries to go. About all that is probable for the middle future is that what the NSPE Code of Ethics calls "environmental quality and the natural resource base essential for future development" will not change much. Humans will continue to need clean air, clean water, sources of energy, a moderate climate, shelter, and so on. Planning for the middle future may, then, consist in large part in "conserving and protecting" environmental quality and natural resources. Even so, planning for the middle future does not seem to be the kind of planning that should be done today and never again. Successful planning for the middle future seems to require regularly revising the plan in response to unpredicted change ("mid-course corrections"). This is a kind of planning that many people do successfully, not only engineers but also architects, economists, politicians, and the like. Of course, who should work on this or that plan depends on the sort of plan it is.

The third subsidiary question concerns the "far future," the time from the end of the middle future until the end of time. When planning for the far future, all that humans, engineers or not, can be reasonably sure of is that that future will probably differ from today in many unpredicted ways, some quite important for planners. There are many causes of unpredicted change, from earthquakes to asteroids, but perhaps the most likely and important is human inventiveness, not only the inventiveness of engineers, architects, economists, and politicians, but also of biologists, business managers, investors, lawyers, musicians, and so on. Anyone who claims to know much about the far future, even just its needs or resources, will probably turn out to be mistaken on many important points. For example: What futurist of 1820 would have guessed that just two centuries later, humans would need the vast quantities of coal, petroleum, methane, nuclear energy, and the like now used? Or that half the world would live in cities? Or that plastics would pollute the oceans? There is not yet a discipline with much success predicting the far future, much less successfully planning for it. Most predictions consist of projecting current trends into the future, not anticipating inventions that inaugurate new trends. That, perhaps, is why not much ages faster than science fiction.

CONCLUDING REMARKS

One conclusion that may seem to follow from what has been said so far is that engineers should have objected to designing, building, or overseeing nuclear power plants when that source of electricity was first proposed seventy or so years ago. Much of nuclear power, especially the disposal of its waste, was then well outside engineering's "data base." There are, however, at least five reasons to doubt this conclusion.

First, seventy years ago engineers did not typically do life cycle analysis of the artifacts (or processes) they designed. Without such an analysis, they would have had much less reason to object to nuclear power than they do today. Life cycle analysis has given engineers information they once lacked, alerting them to problems they used to miss. Life cycle analysis has changed "the state of the art" for engineers.

Second, the chief ground for engineers to object to nuclear power seventy years ago would have had to have been the "public health, safety, and welfare." But the profession did not then agree that "the public" included future generations. Indeed, one reason for the recent addition of "the environment" and "sustainable development" to codes of engineering ethics seems to have been to protect future generations because the earlier references to "the public" arguably did not. Engineering ethics (the special morally permissible standards of conduct that govern engineers just because they are engineers) is not static.

Third, another reason that codes of engineering ethics now include references to "the environment" and "sustainable development" may be that engineers realized that they lacked enough guidance on how to mediate between the current public's need for polluting resources, such as coal, petroleum, and methane, and the competing need for clean air, clean water, healthy forests, and so on. Energy both contributes to human welfare, because it helps to raise the standard of living, and damages that welfare, because production of energy tends to pollute the air and water, destroy forests, and so on. Furthermore, there is a question of justice here. The distribution of benefits and burdens, even among the living at any one time, is not automatically just. Those who suffer most from pollution are rarely, if ever, those using the most energy. Engineers still lack professional standards to guide them in the distribution of such benefits and burdens among those alive today.

Fourth, it is therefore not clear how engineers should mediate between the interests of those alive today and the interests of those born in the next few decades, much less those born a hundred or a thousand years from now. For example, how much, if at all, should engineers discount the interests of future generations as those generations become more likely to have interests hard to predict? Should engineers ignore the interests of those generations that might not exist at all if preceding generations destroy the resource base?

Fifth, engineers seventy years ago may have had good reason to believe that they could solve the problem of safely disposing of nuclear waste without entombing it for 10,000 years, for example, by burning the waste in thermonuclear reactors or rocketing it into the sun. Some options that once looked promising do not look so now.

On the other hand, one conclusion that may seem to follow from the argument made here is that today's engineers, and perhaps tomorrow's, should not build nuclear power plants until they have found a safe (and politically acceptable) way quickly to dispose of the waste such power plants generate (dispose of it, say, in three generations or less).[8]

Of course, drawing that conclusion is beyond the range of this chapter since (among other things) drawing that conclusion would require a detailed assessment of the alternatives to nuclear power—and that is another article, maybe even another long book, not just a few more paragraphs.

NOTES

An earlier version of this chapter was presented to the conference on "Science, Technology, and Future Generations," Center for Applied Philosophy and Ethics, Graduate School of Letters, Kyoto University, Kyoto, Japan, December 15–16, 2018. Thanks to those in the audience for several useful comments. Thanks also to three reviewers for their detailed suggestions for the article originally published as: "Temporal Limits on What Engineers Can Plan," *Science and Engineering Ethics* 25 (October 2019): 1609–1624.

1. For more than a decade now, the NSPE Code has encouraged engineers to "adhere to the principles of sustainable development in order to protect the environment for future generations" (NSPE Code 2007, III.2.d).

2. The design criterion for Repository should have been one million years, not 10,000, because the deadly radiation coming from the waste would not have degraded to something like the environment's natural background radiation until about a million years had passed. At 10,000 years, the nuclear waste would still be "hot" enough to constitute a serious health hazard to nearby humans and animals. 10,000 years seems to have been a compromise, making the design problem easier and the project more plausible. (Wikipedia 2019c).

3. Ratliff (1997), Endres (2009), and Wikipedia (2019c). The Finns are now building a similar repository at Onkalo. For details, see Wikipedia (2019b). The argument made here seems to apply to the Onkalo repository as well. Thanks to Hidekazu Kanemitsu for calling attention to what the Finns are doing.

4. Western Michigan University, 2018. There are a lot of references to this definition on the web, all crediting ABET, but it does not seem to be on ABET's website.

5. That is, if carpentry is a profession. Carpentry certainly is a profession if "profession" just means lawful occupation (which might be all that ABET means here). If, however, "profession" includes other features, such as college education or high social status, carpentry is not a profession.

6. That perhaps is why so many of the surviving programs in nuclear engineering now focus on medical uses of radioactive materials, updating existing nuclear power plants, disposal of nuclear waste, and so on. Not many jobs are in design or construction of new nuclear power stations. If they ever are again, programs in design and construction may have trouble finding suitably qualified engineers to teach the necessary courses.

7. While including some social engineering in tomorrow's curriculum may seem a good idea, engineers have yet to invent a way to do that likely to be useful. A little training in social engineering may not be much better than none; enough training in social engineering to be useful may turn out to demand a curriculum as distinct from engineering's as that of computer science or operations management now is. The resulting "social engineers" would, then, not be engineers strictly so called, however useful they might be.

8. Note that the following articles reached a similar conclusion by different routes: Lemons et al. (1990) and Shrader-Frechette (1993).

REFERENCES

Davis, M. (2009). Is engineering a profession everywhere? *Philosophia*, 37(June), 211–225.

Davis, M. (2010a). Distinguishing architects from engineers: A pilot study in differences between engineers and other technologists. In I. van de Poel & D. Goldberg (eds), Philosophy and engineering: An emerging agenda (pp. 15–30). Berlin: Springer.

Davis, M. (2010b). Engineers and sustainability: An inquiry into the elusive distinction between Macro-, micro-, and meso-ethics. *Journal of Applied Ethics and Philosophy*, 2, 12–20.

Endres, D. (2009). Science and public participation: An analysis of public scientific argument in the Yucca Mountain controversy. *Environmental Communication: A Journal of Nature and Culture*, 3(1), 49–75.

Lemons, J., Brow, D. A., & Varner, G. E. (1990). Congress, consistency, and environmental law: Nuclear Waste at Yucca Mountain Nevada. *Environmental Ethics*, 12(Winter), 311–327.

National Society of Professional Ethics. (2007). Code of ethics for engineers, http://ethics.iit.edu/ecode s/node/4098. Accessed June 11, 2019.

Ratliff, J. N. (1997). The politics of nuclear waste: An analysis of a public hearing on the proposed Yucca Mountain nuclear waste repository. *Communication Studies*, 48(4), 359–380.

Shrader-Frechette, K. S. (1993). Consent and nuclear waste disposal. *Public Affairs Quarterly*, 7 (October), 363–377.

Wikipedia. (2019a). Izumo-taisha, https ://en.wikip edia.org/wiki/Izumo-taisha. Accessed June 11, 2019.

Wikipedia. (2019b). Onkalo spent nuclear fuel repository https://en.wikipedia.org/wiki/Onkalo_spent_nuclear_fuel_repository Accessed June 11, 2019.

Wikipedia. (2019c). Paromeos monastery, https ://en.wikip edia.org/wiki/Parom eos_Monas tery. Accessed June 11, 2019.

Wikipedia. (2019d). Yucca Mountain nuclear waste repository, https://en.wikipedia .org/wiki/Yucca _Mountain_nuclear_waste repository. Accessed June 11, 2019.

YuccaMountain.org. (2019). https ://www.yucca mount ain.org/new.htm. Accessed June 11, 2019.

Chapter 20

Epilogue
A Research Agenda

This chapter has five sections. The first distinguishes three senses of "study" in common use among engineering students, engineers, and philosophers. The second section adds a fourth sense of "study engineering," the empirical investigation of engineering that seems characteristic of the social sciences. The third section briefly explains why the philosophy of engineering (strictly so called) is philosophy strictly speaking. The fourth section discusses three articles that are both empirical research into engineering and the work of philosophers (but not philosophy). It shows there is one empirical work philosophers can do, how they can do it, and why they should do it now and then. It also notes ways in which social scientists might do it better. The fifth section, the last, presents a series of studies that philosophers or social scientists might perform to settle whether engineering is today a single global profession.

1. STUDY OF ENGINEERING: FOUR KINDS

This book has been a study of engineering—but in what sense of "study?" There are at least four senses of "study" worth distinguishing before proposing the research agenda the Preface promised.

First, "study engineering" can mean learning the discipline of engineering itself, for example, the activity that goes on in engineering classes like *Statics* or *Advanced Circuits*, or in an engineering internship. Engineering students study engineering. The study of engineering in this sense passes knowledge of engineering (along with appropriate skills and judgment) from one generation of engineers to the next; it does not add to what engineers, as engineers, are supposed to know (or what engineers as an ordered community actually know). This book is not a study of engineering in that sense.

Second, "the study of engineering" can refer to an activity typical of engineers proper, not of students, not even of engineering students strictly speaking. It is an activity that *adds* to what engineers, as engineers, know or at least can know (engineering's "body of knowledge"). A (rough) synonym for "study of engineering" in this sense is "engineering research." Engineering research leads to formal reports that an engineer might write for a manager or publish as an article in, say, *Computational Mechanics* or the *Annual Review of Bioengineering*. That is another sense in which this book is not a "study of engineering."

Third, "study of engineering" can mean what we have been doing in this book, what philosophers might call "the philosophy of engineering." As noted in the Preface, the philosophy of engineering does not necessarily add to what engineers, as engineers, know or should know, nor does it necessarily add to what the rest of us know or should know. Instead, it chiefly arranges what we know about engineering, or think we know, into a more informative whole. It adds to our understanding of engineering rather than to our knowledge of it.

By this page, it should be clear what I mean by "arrange into a more informative whole," since that is what I have been doing all along (or at least trying to do) for certain central activities of engineering. Yet, many readers, especially philosophers, may find what I have done so far, an odd collation of ordinary philosophy (such as the discussion of definition in Chapter 1), history (such as that reconstruction of engineering's biography in Chapter 2), or the sociology of certain disciplines in Chapter 3 through 6 (architects, chemists, software engineers, and managers).

Of course, many readers may accept that collation because, though not "true philosophy," it is still "interdisciplinary"; interdisciplinary research is now okay, indeed, popular; and philosophers have always had a large enough part in the founding of new sciences or the global revision of old ones that interdisciplinary work has come to many to seem an organic part of philosophy. But I cannot accept what I have done here merely because it is interdisciplinary. Like many of my fellow philosophers, I try to respect disciplinary boundaries. It is too easy to make a fool of oneself outside one's own discipline. More importantly, I do not think that this book has left my discipline behind, though I have left it behind elsewhere (see, e.g., Davis, 1997; Davis, 2009b). So, I owe the reader an explanation of why this book is (more or less) simply philosophy—and why the research program I shall propose in the final section of this chapter is not.

The arranging of what we know (or at least believe we know) that we are calling "philosophy of engineering" must be either *a priori* (not including any empirical knowledge of engineering), *a posteriori* (including at least some such empirical knowledge), or mixed. *Unmixed a priori* philosophy of engineering would follow (something like) this advice: *First, do the philosophy;*

then seek empirical knowledge. The assumption this advice seems to make is that there is no point to trying to understand engineering until we have the concepts with which to express what we know of it (or, at least, think we know). Seeking understanding of engineering belongs to philosophy. Seeking knowledge of engineering is an activity best left to social scientists, historians, journalists, and the like trained in (one sort or another of) empirical research, empirical research that can only be reliably, efficiently, and usefully performed after philosophers have collected enough of the chief concepts in question, clarified them, and used them to understand what we know about engineering.

This way of understanding the philosophy of engineering may seem Cartesian (in the sense of "Cartesian" adopted in Chapter 2), but it is not. It is only close. The chief difference between this way of understanding philosophy of engineering and the standard Cartesian way is that this way allows philosophers to *exchange* what they know of engineering with other philosophers. This way of understanding engineering avoids the charge of solipsism. Nonetheless, this way of understanding the philosophy of engineering seems to share at least one serious flaw with Cartesianism strictly so called. The definition of engineering defended in Chapter 2 seems to have taken several centuries to develop. What developed seemed (in large part at least) to depend on historical circumstance (especially on who happened to be members of the nascent profession and what was asked of them). Philosophers must learn something of such contingencies *before* they can hope to contribute to the philosophy of engineering (e.g., explaining why the engineering curriculum typically ignores aesthetics while the architecture curriculum typically stresses it). Philosophers need not study engineering in our first sense of that term (Statics, Advanced Circuits, and so on) before doing philosophy of engineering, though submitting to the discipline might be helpful. But a philosopher of engineering must learn something of what the term "engineering" refers to, for example, by asking engineers what they do or by reading what others have observed them doing. Otherwise, the philosophy of engineering would be a mere shuffling of words or abstract concepts, an activity not likely to have much connection with understanding engineering as history has made it, that is, for example, as including applied physics but not architecture, naval architecture but not software engineering.

The philosophy of engineering is, then, necessarily (in part a least) *a posteriori*. We must begin by learning something of engineering in the way we ordinarily begin learning other contingent concepts. We must listen, read, observe, discuss, ask questions, and so on. Yet, it would be wrong to say that philosophy of engineering should follow the rule: *First do the science; then the philosophy.* The learning that precedes philosophy of engineering is not science (or, at least, not typically science). Science is a family of *disciplines* ("the sciences"): physics, chemistry, biology, economics, sociology, history, and so on. Each science has its own discipline, that is, certain standard ways of defining problems, collecting relevant information, formulating answers, and defending or criticizing those answers. The cognitive power of each science comes from its distinctive discipline. The supposed knowledge that precedes philosophy of engineering arises, in large part at least, not from any discipline but from ordinary observation, court proceedings, legislative hearings, gossip, and other organs of common sense. Much of this supposed knowledge may eventually turn out to be false and discarded. That is at least part of what makes science interesting, its regular overtaking of common sense. But common sense is what philosophy in general, and philosophy of engineering in particular, must begin.

So, the philosophy of engineering is logically prior to science but logically posterior to common sense. Is philosophy of engineering therefore permanently fixed by the definition of philosophy or engineering once history has spoken, or is the relation of philosophy to engineering (at least in large part) more dynamic? My answer is: More dynamic. Philosophers learn about engineering from engineers themselves, from engineering educators, from engineering students, from the clients or employers of engineers, from novelists, and perhaps from others in a position to know, or think they know, something about engineering in the undisciplined way of common sense.

Philosophers can help these informants, whether an individual, crowd, or community interested in understanding engineering, to sharpen old questions about engineering or frame new ones—that is, by doing what philosophers typically do. Sometimes they identify an existing discipline capable of answering some of those questions, a "science"; sometimes they design a new science for that purpose. The social scientists, engineers, journalists, engineering educators, and so on, then use the new or newly refined science to answer questions the philosophy of engineering could not.

Once the answers are reported, the continuing dialogue (philosophy of engineering) may just accept those answers or—more often—again sharpen the questions or re-frame them, as the individual, crowd, or community try to fit what they have just learned into what they already know or at least believe they know. And so on until everyone still involved is (more or less) satisfied with their understanding of engineering. (Some participants in the dialogue may have dropped out after concluding that their interest is not engineering but architecture, industrial design, ethical management, synthetic chemistry, climate management, or just retirement to a sunny beach in Bali.)

2. EMPIRICAL RESEARCH: THE FOURTH SENSE OF "STUDY"

Philosophers may be inclined to think of this iterative process as what Rawls called "reflective equilibrium"—the process by which a state of coherence among certain beliefs is arrived at by deliberative mutual adjustment among general principles and particular judgments. (Rawls, 1971, 48–51) So, let me briefly explain why I think the process described here is not quite Rawls' reflective equilibrium. For Rawls, what is important about reflective equilibrium is reaching coherence, a certain stasis among abstract entities (judgments, principles, and so on), since that is what makes possible *justifying* moral judgments or principles (at a certain time within a certain society). In contrast, the iterative process I just described is what makes possible the *study* of engineering in our fourth sense. That process explains how common sense, philosophy, and social science, can each have a part in a study of engineering that is empirical without being engineering research.

We might call this fourth sense "empirical research *into* engineering" to distinguish it from "engineering research" (our second sense). This fourth sense recognizes research into engineering as the work of non-engineering disciplines (philosophically refined) rather than of philosophy or research according to engineering's discipline. It is a catchall for the study of engineering according to such historically defined disciplines as anthropology, history, political science, sociology, and so on. What distinguishes these disciplines from philosophy is that they seek new observations or experience rather than seeking to understand the observations and experiences already known to all reasonable persons (mere common sense).

The sciences are *not* activities one can always perform in one's favorite armchair (as a Cartesian performs philosophy) or even in dialogue with others in other armchairs (as Socratics may perform philosophy). The sciences are supposed to advance knowledge beyond common sense and common-sense philosophy. Even Einstein had to submit his theories to the non-philosophical tests that physics suggested.

This distinction between philosophy of engineering and empirical research into engineering is rough, I admit. For example, it may seem to suggest to some that the research of scientists (and perhaps even of engineers) typically ends when they have gathered enough new information, that their empirical research does not require a favorite armchair or dialogue with others outside their discipline or within). Yet, much scientific research is typically done "in an armchair," for example, by reading relevant work of other scientists, formulating research questions, designing research instruments, and drawing conclusions from the data that empirical research has generated. Among the important activities that scientists can perform in an armchair is "theory," that is, designing a framework for further valuable empirical research.

This distinction between philosophy of engineering and empirical research into engineering may also suggest a "unity of the sciences" with each science touching the same subject is likely to draw the same conclusion as the others or at least typically producing data about engineering that will necessarily add a new piece to the puzzle. Yet, in practice, each science seems limited by its distinctive discipline. For example, anthropologists typically rely on in-depth interviews with a few informants while historians typically rely on libraries or files of documents privately held. Anthropologists who study engineering (in our fourth sense of study) may, then, expect to learn about engineering primarily from the few living engineers who can afford to take the time to give extensive interviews. Historians, in contrast, may suffer two other limitation. First, their informants will typically be behind the times, reporting (accurately or not) how things *were* in their time. (As the t-shirt says: "Historians tell it like it *was*.") Second, and perhaps more important, historians typically only have access to informants through those documents (and artifacts) that happen to survive. Most engineering documents seem to be lost within a few years of being prepared. Those that survive long enough to come into the hands of an historian are likely to introduce certain biases into the historical record. The rich are more likely than the poor to preserve documents; the winners in a "technology race" more likely to write about what they did than the losers; and so on. The methods of historians tend to overlook ordinary engineers and their small ever-day work.

We may, then, find the dialogue we call philosophy of engineering more complicated than expected. It may begin with philosophy asking questions of common sense until certain concepts of common sense are exact enough to permit scientific research within one or more non-engineering disciplines. All else equal, the more exact the concepts, the easier it is to disprove claims inherited from common sense. All else equal, the more objective a science, the easier it is to reduce the effects of bias. On questions to which more than one scientific discipline has something to contribute, the contributions may seem incompatible at first or, indeed, be incompatible, because of inherent bias. The philosophy of engineering may have to mediate between one science and another, as well as between the sciences in general and common sense. The mediation may require the philosophy of engineering to imagine research none of the existing sciences can afford to perform.

So, the case studies that constitute Part IV of this book belong to philosophy, not to any social science. As the references demonstrate, much of the information on which the three chapters rely comes from the websites of newspapers, works of social science, and what engineers have told my colleagues or me. None of the information relied on for the chapters of Part IV came from my own empirical research into engineering (Davis, 1997; Davis, 2009b). Since, in principle at least, I did not have to leave my armchair to write those chapters, they belong to philosophy of engineering, not to one of the social sciences. I have tried to resolve the question of engineering's status as a global profession in the same way—defining "profession," "engineering," and "global"—and then using those definitions to interpret a certain data set drawn (largely) from common sense (see, especially, Chapter 2, Chapter 14, and Chapters 16 through 18).

But I cannot claim success. I made arguments based on what I then knew, or at least believed I knew. I may have won over some engineers, philosophers, and others interested in the question of engineering's status as a global profession but, mostly, I seem so far to have only provoked a few social scientists to try to transform elements of the philosophical dialogue into a theory they could refute. Most of these (attempted) refutations relied on examination of official documents (see, e.g., Cao, Su, Hu, 2013; alZahir and Kombo, 2014).

Why (it may be asked) do I count research relying on official documents as social science rather than philosophy (or common sense)? Whether we should count research relying on official documents as social science rather than philosophy (or philosophy rather than social science) depends on how much such research relies on one or more disciplines. The more common sense is all that we need for the research, the more the research is philosophical. The more we need special skills to find the documents or understand them—for example, a knowledge of the terms and indexes typically used in legal research—the more it makes sense to count the research as social science (i.e., in this example, the study of law). The study of official documents, especially codes, is typically the study of law.

Other refutations relied on long interviews of a few engineers (see, e.g., Zhu, 2017). The conclusion of each of these studies was that engineers in one country, region, or "culture" differed from those in other countries, regions, or "cultures" in certain ways (while having much in common). What none of these studies showed was that *engineers* consider the discovered differences important enough to prevent engineering today from being the same profession everywhere despite those differences. Indeed, many of the studies did not even imagine comparisons across international, regional, or cultural boundaries using the same methods. They relied instead on a common sense understanding of their own country, region, or culture, comparing that to what they found in the official documents or documented practices of other countries, regions, or "cultures."

3. PHILOSOPHERS DOING EMPIRICAL RESEARCH

I now want to discuss three papers a few philosophers and I published between 2017 and 2020. They are not included in this volume because I consider them to be (primarily) works of social science and wanted this book to be a work of philosophy. I nonetheless want to discuss them here because they are examples of the research that social scientists might do if they adopted the research agenda I will be describing. That philosophers, not social scientists, did the research is relevant now only insofar as it suggests how easy it would be for social scientists to do such research better—alone or in teams including a philosopher.

I undertook this empirical research because the discussion of engineering as a global profession has gotten ahead of empirical research and I could not get social scientists to do the needed empirical research themselves or in cooperation with me. When I asked social scientists whom I knew why they had not done the research on their own or with me, they offered at least one of these six (unsatisfactory) explanations (and perhaps others):

First, profession is no longer a subject of interest in their field (meaning by "profession" one of the senses that Chapter 2 labeled "sociological"). The "hot topic" for research today is (they said) "professionalization," that is, "the process" of becoming a professional in one or more of the sociological definitions identified in Chapter 2. (They had lost interest in profession in part because they recognized the inadequacy of the social sociological definitions.)

Second, the social scientists I questioned did not think they could publish their research into engineering if they used my definition of "profession." "Profession" in that sense is not empirical enough for their journals (though it is the only definition of profession that has much empirical support).

Third, some of the social scientists I questioned thought they lacked the preparation necessary to interview engineers effectively. Their knowledge of social science may have prepared them to study a natural science because there is considerable overlap in method, though not in subject matter, between the natural sciences and the social sciences, especially the "scientific method." There is, however, not enough overlap with engineering's methods. Engineers do not study any part of nature (not even society); they just (sic) study how to design things.

Fourth, the social scientists I talked with did not know how to find the research subjects (the engineers) needed for the research that I suggested. They lacked contacts on the engineering side of the big corporations for which engineers typically work. Most of those social scientists even doubted that they could find an engineering class related enough to their research to justify a professor giving class time to a survey.

Fifth, they did not know how to attract sufficient practicing engineers into such a study. The usual enticements, such as free pizza or extra credit, that work with students seemed unlikely to work with practicing engineers, especially those long out of school or high in a corporate hierarchy.

Sixth, much of the discussion of globalism in engineering tends to contrast Asian engineers with Western engineers (or Chinese engineers with American engineers). Yet, few social scientists have the resources to compare a large sample of Asian engineers with a large sample of Western engineers. So, the best they can do is one or two small studies comparing the results of one empirical study of the engineers of one country with the results of a *presumed* empirical study of another (common sense underwriting the presumption).

4. THREE EMPIRICAL STUDIES OF ENGINEERING

That, anyway, is how matters stood when a mainland Chinese publication, *Engineering Studies*, invited a paper about whether Chinese engineers were members of the same profession as Western engineers. My response was philosophical, an attempt to understand the question well enough for social scientists to do the empirical work to answer it (Davis, 2007). Since that paper was published in Chinese without its original English version, I also published an expanded English version clarifying the more general question, "Is Engineering a Profession Everywhere?" (Davis, 2009a)

I then waited for a rush of publications in the social sciences reporting empirical work of the sort I invited. There was *none*. Instead, there was a trickle of publications seeking to show how different engineering in one country, region, or culture could be from engineering in another (see, e.g., Iseda, 2008). I also received a few emails from philosophers asking specific questions about my arguments, helping me tighten them but not forcing any substantial change.

Then, early in 2014, I received an inquiry from a Chinese philosopher at Beijing Institute of Technology, Hengli Zhang, asking whether he could visit IIT's Center for the Study of Ethics in the Professions for a year. The inquiry made it clear that he had read much of my work in philosophy of engineering and was interested in its application to teaching engineering ethics in China. He was especially interested in my concept of profession. He would be fully funded if I invited him. So, of course, I invited him. He arrived August 2015.

Discussion about projects he might undertake during the year revealed that he had a pool of former students and engineer friends who might serve as research subjects. He also had up-to-date contact information and an inexpensive way to contact them (email). Instead of waiting any longer for social scientists to take up my research suggestions, I could—with Zhang's help—start on the research myself.

Zhang and I took most of the fall semester to prepare a questionnaire. There were at least two reasons why preparing the questionnaire took so long. One was that there was no model for the questionnaire we needed. We wanted to determine whether Chinese engineers had the same concept of profession as American (or Western) engineers (one of those concepts extant in Western literature on the professions). We wanted to be sure we did not beg any question. Second, there was the language problem. I did not read, write, or speak Chinese at all; Zhang's English was uneven. So, when Zhang produced a first draft of the questionnaire, much of the effort of revising, especially at first, consisted of putting the questionnaire into Standard English. Zhang objected to many of my "corrections" because they changed the meaning he assigned to the question. Zhang's objections often led to long discussions of the etymology, connotations, and subtexts of the English word or phrase. Sometimes Zhang had a single Chinese word or phrase in mind for a translation of the English. At these times, he had to explain the Chinese word or phrase to me. Usually, we found a less treacherous English term. But, occasionally, Zhang ended up choosing a less treacherous Chinese term. Explaining English to Zhang made me realize how poorly even English, much less Chinese, mapped onto the concepts that philosophy of engineering seemed to take for granted.

When we were both satisfied with the English questionnaire, Zhang began translating it into Chinese. That turned out to be harder than we expected. For example, we had agreed to avoid "profession" (and its transforms) in the questionnaire to avoid seeming to beg the question whether Chinese engineers had a Western concept of profession. But one question (7) seemed to require the Chinese term for "profession" rather than the Chinese term for "occupation." Eventually, we decided to revise the *English* version of the questionnaire to read "professional and technical titles of engineer" as closer to what Zhang thought was the best he could do in Chinese.

We then carried out the survey, enrolling seventy-one practicing engineers. Among our results, three are worth mention here. First, most of the engineers we surveyed seemed to reject the common sociological definitions of profession. This was not surprising given that engineers in China (including our sample) seem to be paid about as much as skilled workers, seemed to lack high social status, are mostly unlicensed, and so on. One side of the "social contract" seemed to be missing. Second, most of the engineers we surveyed nonetheless held themselves and each other to engineering's special standards of conduct (the standards typically present in American and European codes of engineering ethics). Our research subjects thought of engineers as a group working together rather than as a category of individuals who merely share skills. Third, the primary reason the engineers we surveyed gave for adhering to engineering's standards, technical as well as ethical, was that the resulting products or services would be better for society. In short, the responses of the engineers we surveyed seemed (roughly) to track the definition of profession defended here—even though the Chinese language does not yet have an obvious term for that sense. There seems to be no empirical support for the supposedly empirical definitions of sociology (Davis and Zhang, 2017).

This study may be unsophisticated by the standards of the social sciences. Among the technical changes that would make it better would be a larger sample of Chinese engineers, a more sophisticated statistical analysis, and a more varied sample (especially one in which a crucial factor for inclusion of research subjects in the study was not acquaintanceship with one of the researchers, Zhang). But perhaps the most important improvement would be a control group consisting of American (or Western) engineers.

Nonetheless, I think this study did what it was designed to do. First, it showed that some Chinese have the concept of profession defended here (a "Western" concept) even though they lack a term for it. It showed that "Western" concepts cannot be dismissed from "Eastern" discussions of professional ethics just because they are "Western" or because the East does not yet have a term for it. Second, insofar as the survey showed that, it also showed that concepts can cross cultural divides that might seem to make the professional ethics of engineers in one country incomprehensible to others. It makes plausible a cosmopolitan approach to professional ethics in general and engineering ethics in particular.

About the time this paper was published (2017), I received an email from Lina Wei, a doctoral student in the Department of Philosophy, School of Humanities, Zhejiang University, Hangzhou, China, asking to spend a full academic year with me while she worked on a dissertation concerned with the professional competence of engineers. We soon worked out an arrangement for her like Zhang's. She arrived in August 2017.

After a month of readings that I suggested, and corresponding weekly discussions, Wei identified three (related) questions she might like to answer:

First, *what is professional competence in engineering?* Wei had become interested in this question because of several recent engineering disasters in China, including the failure of eight large highway bridges since 2011. These disasters seemed to her to involve ethical lapses, especially bribery, rather than technical incompetence. She wondered whether such lapses could count as evidence of professional *incompetence*. That seemed to her a philosophical question.

The second question she was interested in arose in part from my definition of profession, but not from the sociological definitions dismissed in Chapter 2. According to my definition, there could be no profession of engineering if engineers did not consider themselves to be working together to maintain common standards of conduct beyond those set by law, market, morality, and public opinion. Wei thought she knew of such standards in China (the government's official code of engineering ethics), but not what attitude engineers practicing in China took toward them. Since what the attitude of China's engineers is seems to be an empirical question, she wondered how to go about determining their attitude. *How is one to determine what attitude, if any, most engineers in China take to various candidates for their code of ethics?* This was another philosophical question.

Third, there was little empirical research on the professionalism of Chinese engineers, but what there was seemed to show that Chinese engineers regarded the standards of engineering as personal morality or social ethics rather than engineering ethics (see, e.g., Zhu and Jesiek, 2015). Wei noted that none of this empirical research tested "the Socratic definition" of profession against the others. She wondered whether using the Socratic definition would yield a different answer. That seemed to her to be an empirical question best left to sociologists—until she read Davis and Zhang, 2017, and heard why we had done the research that social scientists should have done.

She had thought her dissertation should answer all three of these questions. That was when she informed me that she could get access to several hundred practicing engineers (students in a professional master's program) who might be willing to fill out a questionnaire asking the questions that interested her. She soon did a draft of the questionnaire in English, using the Davis-Zhang questionnaire as a model. Once she had a satisfactory draft of a questionnaire in English, we set about preparing the Chinese version. We encountered many questions of translation like those Zhang and I had to work through. Though Wei's English was better than Zhang's, she asked her mentor, Hangqing Cong, to help with the Chinese version, since his graduate class in engineering ethics would constitute her research subjects. Cong's participation eventually became important enough that both Wei and I thought he should be listed as an author. He agreed.

Wei and Cong carried out the survey in December 2017, sending an explanation of the research, with a link to the questionnaire. When all 229 potential respondents had responded (yes, all), Wei began analyzing the data that had been collected. Early in spring 2019, we began preparing a paper. Part of the survey confirmed the conclusion that Davis and Zhang, 2017 had drawn: Chinese engineers seemed to have a profession if "profession" was understood as I claimed it should be, but not if any of the popular sociological definitions were used. The survey also showed that the engineers surveyed treated standards of engineering ethics much as they treated its technical standards: both were professional standards—and the ability to follow them was a measure of professional competence (Wei, Cong, and Davis, 2019).

Once our first paper had been accepted for publication, Wei proposed a second, one that would take advantage of some answers the first had ignored. For over a decade, Nanyan Cao (and some other scholars) have implicitly or explicitly claimed that engineering ethics in China is importantly different from engineering ethics in the United States (even if the United States and China share a common concept of profession). The evidence for that claim came from official documents (e.g., "codes of ethics" that the government imposed) or certain large features of Chinese society (e.g., Taoism, Confucianism, or Buddhism). Though neither the study of official documents nor the study of large features of Chinese culture is an uncommon approach to studying engineering ethics in China, neither approach actually studies the implicit concepts of engineering *practice* in China ("China's unwritten code of engineering ethics"). The paper Wei proposed would—or at least would get much closer. Wei had asked more than two hundred (Mainland) Chinese engineers about what they do and why they do it. Their responses suggested that Chinese engineers, or at least those surveyed, think about engineering ethics much as American engineers seem to (Compare Wei and Davis, 2020, with Davis, 1997). Chinese engineers had an unofficial ethics in addition to the state-imposed official ethics that empirical studies had so far focused on. At least some central features of Chinese engineering ethics did not seem to have much to do with "Chinese culture."

5. CONCLUSIONS: A RESEARCH AGENDA

I now want to propose an agenda of empirical research for philosophers of engineering to follow until social scientists, including engineering educators, are ready to take over, a research agenda that they can follow now.

The first item on the agenda is to identify empirical questions that come up in philosophy of engineering, especially engineering ethics, that stand in the way of studying engineering ethics in a particular country, region, or culture (in our fourth sense of "study"). For example, once philosophers in a certain country accept a definition of engineering, they are left with the empirical question: do engineers in that country satisfy that definition? If philosophers in that country have access to many practicing engineers, they can carry out a survey such as Wei, Zhang, Cong, I carried out. They could use the questionnaire developed for China, amended to suit their own country and research interests. The results might show that there is more than one profession of engineering in their country, just one, or none. Any of these results would be important by itself. But the research would be stronger if it included a control group, for example, ordinary engineering classes (rather than engineering ethics classes) with the questionnaire revised just enough to fit the full range of professions likely to be nascent in those classes.

Which brings me to the second item on my agenda. If, as I believe, these surveys will, country by country, build a case for the profession of engineering being much the same in all the countries surveyed, we will be entitled to conclude that there is only one engineering profession. Furthermore, if such surveys are done with similar results in all the countries of the world having engineers, we will be entitled to conclude that in every country, most or all engineers accept much the same standards as the others do or, in other words, that engineering is a single global profession.

However, this collection of empirical studies turns out, a related philosophical question will remain: *Should* engineering be one global profession? Though the empirical research suggested here cannot settle that philosophical question, it might help to strengthen one side or the other of the argument. For example, if the empirical research seems to show that there are several professions of engineering, all global, that is a reason not to require one global profession of engineering. What is to be gained by having one profession rather than several? Without a good answer to that question, several professions of engineering is, all else equal, the more attractive arrangement simply because it is the arrangement that already exists. If, on the other hand, the empirical research seems to show that engineering is one global profession, we are invited to ask: what is to be gained by allowing for more than one? All else equal, we should stick to what works—unless the benefits of changing over to another system will at least repay the transition costs. Not only would the philosophy of engineering enrich the empirical research, but the empirical research might also enrich the philosophy of engineering, providing new arguments for discussion.

So, we now have a definition of a global profession. A profession is global if it has many practitioners in most or all the countries in the world. A profession is a number of individuals in the same occupation voluntarily organized to earn a living by openly serving a moral ideal in a morally permissible way beyond what law, market, morality, and public opinion would otherwise require.

REFERENCES

alZahir, S. and Kombo, L. (2014). Towards a Global Code of Ethics for Engineers. IEEE DOI: 10.1109/ETHICS.2014.6893407.

Cao, N. Y., Su, J. B. & Hu, M. Y., (2013) Ethical Awareness in Chinese Professional Engineering Societies: Textual Research on Constitutions of Chinese Engineering Organizations, in D.P. Michelfelder et al. (ed.), *Philosophy and Engineering:*

Reflections on Practice, Principles and Process, Philosophy of Engineering and Technology 15, 203–214.

Calero, T., Davis, M., and Weil, V., Responsible Communication Between Engineers and Managers, in *Responsible Communications: Ethical Issues in Business, Industry, and the Professions*, ed. James A. Jaska and Michael S. Pritchard (Hampton Press, 1996), pp. 307–321.

Davis, M. Better Communications Between Engineers and Managers: Some Ways to Prevent Ethically Hard Choices, *Science and Engineering Ethics* 3 (April 1997): 171–213.

Davis M. How is a Profession of Engineering in China Possible? [in Chinese] *Engineering Studies* [China] 2007: 132–141.

Davis M. Is Engineering a Profession Everywhere? *Philosophia* 37 (June 2009a): 211–225.

Davis, M. *Code Writing: How Software Engineering Became a Profession*, Center for the Study of Ethics in the Professions: Chicago, 2009b (http://ethics.iit.edu/sea/sea.php/9).

Davis M. Collaborating across disciplines, in *The SAGE Handbook of Research Management*, ed. Robert Dingwall and Mary McDonnell (SAGE, 2015), pp. 213–224.

Davis, M. and Zhang, H. Proving that China has a Profession of Engineering: A Case Study in Operationalizing a Concept across a Cultural Divide, *Science and Engineering Ethics* (December 2017): 1581–1596.

Davis, M., Wei, L. and Cong, H. Professionalism among Chinese Engineers: An Empirical Study, *Science and Engineering Ethics* 26 (4): 2121–2139 (2020).

Iseda, T. (2008). How should we foster the professional integrity of engineers in Japan? A pride-based approach. *Science and Engineering Ethics, 14,* 165–176.

McGinn, R.E. (2003). Mind the Gaps: An Empirical Approach to Engineering Ethics, 1997–2001. *Science and Engineering Ethics* 9: 517–542.

Rawls, J. *A Theory of Justice* (Cambridge, Ma.: The Belknap Press of Harvard University Press, 1971).

Su, Junbin and Cao, N. A Study of Professional Code of Ethics Based on an Investigation Into Bylaws of Registered Professional Engineer and Constitution of Engineering Society in China (中国注册工程师制度和工程社团章程的伦理意识考察), *Journal of HuaZhong Science and Technology University* (Social Science edition), 4 (2007): 95–100.

Su, J. B. and Cao, N.Y. (2008) A study of the Evolution of Ethics Considerations of Chinese Engineers: Based on a historic inquiry into the revisions of the ethics code of China Institute of Engineer from 1933 to 1996 (中国工程师伦理意识的变迁——关于《中国工程师信条》1933–1996 年修订的技术与社会考察), *Journal of Dialectics of Nature* 6: 14–19.

Wei, L., Cong, H. and Davis, M. "Professionalism among Chinese Engineers: An Empirical Study," *Science and Engineering Ethics*, November 2019. DOI: 10.1007/s11948-019-00158-4.

Wei, L. and Davis, M. China's unwritten code of engineering ethics, *Business & Professional Ethics Journal* 39 (Summer 2020): 169–206.

Zhu, Q. (2017) *Working Effectively with Confucian Engineers: Sociopolitical Contexts, Cultures of Engineering Practice, and Collaborative Strategies*, Dissertation, Purdue University.

Zhu, Q. and Jesiek, B. K. (2015) Confucianism, Marxism, and Pragmatism: The Intellectual Contexts of Engineering Education in China, S.H. Christensen et al. (eds), International Perspectives on Engineering Education, *Philosophy of Engineering and Technology* 20: 151–170.

Index

ABET (formerly the Accreditation Board of Engineering and Technology), 25n4, 26, 35, 39, 61–62, 70, 92, 103, 274
ACM (formerly the Association for Computing Machinery), 67, 69
American Association of Engineers (AAE), 62, 66n4
American Institute of Chemists (AIC), 60–61
American Society of Civil Engineers (ASCE), 26, 40, 103, 139
American Society of Mechanical Engineers (ASME), 139, 155, 172, 177–80, 195–96, 263
applied ethics, 168. *See also* practical ethics
artist, 10, 44–47, 50, 52–53n7, 147n36, 271, 275
Australian Institute of Mining and Metallurgy (AIMM), 63

Broome, Taft, xv, xvii
Builder, chief, 4, 39, 44, 115

calling, 77, 98–100, 108
care, 6, 14, 43, 50, 62, 81, 111, 124, 128–29, 132, 138, 153, 178, 185–89, 208–9, 212, 219, 229, 258, 260

Chemical Institute of Canada (CIC), 62–63
Chinese, xvii, xix, 25, 27–33, 111, 295–99
code, xix, 10, 15–18, 19n11, 26–28, 30–33, 43–44, 51, 58–63, 65, 69, 77–78, 80, 100–5, 111, 117–19, 136, 138–39, 146, 155–56, 162–63, 172, 174–76, 179, 186–87, 190, 191, 195–96, 198, 213, 221–30, 262, 264, 267, 269–71, 283–84, 293, 296
contract, xix, 12, 25, 106, 110, 126, 142, 152–53, 156, 177, 186–88, 193–94, 196, 223, 228–29, 259, 261, 296
cooperative practice, 14, 81, 84, 108–10, 118, 139, 172, 282
Corps de génie, xv
culture, 28–30, 32–34, 78, 82, 123, 137, 151–55, 191–98, 202, 295, 210, 214, 219, 221, 224, 237, 259n3, 264, 293, 295, 299

da Vinci, Leonardo, xv, 39, 47
DeGeorge, Richard T., 76
Delft Technological University (TUDeft), xvii, 156–57, 159n8
discipline, 3–5, 10, 14, 18, 19n8, 23, 34, 38–43, 47, 50–51, 59, 68–69, 74, 80–82, 92, 94–97, 100–1, 105, 108–9,

112–14, 117, 156, 192–94, 272–78, 280–81, 287–93
Durbin, Paul, xiv, 87
Durkheim, Emile, 5, 19n4

employer, xv, 5, 13–14, 17, 25, 27–28, 30, 38, 60–62, 69, 76–78, 80–86, 93, 98, 102, 104, 118, 126, 136–39, 151, 155, 163, 172, 174, 176, 180, 185–88, 190, 193–94, 196–97, 208–9, 227–29, 247, 263, 265, 272–73, 278–79, 281–82, 290
Erie Canal, 50–51, 53n13, 113–14
ethical theory. *See* moral theory
ethics, xiii, xvii–xv, xix, xx, 3, 5–6, 10, 15–18, 24, 26, 30–33, 35n2, 36n5, 43–44, 51, 55–63, 65–66n2, 69, 76–80, 86, 93, 100–6n2, 107–13, 115–16, 118–19, 124, 136, 138–39, 146n23, 147n31, 151–60, 162–63, 167–81, 185–91, 195–99, 202, 211, 213, 219, 225–30, 237, 257–58, 262, 264, 267, 269, 281–84, 295–300
experience, xx, 9, 29, 31, 34, 39, 51, 53n13, 76, 94, 113–14, 117, 132, 157–58, 169, 194, 212, 214, 223, 242, 251, 261, 263, 272–74, 278–79, 291
experts, xvi, 10, 12, 28, 46, 49–50, 75, 85, 108, 140, 147n32, 214, 245, 247, 266

FEANI (European Federation of National Engineering Associations), 100, 163, 227–28, 230
France, xv, 3, 17, 30, 47–49, 118
function, xvi, 5–6, 18, 34, 38–39, 42, 46, 60–61, 67–68, 80–81, 85, 91–92, 94–95, 105, 108, 113–14, 117, 129, 143n4, 209, 228, 244, 247, 273

guild, 5, 45, 47, 94

Hegel, Georg Wilhelm Friedrich, 92
Herkert, Joseph R., 169–70, 173, 176–80
history of engineering, xiv–xvi, 40–41, 49, 51, 94–95, 114, 153, 278

IEEE (formerly the Institute for Electrical and Electronic Engineering), 35n2, 69, 103–5, 139, 155, 169–70, 180, 195–96, 263
Illinois Institute of Technology (IIT), xvii, 40–42, 93, 156, 295
International Organization for Standardization (ISO), 139, 165, 178, 189, 195, 263, 265

jurisprudence, xix. *See also* philosophy of law
justification, xx, 12, 125–26, 128, 142, 206–7, 211, 258, 265

Kant, Immanuel, 145n18

labor union, 5, 16, 225
license, xvi, 5–6, 14, 16, 24, 26, 35n1, 37, 59, 105, 195, 265, 273

Mafia, 7, 19n6, 25, 33
malfunction, 131–32, 135, 142
manager, 14–15, 28, 32–33, 41, 53, 71, 73–78, 82–86, 93–94, 141, 152–53, 193–94, 214, 223–24, 236, 240, 242, 247, 250, 260, 266–67, 281, 283, 288
method, xx, 7–9, 12, 24, 39, 41, 45, 56, 69, 93–94, 113, 115, 151, 179, 241, 249–50, 270–71, 292–94
Michelangelo, 40, 45–47
Mitcham, Carl, xv, xvii, xxi
MIT (Massachusetts Institute of Technology), xvii, 53n12
monopoly, 5–6, 59, 82, 165
moral ideal, 9, 11–15, 17, 19n9, 20, 34, 59–61, 65, 69, 76, 82, 100, 108–9, 111–12, 170, 300
morality, xix, 5, 7, 9–13, 15–17, 57–60, 108–10, 113, 126, 144n8, 170–71, 176–77, 186–87, 251, 257, 282, 298, 300
moral theory, 66n2, 158–60, 170, 257

National Society of Professional Engineers (NSPE), 26, 103, 269

nature, 11–12, 34, 43, 55–56, 63, 213, 227–28, 274, 294
Nuclear Regulatory Commission (NRC), 239

occupation, xv, xvi, 3–7, 9–18, 30, 39, 59, 60, 63, 69, 93, 97–99, 105, 108, 110–11, 113, 119, 143, 296, 300
officer, military, 4, 49–50, 97

Parson, Talcott, 19n4
Paxton, Joseph, 95–96
philosophy-of, xiv, xix, 158
philosophy of engineering, xiii–xviii, 24, 37–40, 51, 287–93, 295–96, 299–300
philosophy of law, xiii, xiv, xviii–xix, 159
philosophy of medicine, xiii–xiv, xix
philosophy of science, xiv, xviii
positive morality, 58, 170
practical ethics, 154, 182n1
profession, xx, 3–4, 9, 12, 16, 17, 57, 59, 76, 79, 98–99, 108, 110–11, 285, 294, 296, 298
profit, xiii, 55, 75–77, 81, 83, 85, 110, 151, 163, 208, 223, 228
public service, 11
pyramid, xv, 39, 95, 271–72, 278–80

rational, 36, 126–28, 142, 215n5
reasonable, xiii, xviii, 11–13, 24, 27, 32, 58, 60–61, 69, 76, 80, 100–1, 104, 109–10, 118, 123, 128–29, 132, 135, 138, 142, 145, 163, 170–72, 178, 185–89, 195, 203–4, 206, 209–14, 215n5, 248–49, 280, 291

Rensselear Polytechnic Institute (RPI), 40–42

Siege master, 4, 44
social science, xx, 4, 113, 154, 158, 287, 291–95, 297
social scientist, xiii, xvii, xx, 17, 33, 78, 93, 107, 108, 112, 119, 146, 287, 289, 290, 293–96, 298, 299
sociology, xx, 6, 153–54, 156, 158, 290, 291
sociology of engineering, xiv–xv, xviii, 177
sociology of professions, 4, 19n6, 153
sociology of science, xiv, xviii
Socratic definition of profession, 9–16, 298
standard, xxi, 5, 7, 11–17, 25–28, 31–32, 41, 43, 46, 55, 57–60, 62–63, 65, 70, 78–80, 83, 86, 92, 94, 100–5, 108–11, 113, 118–19, 125, 136, 139, 144, 151, 155–56, 160, 164, 168, 170–76, 178–81, 185–86, 188–89, 191–95, 197–99, 208, 211–13, 222–24, 240, 247, 252, 257–59, 262–67, 270, 272, 275, 277, 282, 284, 289–90, 296–98, 300
Stonehenge, xv
Stonemason, 4, 45–47, 50, 271, 275, 279
STS (science and technology studies), 157

van de Poel, Ebo, xvii
Veblen, Thorstein, 74–75, 79, 82, 85

Yucca Mountain Nuclear Repository, 277, 278, 281–82

www.ingramcontent.com/pod-product-compliance
Lightning Source LLC
Chambersburg PA
CBHW022009300426
44117CB00005B/101